Slope Stability Analysis and Stabilization

New Methods and Insight

Second Edition

Slope Stability Analysis and Stabilization

New Methods and Insight

Second Edition

Y. M. Cheng
C. K. Lau

CRC Press
Taylor & Francis Group
Boca Raton London New York

CRC Press is an imprint of the
Taylor & Francis Group, an **informa** business
A SPON BOOK

CRC Press
Taylor & Francis Group
6000 Broken Sound Parkway NW, Suite 300
Boca Raton, FL 33487-2742

First issued in paperback 2017

Version Date: 20140319

ISBN 13: 978-1-4665-8283-5 (hbk)
ISBN 13: 978-1-138-07550-4 (pbk)

Library of Congress Cataloging-in-Publication Data

Cheng, Y. M.
Slope stability analysis and stabilization : new methods and insight / Y. M. Cheng, C. K. Lau. -- Second edition.
pages cm
Includes bibliographical references and index.
ISBN 978-1-4665-8283-5 (hardback)
1. Slopes (Soil mechanics) 2. Soil stabilization. I. Lau, C. K. II. Title.

TA749.C44 2014
624.1'51363--dc23 2014002523

Visit the Taylor & Francis Web site at
http://www.taylorandfrancis.com

and the CRC Press Web site at
http://www.crcpress.com

Contents

Preface

After the publication of the first edition, we received several e-mails regarding some of the background material of slope stability analysis discussed in the book. We also received e-mails from research students enquiring about the procedure of numerical implementation in some of the stability analysis methods.

In the second edition, the more advanced concepts and case studies involved in slope stability analysis have been covered in greater detail based on our research work. In particular, we have added more examples and illustrations on the distinct element of slope, the relation between limit equilibrium and plasticity theory, the fundamental relation between slope stability analysis and bearing capacity problem, as well as three-dimensional slope stability under patch load conditions. The results of some of the laboratory tests that we have conducted are also included for illustration. Most importantly, we have added a chapter detailing the procedures involved in performing limit equilibrium analysis. This should help engineers carry out calculations or develop simple programs to carry out the analysis. Another new chapter deals with the design and construction practice in Hong Kong. This will be useful for those who are interested in slope stabilization works in Hong Kong.

The central core of SLOPE 2000 and SLOPE 3D has been developed mainly by Cheng, while many research students have helped in various works associated with the research results and the programs. We would like to thank Yip C.J., Wei W.B., Li N., Ling C.W., Li L. and Chen J. for helping with the preparation of the book.

Chapter 1

Introduction

1.1 OVERVIEW

The motive of writing this book is to address a number of issues of the current design and construction of engineered slopes. The book sets out to critically review the current situation and offer alternative and, in our view, more appropriate approaches for the establishment of a suitable design model, enhancement of the basic theory, locating the critical failure surfaces and overcoming numerical convergence problems. The latest developments in three-dimensional (3D) stability analysis and finite-element method will also be covered. It will provide helpful practical advice in ground investigation, design and implementation on site. The objective is to contribute towards the establishment of best practice in the design and construction of engineered slopes. In particular, the book will consider the fundamental assumptions of both limit equilibrium and finite-element methods in assessing the stability of a slope, and provide guidance in assessing their limitations. Some of the more up-to-date developments in slope stability analysis methods based on the author's works will also be covered in this book.

Some salient case histories to illustrate how adverse geological conditions can have serious implication on slope design and how these problems could be dealt with will also be given. Chapter 6 touches on the implementation of design on site. Emphasis is on how to translate the conceptual design conceived in the design office into physical implementation on site in a holistic way taking into account the latest developments in construction technology. Because of our background, a lot of cases and construction practices referred to in the book are related to the experience gained in Hong Kong, but the engineering principles should nevertheless be applicable to other regions as well.

1.2 BACKGROUND

Planet earth has an undulating surface and landslides occur regularly. Early humans would try to select relatively stable ground for settlement. As population grows and human life becomes more urbanized, there is a necessity to create terraces and corridors to make room for buildings and infrastructures like quays, canals, railways and roads. Man-made cut and fill slopes would have to be formed to facilitate such developments. Attempts have been made to improve the then rules of thumb approach by mathematically calculating the stability of such cut and fill slopes. One of the earliest attempts was by a French engineer, Alexander Collin (Collin, 1846). In 1916, K.E. Petterson (1955) used the limit equilibrium method to back-calculate mathematically the rotational stability of the Stigberg quay failure in Gothenburg, Sweden. A series of quay failures in Sweden motivated Swedes to make one of the earliest attempts to quantify slope stability by using the method of slices and the limit equilibrium method. The systematical method has culminated in the establishment of the Swedish Method (or the Ordinary Method) of Slices (Fellenius, 1927). A number of subsequent refinements to the method were made: Taylor's stability chart (Taylor, 1937); Bishop's Simplified Method of Slices (Bishop, 1955), which ensures that the moments are in equilibrium; Janbu extending the circular slip to a generalized slip surface (Janbu, 1973); Morgenstern and Price (1965) ensuring that moments and forces equilibrium are achieved simultaneously; Spencer's parallel inter-slice forces (1967); and Sarma's imposed horizontal earthquake approach (1973). These methods have resulted in the Modern Generalized Method of Slices (e.g. Low et al., 1998).

In the classical limit equilibrium approach, the user has to a priori define a slip surface before working out the stability. There are different techniques to ensure a critical slip surface can indeed be identified. A detailed discussion will be presented in Chapter 3. As expected, the ubiquitous finite-element method (Griffiths and Lane, 1999) or the equivalent finite-difference method (Cundall and Strack, 1979), namely, fast Lagrangian analysis of continua (FLAC), can also be used to evaluate the stability directly using the strength reduction algorithm (Dawson et al., 1999). Zhang (1999) has proposed a rigid finite-element method to work out the factor of safety (FOS). The advantage of these methods is that there is no need to assume any inter-slice forces or slip surface, but there also are limitations of these methods, which are covered in Chapter 4. Conversely, other assumptions will be required for the classical limit equilibrium method, which will be discussed in Chapter 2.

In the early days when computer was not available as widely, engineers may have preferred to use the stability charts developed by Taylor (1937), for example. Now that powerful and affordable computers are readily available, practitioners invariably use computer software to evaluate the stability in a design. However, every numerical method has its own postulations and thus limitations. It is therefore necessary for practitioners to be fully aware of them so that the method can be used within its limitations in a real design situation. Apart from the numerical method, it is equally important for engineers to have an appropriate design model for the design situation.

There is, however, one fundamental issue that has been bothering us for a long time: all observed failures are invariably 3D in nature, but virtually all calculations for routine design always assume the failure is in plane strain. Shear strengths in 3D and 2D (plane strain) are significantly different from each other. For example, typical sand can mobilize in plane strain up to 6° higher in frictional angle when compared with the shear strength in 3D or axisymmetric strain (Bishop, 1972). It seems we have been conflating the two key issues: using 3D strength data but a 2D model, and thus rendering the existing practice highly dubious. However, the increase in shear strength in plane strain usually far outweighs the inherent higher FOS in a 3D analysis. This is probably the reason why in nature all slopes fail in 3D as it is easier for slopes to fail this way. Now that 3D slope stability analysis has become well established, practitioners would no longer have any excuses to not be able to perform the analysis correctly, or at least, take the 3D effect into account.

1.3 CLOSED-FORM SOLUTIONS

For some simple and special cases, closed-form but non-trivial solutions do exist. These are very important results because apart from being academically pleasing, these should form the backbone of our other works presented in this book. Engineers, particularly younger ones, tend to rely heavily on code calculation using a computer and find it increasingly difficult to have a good feel of the engineering problems they are facing in their work. We hope that by looking at some of the closed-form solutions, we can put into our toolbox some very simple and reliable back-of-the-envelope-type calculations to help us develop a good feel of the stability of a slope and whether the computer code calculation is giving us a sensible answer. We hope we can offer a little bit of help in avoiding the current phenomenon where engineers tend to over-rely on a readymade black box–type solution and more on simple but reliable engineering sense in their daily work so that design can proceed with more understanding and less arbitrary leap into the dark.

For a circular slip failure with $c \neq 0$ and $\phi = 0$, if we take moment at the centre of rotation, the FOS will be obtained easily, which is the classical Swedish method, which will be covered in Chapter 2. The FOS from the Swedish method should be exactly equal to that from Bishop's method for this case. On the other hand, the Morgenstern–Price method will fail to converge easily for this case, whereas the method of Sarma will give a result very close to that of the Swedish method. Apart from the closed-form solutions for circular slip for the $c \neq 0$ and $\phi = 0$ case, which should already be very testing for the computer code to handle, the classical bearing capacity and earth pressure problem where closed-form solutions also exist may also be used to calibrate and verify a code calculation. A bearing capacity problem can be seen as a slope with a very gentle slope angle but with substantial surcharge loading. The beauty of this classical problem is that it is relatively easy to extend the problem to the 3D or at least the axisymmetric case where a closed-form solution also exists. For example, for an applied pressure of 5.14 Cu for the 2D case and 5.69 Cu for the axisymmetric case (Shield, 1955), where Cu is the undrained shear strength of the soil, the ultimate bearing capacity will be motivated. The computer code should yield FOS = 1.0 if the surcharge loadings are set to 5.14 Cu and 5.69 Cu, respectively. Likewise, similar bearing capacity solutions also exist for frictional material in both plane strain and axisymmetric strain (Cox, 1962; Bolton and Lau, 1993). It is surprising to find that many commercial programs have difficulties in reproducing these classical solutions, and the limit of application of each computer program should be assessed by the engineers.

Similarly, earth pressure problems, both active and passive, would also be a suitable check for the computer code. Here, the slope has an angle of 90°. By applying an active or passive pressure at the vertical face, the computer should yield FOS = 1.0 for both cases, which will be illustrated in Section 3.9. Likewise, the problem can be extended to 3D, or more precisely the axisymmetric case, for a shaft stability problem (Kwong, 1991).

Our argument is that all codes should be benchmarked and validated by subjecting them to solving the classical problems where *closed-form* solutions exist for comparison. Hopefully, the comparison would reveal both their strengths and limitations so that users can put things into perspective when using the code for design in real life. More on this topic can be found in Chapter 2.

1.4 ENGINEERING JUDGEMENT

We all agree that engineering judgement is one of the most valuable assets of an engineer because engineering is very much an art as it is a science. In our view, however, the best engineers always use their engineering judgement sparingly. To us, engineering judgement is really a euphemism for a leap into the dark. So in reality, the less we leap, the more comfortable we would be. We would therefore like to be able to use simple and understandable tools in our toolbox so that we can routinely do some back-of-the-envelope-type calculations that would help us to assess and evaluate the design situations we are facing so that we can develop a good feel of the problem, which will enable to do slope stabilization on a more rational basis.

1.5 GROUND MODEL

Before we can set out to check the stability of a slope, we need to find out what it is like and what it consists of. From a topographical survey, or more usually an aerial photograph interpretation and subsequent ground-truthing, we can determine its height, slope angle, and whether it has berms and is served by a drainage system or not. In addition, we also need to know its history, both in terms of its geological past, whether it has suffered failure or distress, and whether it has been engineered before. In a nutshell, we need to build a geological model of the slope featuring the key geological formations and characteristics. After some simplification and idealization in the context of the intended purpose of the site, a ground model can then be set up. When the design parameters and boundary conditions are delineated, a design model as defined by the Geotechnical Engineering Office in Hong Kong (GEO, 2007) should be established.

1.6 STATUS QUO

Despite being *properly* designed and implemented, slopes would still become unstable and collapse at an alarming rate. Wong's (2001) study suggests that the probability of a major failure (defined as >50 m^3) of an engineered slope is only about 50% better than that of a non-engineered slope. Martin (2000) pointed out that the most important factor with regard to major failures is the adoption of an inadequate geological or hydrogeological model in the design of slopes. In Hong Kong, it is an established practice for the Geotechnical Engineering Office to carry out landslip investigation whenever there is a significant failure or fatality. It is of interest to note that past failure investigations also suggest that the most usual causes of failure are some *unforeseen* adverse ground conditions and geological features in the slope. It is, however, widely believed that such adverse geological features, though *unforeseen*, should really be foreseeable if we set out to identify them at the outset. Typical unforeseen ground conditions are the presence of adverse geological features and adverse groundwater conditions.

1. Examples of adverse geological features in terms of strength are as follows:
    a. Adverse discontinuities, for example, relict joints
    b. Relict instability caused by discontinuities: dilation of discontinuities, with secondary infilling of low-friction materials, that is, soft bands, sometimes in the form of kaolin infill
    c. Re-activation of a pre-existing (relict) landslide, for example, a slickensided joint
    d. Faults

2. Examples of complex and unfavourable hydrogeological conditions are as follows:
   a. Drainage lines.
   b. Recharge zones, for example, open discontinuities, dilated relict joints.
   c. Zones with a large difference in hydraulic conductivity resulting in a perched groundwater table.
   d. A network of soil pipes and sinkholes.
   e. Damming of the drainage path of groundwater.
   f. Aquifer, for example, relict discontinuities.
   g. Aquitard, for example, basalt dyke.
   h. Tension cracks.
   i. Local depression.
   j. Depression of the rockhead.
   k. Blockage of soil pipes.
   l. Artesian conditions – Jiao et al. (2006) have pointed out that the normally assumed unconfined groundwater condition in Hong Kong is questionable. They have evidence to suggest that it is not uncommon for a zone near the rockhead to have a significantly higher hydraulic conductivity resulting in artesian conditions.
   m. Time delay in the rise of the groundwater table.
   n. Faults.

It is not too difficult to set up a realistic and accurate ground model for design purpose using routine ground investigation techniques but for the features mentioned earlier. In other words, it is actually very difficult to identify and quantify the highlighted adverse geological conditions. If we want to address the *so what* question, the adverse geological conditions may have two types of quite distinct impacts when it comes to slope design. We have to remember we do not want to be pedantic, but we still have a real engineering situation to deal with. The impacts would boil down to two types: (1) the presence of narrow bands of weakness and (2) the existence of complex and unfavourable hydrogeological conditions, that is, the transient ground pore water pressure may be high and may even be artesian.

While there is no hard-and-fast rule on how to identify the adverse geological conditions, the mapping of relict joints at the outcrops and the split continuous triple tube core (e.g. Mazier) samples may help identify the existence of zones and planes of weakness so that these can be incorporated properly in the slope design. The existence of complex and unfavourable hydrogeological conditions may be a lot more difficult to identify as the impact would be more complicated and indirect. Detailed geomorphological mapping may be able to identify most of the surface features like drainage lines, open discontinuities, tension cracks, local depression, etc. More subtle features would be recharge zones, soil pipes, aquifers, aquitards, depression of the rockhead and faults. Such features may manifest as extremely high perched groundwater table and artesian conditions. While it would be ideal to be able to identify all such hydrogeological features so that a proper hydrogeological model can be built for some very special cases, under normal design situations, we would suggest installing a redundant number of piezometers in the ground, so that the transient perched groundwater table and the artesian groundwater pressure, including any time delay in the rise of the groundwater table, can be measured directly using compact and robust electronic proprietary groundwater pressure–monitoring devices, for example, DIVER developed by Van Essen. Such devices may cost a lot more than the traditional Halcrow buckets, but can potentially provide the designer with the much needed transient groundwater pressure so that a realistic design event can be built for the slope design.

Although ground investigation should be planned with the identification of the adverse geological features firmly in mind, as a matter of course, one must also bear in mind that engineers have to deal with a huge number of slopes and it may not be feasible to screen each and every one of the slopes thoroughly. One must accept that no matter what one does, some of them would certainly be missed in our design. It is nevertheless still the best practice to attempt to identify all potential adverse geological features so that these can be properly dealt with in the slope design.

As an example, a geological model could be rock at various degrees of weathering, resulting in the following geological sequence in a slope, that is, completely decomposed rock (saprolite) overlying moderately to slightly weathered rock. The slope may be mantled by a layer of colluvium. In order to get this far, the engineer has to spend a lot of time and resources. But this is probably still not enough. We know rock mass behaviour is strongly influenced by discontinuities. Likewise, when rock mass decomposes, it would still be heavily influenced by relict joints. An engineer has no choice but to be able to build a geological model with all salient details for the design. It helps a lot if he also has a good understanding of the geological processes that may assist him to find out the existence of any adverse geological features. Typically, such adverse features are soft bands, internal erosion soil pipes, fault zones, etc., as listed previously. Such features may result in planes of weakness or create a very complicated hydrogeological system. Slopes often fail along such zones of weakness, or as a result of the very high water table or even artesian water pressure if these are not dealt with properly in the design, like the installation of relief wells and sub-horizontal drains. With the assistance of a professional engineering geologist, if necessary, the engineer should be able to construct a realistic geological model for his design. A comprehensive treatment of engineering geological practice in Hong Kong can be found in GEO publication no. 1/2007 (2007). This document may assist the engineer to recognize when specialist engineering geological expertise should be sought.

1.7 GROUND INVESTIGATION

Ground investigation, defined here in the broadest possible sense involving desk study, site reconnaissance, exploratory drilling, trenching and trial pitting, in situ testing, detailed examination during construction when the ground is opened up and supplementary investigation during construction, should be planned, supervised and interpreted by a geotechnical specialist appointed at the inception of a project. It should be instilled into the minds of the practitioners that a ground investigation does not stop when the ground investigation contract is completed, but should be conducted as an operation of discovery throughout the construction period. For example, mapping of the excavation during construction should be treated as an integral part of ground investigation. More use of new monitoring techniques like differential global positioning systems (GPSs) (Yin et al., 2002) to detect ground movements should also be considered.

In Hong Kong, ground investigation typically constitutes less than 1% of the total construction costs of foundation projects, but is mainly responsible for overrun in time (85%) and budget (30%) (Lau and Lau, 1998). The adage is that one pays for a ground investigation irrespective of whether it happens or not. That is, you either pay up front or at the bitter end when things go wrong. So it makes good commercial sense to invest in a thorough ground investigation at the outset.

The geological model can be established by mapping the outcrops in the vicinity and by sinking exploratory boreholes, trial pitting and trenching. For the identification of pre-existing slip surface of an old landslide where only residual shear strength is mobilized, it can be mapped through splitting and logging of a continuous Mazier sample (undisturbed sample) or even the sinking of an exploratory shaft.

In particular, Martin (2000) advocated the need to appraise relict discontinuities in saprolite and for more reliable prediction of transient rise in the perched groundwater table by using the following methods:

1. More frequent use of shallow standpipe piezometers sited at potential perching horizons.
2. Splitting and examining continuous triple-tube drill hole samples, in preference to alternative sampling and standard penetration testing.
3. More extensive and detailed walkover surveys during ground investigation and engineering inspection especially of the natural terrain beyond the crest of cut slopes. Particular attention should be paid to drainage lines and potential recharge zones.

1.8 DESIGN PARAMETERS

The next step would be to assign appropriate design parameters for the geological materials encountered. The key parameters for the geological materials are shear strength, hydraulic conductivity, density, stiffness and in situ stress. The latter two are probably of less importance compared with the others. The boundary conditions would also be important. The parameters can be obtained by index, triaxial, shear box and other in situ tests.

1.9 GROUNDWATER REGIME

The groundwater regime would be one of the most important aspects for any slope design. As mentioned earlier, slope stability is very sensitive to the groundwater regime. Likewise, the groundwater regime is also heavily influenced by the intensity and duration of local rainfall and the drainage provision. Rainfall intensity is usually measured by rain gauges and groundwater pressure is measured using standpipe piezometers installed in boreholes. Halcrow buckets or proprietary electronic groundwater-monitoring devices, for example, DIVER by Van Essen, should be used to monitor the groundwater conditions. The latter devices essentially are miniature pressure transducers (18 mm outer diameter) complete with a data logger and multiyear battery power supply so that they can be inserted into a standard standpipe piezometer (19 mm inner diameter). They usually measure the total water pressure so that a barometric correction could be made locally to account for changes in atmospheric pressure. A typical device can measure groundwater pressure once every 10 min for 1 year with a battery lasting for a few years. The device has to be retrieved from the ground and connected to a computer in order to download the data. The device, for example, DIVER, is housed in a strong, watertight stainless steel housing. As the metallic housing acts as a Faraday cage, the device is protected from strayed electricity and lightning. More details on such devices can be found at the manufacturer's website (http://www.swstechnology.com). One should also be wary of any potential damming of the groundwater flow as a result of the underground construction works.

1.10 DESIGN METHODOLOGY

We have to tackle the problem from both ends: the probability of a design event occurring and the consequence should such a design event occur. Much more engineering input has to be given to cases with high chance of occurrence and with significant consequences should such an event occur. For such sensitive cases, the engineer has to be more thorough in his identification of adverse geological features. In other words, he has to follow the best practice for such cases.

1.11 CASE HISTORIES

Engineering is both a science and an art. Engineers cannot afford to defer making design decisions until everything is clarified and understood. They would have to make provisional decisions so that progress can be made on site. It is expected that failures would occur whenever we are moving from the comfort zone. Precedence would be extremely important so that we know what our comfort zone is. Past success obviously would be good for morale. Ironically, past failures are what should be equally, if not more, important. Past failures usually are associated with working at the frontiers of technology or design based on extrapolation of past experience. Therefore, studying past mistakes and failures would be extremely instructive and valuable. In Hong Kong, the Geotechnical Engineering Office carries out detailed landslide investigation whenever there is a major landslide or landslide with fatality. We have selected some typical studies to illustrate some of the controlling adverse geological features mentioned in Section 1.6.

Case 1: The Shum Wan Road landslide occurred on 13 August 1995. Figure 1.1 shows a simplified geological section through the landslide. There is a thin mantle of colluvium overlying partially weathered fine-ash to coarse-ash crystallized tuff. Joints within the partially weathered tuff were commonly coated with manganese oxide and infilled with white clay to about 15 mm in thickness. An extensive soft yellowish brown clay seam

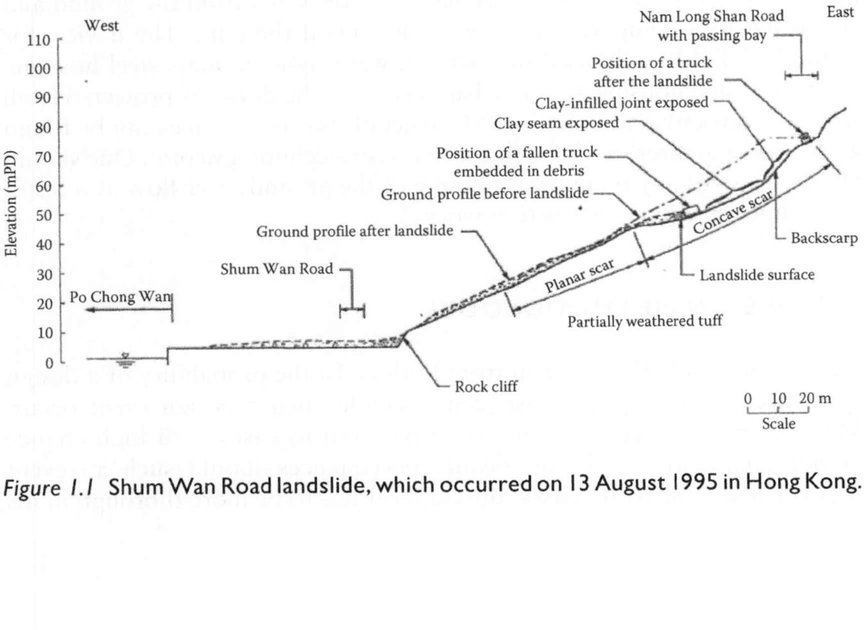

Figure 1.1 Shum Wan Road landslide, which occurred on 13 August 1995 in Hong Kong.

typically 100–350 mm thick formed part of the base of the concave scar. Laboratory tests suggest the following shear strength of the materials:

CDT: $c' = 5$ kPa; $\phi' = 38°$

Clay seam: $c' = 8$ kPa; $\phi' = 26°$

Clay seam (slickensided): $c' = 0$; $\phi' = 21°$

One of the principal causes of failure is the presence of weak layers, that is, clay seams and clay-infilled joints, in the ground. A comprehensive report on the landslide can be found in the GEO's report (GEO, 1996b).

Case 2: The Cheung Shan Estate landslip occurred on 16 July 1993. Figure 1.2 shows the cross section of the failed slope. The ground at the location of the landslip comprised colluvium to about 1 m thickness over partially weathered granodiorite. The landslip appears to have taken place entirely within the colluvium. When rainwater percolated the colluvium and reached the less permeable partially weathered granodiorite, a *perched water table* could have been developed and caused the landslip. More details on the failure can be found in the GEO's report (GEO, 1976).

Case 3: The three sequential landslides at milestone 14 1/2 Castle Peak Road occurred twice on 23 July 1994 and once on 7 August 1994. The cross

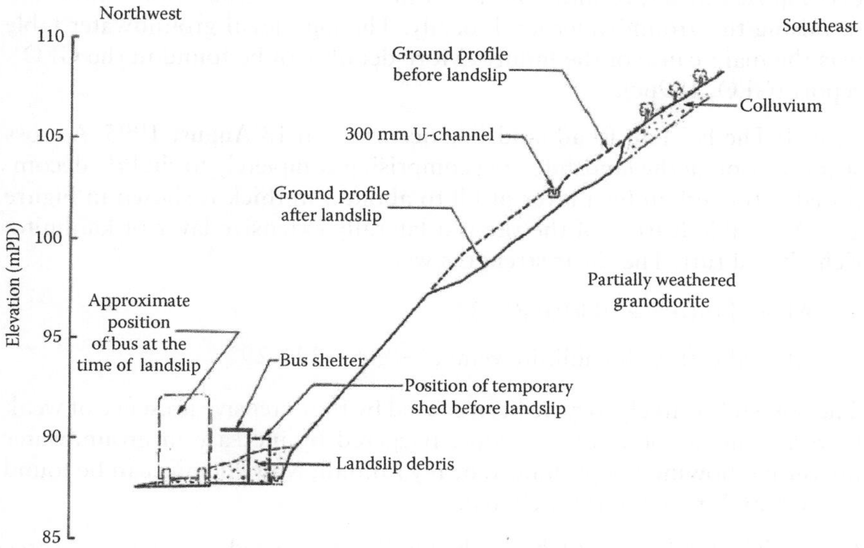

Figure 1.2 Cheung Shan Estate landslip, which occurred on 16 July 1993 in Hong Kong.

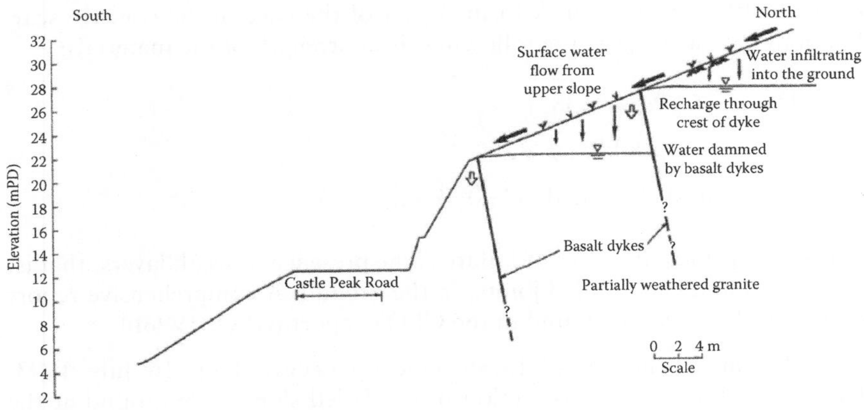

Figure 1.3 Landslide at Castle Peak Road, which occurred twice on 23 July and once on 7 August 1994 in Hong Kong.

section of the failed slope is shown in Figure 1.3. The granite at the site was intruded by sub-vertical basalt dykes of about 800 mm thick. The dykes were exposed within the landslide scar. When completely decomposed, the basalt dykes become rich in clay and silt and much less permeable than the partially weathered granite. Hence, the dykes act as barriers to water flow. The groundwater regime is thus likely to be controlled by a number of decomposed dykes, resulting in the damming of the groundwater flow and in raising the groundwater level locally. The high local groundwater table was the main cause of the failure. More details can be found in the GEO's report (GEO, 1996c).

Case 4: The Fei Tsui Road landslide occurred on 13 August 1995. A cross section through the landslide area comprising completely to slightly decomposed tuff overlain by a layer of fill to about 3 m thick is shown in Figure 1.4. A notable feature of the site is a laterally extensive layer of kaolinite-rich altered tuff. The shear strengths were

Altered tuff: $c' = 10$ kPa; $\phi' = 34°$

Altered tuff with kaolinite vein: $c' = 0$; $\phi' = 22°–29°$

The landslide is likely to have been caused by the extensive presence of weak materials in the body of the slope, triggered by increase in groundwater pressure following the prolonged heavy rainfall. More details can be found in the GEO's report (GEO, 1996a).

Case 5: The landslides at Ching Cheung Road involved a sequence of three successively larger progressive failures that occurred on 7 July (500 m^3), 17 July (700 m^3) and 3 August (2000 m^3) of 1997 (Figure 1.5). The cut slope

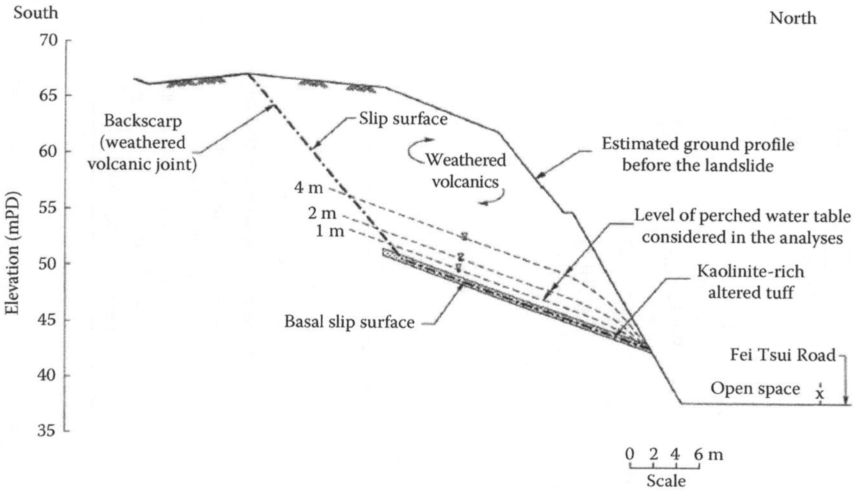

Figure 1.4 Fei Tsui Road landslide, which occurred on 13 August 1995 in Hong Kong.

was formed in 1967. Before its construction, the site was a borrow area and had suffered two failures in 1953 and 1963. A major landslide occurred during the widening of Ching Cheung Road (7500 m^3). Remedial works involved cutting back the slope and installation of raking drains. Under the Landslip Preventive Measures (LPM) programme, the slope was trimmed back further between 1990 and 1992. While this may help improve stability against any shallow failures, there would be significant reduction in the FOS of more deep-seated failures. In 1993, a minor failure occurred. In terms of geology, there are a series of intrusion of basalt dykes up to 1.2 m thickness, occasionally weathered to clayey silt. The hydraulic conductivity of the dykes would be notably lower than that of the surrounding granite and, therefore, the dykes probably locally act as aquitards, inhibiting the downward flow of the groundwater. There are also a series of erosion pipes of about 250 mm diameter at 6 m spacing. It seems likely that the first landslide that occurred on 7 July 1997 caused the blockage of natural pipes. The fact that the drainage line at the slope crest remained dry despite heavy rainfall may suggest that stormwater recharged the ground upstream rather than runoff along the surface. There was gradual building up of the groundwater table as a dual effect of recharging at the back and damming at the slope toe. The causes are likely to be reactivation of the pre-existing slip surface. Also, it is likely that the initial failure had caused blockage of the raking drains and the natural pipe system. The subsequent recharging from upstream and the blockage of the sub-soil drains, both natural and artificial, caused the final and most deep-seated third landslide. It is of interest to note that after the multiple failures at Ching Cheung Road, as a

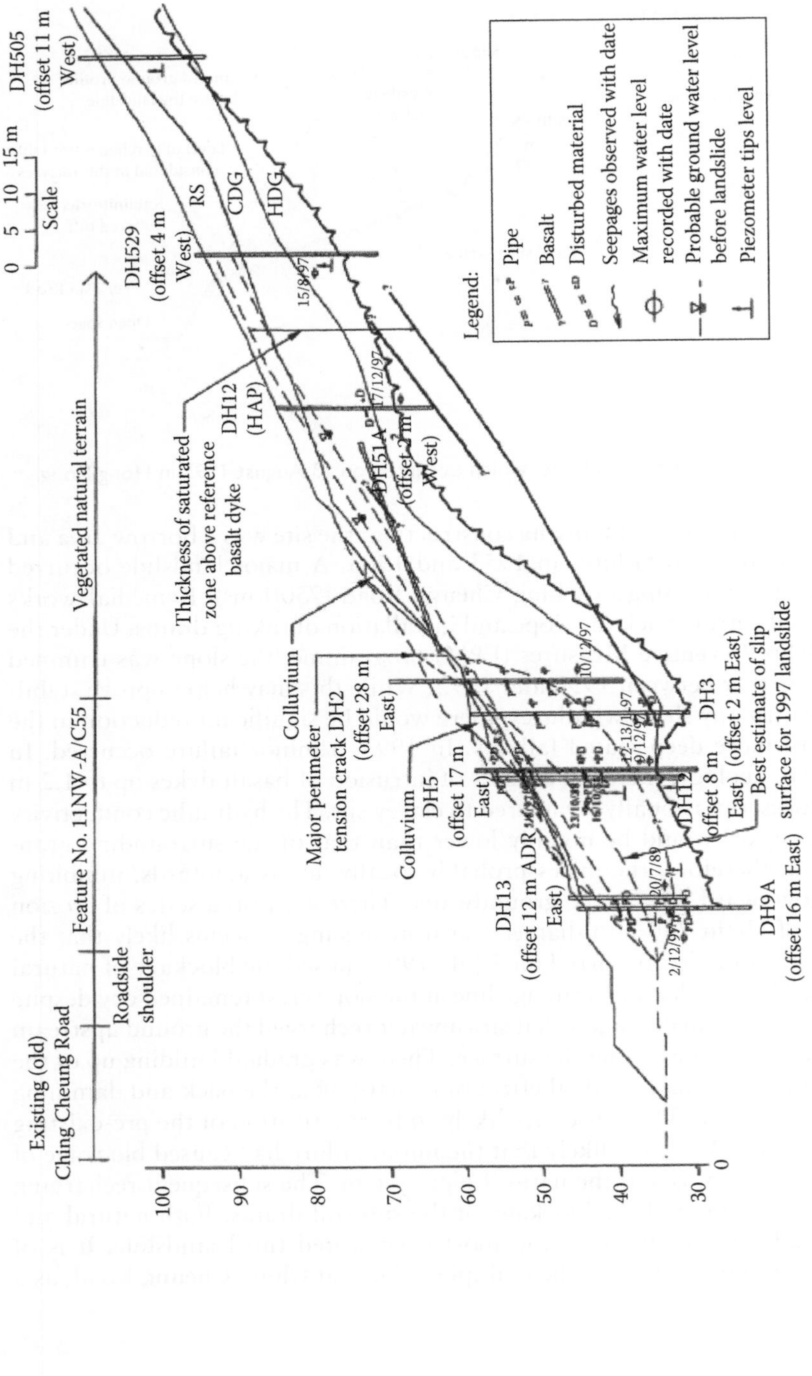

Figure 1.5 Landslides at Ching Cheung Road in 1997 (Hong Kong).

result of flattening of the slope and the complex hydrogeology, the engineer placed the ballast back at the toe as an emergency measure to stabilize the slope. It seems the engineer knew intuitively that removing the toe weight would reduce the stability of the slope against deep-seated failure and likewise the first solution that would come to mind for stabilizing the slope is to put dead weight back on the slope toe. More details on the landslide at Ching Cheung Road can be found in the GEO's Report (GEO, 1998).

Case 6: The Kwun Lung landslide occurred on 23 July 1994. One of the key findings was that leakage from the defective buried foul-water and stormwater drains was likely to be the principal source of subsurface seepage flow towards the landslide location, causing the failure.

In retrospect, we should install standpipe piezometers at the interface between the colluvium and the underlying partially weathered granodiorite and within the zone blocked by aquitards, and monitor the groundwater pressure accordingly using devices such as DIVER for at least one wet season. We should also identify weak zones and take them into account in our design. In addition, we need to take good care of the buried water bearing services in the vicinity.

Chapter 2

Basic slope stability analysis methods

2.1 INTRODUCTION

In this chapter, the basic formulation of the two-dimensional (2D) slope stability method will be discussed. More advanced formulation and related problems will be covered in Chapter 3. Currently, the theory and software for 2D slope stability analysis are rather mature, but there are still some important and new findings that will be discussed in this chapter. Most of the methods discussed in this chapter are available in the program SLOPE 2000 developed by Cheng, which can be downloaded at http://www.cse.polyu.edu.hk/~ceymcheng/, and an outline of this program is given in the Appendix.

2.1.1 Types of stability analyses

There are two different ways of carrying out slope stability analyses. The first approach is the *total stress approach*, which corresponds to clayey slopes or slopes with saturated sandy soils under short-term loadings where the pore pressure is not dissipated. The second approach corresponds to the *effective stress approach*, which applies to long-term stability analyses in which drained conditions prevail. Natural slopes and slopes in residual soils should be analyzed with the effective stress method, considering the maximum water level under severe rainstorms. This is particularly important for cities like Hong Kong where intensive rainfall may occur over a long period, and the water table can rise significantly after a rainstorm.

2.1.2 Definition of the factor of safety

The factor of safety for slope stability analysis is usually defined as the ratio of the ultimate shear strength divided by the mobilized shear stress at incipient failure. There are several ways of formulating the factor of safety F. The most common formulation for F assumes the factor of safety to be constant along the slip surface, and is defined with respect to force equilibrium or moment equilibrium, which are as follows:

1. *Moment equilibrium:* generally used for the analysis of rotational landslides. Considering a slip surface, the factor of safety F_m defined with respect to moment is given by

$$F_m = \frac{M_r}{M_d} \tag{2.1}$$

   where
   M_r is the sum of the resisting moments
   M_d is the sum of the driving moments

   For a circular failure surface, the centre of the circle is usually taken as the moment point for convenience. For a non-circular failure surface, an arbitrary point for the moment consideration may be taken in the analysis. It should be noted that for methods that do not satisfy horizontal force equilibrium (e.g. Bishop method), the factor of safety will depend on the choice of the moment point as *true* moment equilibrium requires force equilibrium. Actually, the use of the moment equilibrium equation without enforcing the force equilibrium cannot guarantee *true* moment equilibrium.
2. *Force equilibrium*: generally applied to translational or rotational failures composed of plane or polygonal slip surfaces. The factor of safety F_f defined with respect to force is given by:

$$F_f = \frac{F_r}{F_d} \tag{2.2}$$

   where
   F_r is the sum of the resisting forces
   F_d is the sum of the driving forces

For *simplified methods* that cannot fulfil both force and moment equilibrium simultaneously, these two definitions will be slightly different in value and meaning, although most design codes do not have a clear requirement of these two factors of safety, and a single factor of safety is actually specified in many design codes. A slope may actually possess several factors of safety according to different methods of analysis, which are covered in the later sections, but an extremum method giving a unique factor of safety that satisfies both force and moment equilibrium and the lower bound requirement has been developed by Cheng et al. (2010), which will be discussed in Section 3.12. Under the extremum principle, there is only one factor of safety which can satisfy the force and moment equilibrium (locally and globally) and the maximum strength of the system is fully mobilized. Based on this principle, any system possesses only one factor of safety, and this definition can avoid the difficult dilemma of multiple factors of safety using different methods of analysis, which is illustrated in Section 3.11.

A slope is considered unstable if $F \leq 1.0$. It is, however, common that many natural stable slopes have factors of safety less than 1.0 according to the commonly adopted design practice, and this phenomenon can be attributed to the following reasons:

1. The application of additional factor of safety on the soil parameters is quite common.
2. There is heavy rainfall with long recurrent periods in the analysis.
3. Three-dimensional effects are not considered in the analysis.
4. Additional stabilization due to the presence of vegetation or soil suction is not considered.

An acceptable factor of safety should be based on the consideration of the recurrent period of heavy rainfall, the consequence of the slope failures, the knowledge of the long-term behaviour of geological materials and the accuracy of the design model. Tables 2.1 and 2.2 show the requirements

Table 2.1 Recommended factors of safety F

		Risk of human losses		
		Negligible	*Average*	*High*
Risk of economic losses	Negligible	1.1	1.2	1.4
	Average	1.2	1.3	1.4
	High	1.4	1.4	1.5

Source: Geotechnical Engineering Office, *Geotechnical Manual of Slopes*, Hong Kong Government, HKSAR Government, Hong Kong, 1984.

Table 2.2 Recommended factor of safety for rehabilitation of failed slopes

Risk of human losses		
Negligible	*Average*	*High*
F> 1.1	F> 1.2	F> 1.3

Source: Geotechnical Engineering Office, *Geotechnical Manual of Slopes*, Hong Kong Government, HKSAR Government, Hong Kong, 1984.

Note: *F* for recurrence period of 10 years.

adopted in Hong Kong, where these values are found to be satisfactory. The slopes in the Three Gorges Project in China are very high and steep, and there is a lack of previous experience as well as knowledge of the long-term behaviour of the geological materials; hence, a higher factor of safety is adopted for the design. In this respect, an acceptable factor of safety will fulfil the basic requirement from the soil mechanics principle as well as the long-term performance of the slope.

The geotechnical engineers should consider the current slope conditions as well as future changes, such as the possibility of cuts at the slope toe, deforestation, surcharges and excessive infiltration. For very important slopes, there may be a need to monitor the pore pressure and suction with a tensionmeter and piezometer, and the displacement can be monitored by inclinometers, the Global Positioning System (GPS) or microwave reflection. Use of strain gauges or optical fibres in soil nails to monitor the strain and the nail loads may also be considered if necessary. For large-scale projects, the use of classical monitoring methods is expensive and time consuming, and GPS has become popular in recent years.

2.2 SLOPE STABILITY ANALYSIS: LIMIT EQUILIBRIUM METHOD

The slope stability problem is a statically indeterminate problem, and there are different methods of analysis available to engineers. Slope stability analysis can be carried out by the limit equilibrium method (LEM), the limit analysis method, the finite element method (FEM) or the finite difference method. By far, most engineers still use the LEM, which they are familiar with. The other methods are not commonly adopted in routine design, but will be discussed in the later sections of this chapter and in Chapter 4.

Currently, most slope stability analyses are carried out by the use of computer software. Some of the early LEMs are, however, simple enough and can be computed by hand calculation, such as the infinite slope analysis (Haefili, 1948) and the $\phi_u = 0$ undrained analysis (Fellenius, 1918). With the advent of computers, more advanced methods have been developed. Most of the LEMs are based on the techniques of slices, which can be vertical, horizontal or inclined. The first slice technique (Fellenius, 1927) was based more on engineering intuition than on a rigorous mechanics principle. There was rapid development of the slice methods in the 1950s and 1960s by Bishop (1955), Janbu (1957), Lowe and Karafiath (1960), Morgenstern and Price (1965), Spencer (1967) and Janbu (1973). The various 2D slice methods of LEM have been well surveyed and summarized (Fredlund and Krahn, 1984; Nash, 1987; Morgenstern, 1992; Duncan, 1996). The common features of the methods of slices have been summarized by Zhu et al. (2003):

a. The sliding body over the failure surface is divided into a finite number of slices. The slices are usually cut vertically, but horizontal as well as inclined cuts have also been used by various researchers. In general, the differences between different methods of cutting are not major, and vertical cut is preferred by most engineers at present.
b. The strength of the slip surface is mobilized to the same degree to bring the sliding body into a limit state. This means there is only a single factor of safety which is applied throughout the whole failure mass.
c. Assumptions regarding interslice forces are employed to render the problem determinate.
d. The factor of safety is computed from force and/or moment equilibrium equations.

The classical limit equilibrium analysis considers the ultimate limit state of the system and provides no information on the development of strain that actually occurs. For a natural slope, it is possible that part of the failure mass is heavily stressed so that the residual strength will be mobilized at some locations while the ultimate shear strength may be applied to other parts of the failure mass. This type of progressive failure may occur in overconsolidated or fissured clays or materials having brittle behaviour. The use of FEM or the extremum principle by Cheng et al. (2011b) can provide an estimation of the progressive failure. The use of a distinct element can, however, estimate the progressive failure and the post-failure condition with ease, which is not possible for the classical continuum-based methods.

Whitman and Bailey (1967) presented a very interesting and classical review of the limit equilibrium analysis methods, which can be grouped as follows:

1. *Method of slices*: The unstable soil mass is divided into a series of vertical slices and the slip surface can be circular or polygonal. Methods of analysis which employ circular slip surfaces include those of Fellenius (1936), Taylor (1948) and Bishop (1955). Methods of analysis which employ non-circular slip surfaces include those of Janbu (1957, 1973), Morgenstern and Price (1965), Spencer (1967) and Sarma (1973) and general limit equilibrium (GLE).
2. *Wedge methods*: The soil mass is divided into wedges with inclined interfaces. This method is commonly used for some earth dam (embankment) designs but is less commonly used for slopes. Approaches that employ the wedge method include those of Sultan and Seed (1967) and Sarma (1979).

The shear strength mobilized along a slip surface depends on the effective normal stress σ' acting on the failure surface. Fröhlich (1953) analysed the influence of the σ' distribution on the slip surface on the calculated F.

He suggested an upper and a lower bound for the possible F values. When the analysis is based on the lower bound theorem of plasticity, the following criteria apply: equilibrium equations, failure criterion and boundary conditions in terms of stresses. On the other hand, if one applies the upper bound theorem of plasticity, the following alternate criteria apply: compatibility equations and displacement boundary conditions, in which the external work equals the internal energy dissipation.

Hoek and Bray (1981) suggested that the lower bound assumption gives accurate values of the factor of safety. Taylor (1948), using the friction method, also concluded that the solution using the lower bound assumption leads to an accurate value of F for homogeneous slopes with circular failures. The use of the lower bound method is difficult in most cases, so different assumptions to evaluate the factor of safety have been used classically in the LEM. Cheng et al. (2010, 2011b) have developed a numerical procedure, described in Sections 3.12 and 3.13, which is effectively the lower bound method but is applicable to a general type of problem.

In the conventional LEM, the shear strength τ_m, which can be mobilized along the failure surface is given by

$$\tau_m = \frac{\tau_f}{F} \tag{2.3}$$

where F is the factor of safety (based on force or moment equilibrium in the final form) with respect to the ultimate shear strength τ_f, which is given by the Mohr–Coulomb relation as

$$\tau_f = c' + \sigma_n' \tan\phi' \quad \text{or} \quad c_u \tag{2.4}$$

where
c' is the cohesion
σ_n' is the effective normal stress
ϕ' is the angle of internal friction
c_u is the undrained shear strength

In the classical stability analysis, F is usually assumed to be constant along the entire failure surface. Therefore, an average value of F is obtained along the slip surface instead of the actual factor of safety, which varies along the failure surface if progressive failure is considered. There are some formulations where the factors of safety can vary along the failure surface. These kinds of formulations attempt to model the progressive failure in a simplified way, but the introduction of additional assumptions is not favoured by many engineers. Chugh (1986) presented a procedure for determining a variable factor of safety along the failure surface within the framework of the LEM. Chugh (1986) predefined a characteristic shape for the variation of the factor of safety along a failure surface, and this idea actually follows the idea of the variable interslice shear force function in the Morgenstern–Price method (1965). The suitability of this variable factor of safety distribution is, however, questionable, as the local factor of safety should be mainly controlled by the local soil properties. In view of these limitations, most engineers prefer the concept of a single factor of safety for a slope, which is easy for the design of the slope stabilization measures. Law and Lumb (1978) and Sarma and Tan (2006) have also proposed different methods with varying factors of safety along the failure surface. These methods, however, also suffer from the use of assumptions without a strong theoretical background. Cheng et al. (2011b) have developed another stability method based on the extremum principle as discussed in Section 3.13, which can allow for different factors of safety at different locations without the need of the previous assumptions.

2.2.1 Limit equilibrium formulation of slope stability analysis methods

LEM is the most popular approach in slope stability analysis. This method is well known to be a statically indeterminate problem, and assumptions on the interslice shear forces are required to render the problem statically determinate. Based on the assumptions of the internal forces and force and/or moment equilibrium, there are more than 10 methods developed for slope stability analysis. The famous methods include those by Fellenius (1936), Bishop (1955), Janbu (1957, 1973), Lowe and Karafiath (1960), Spencer (1967) and Morgenstern and Price (1965).

Since most of the existing methods are very similar in their basic formulations with only minor differences in the assumptions of the interslice shear forces, it is possible to group them under a unified formulation. Fredlund and Krahn (1977) and Espinoza and Bourdeau (1994) have proposed slightly different unified formulations of the more commonly used slope stability analysis methods. In this section, the formulation by Cheng and Zhu (2005), which can degenerate to many existing methods of analysis, will be introduced.

Based upon the static equilibrium conditions and the concept of limit equilibrium, the equations and unknown variables are summarized in Tables 2.3 and 2.4.

From these tables, it is clear that the slope stability problem is statically indeterminate on the order of $6n-2-4n=2n-2$. In other words, we have to introduce additional $(2n-2)$ assumptions to solve the problem. The locations of the base normal forces are usually assumed to be in the middle of the slice, which is a reasonable assumption if the width of the slice is limited. This assumption will reduce the unknowns so that there are only $n-2$ equations to be introduced. The most common additional assumptions are either the location of the interslice normal forces or the relation between

Table 2.3 Summary of system of equation

Equations	*Condition*
n	Moment equilibrium for each slice
2*n*	Force equilibrium in *X* and *Y* directions for each slice
n	Mohr–Coulomb failure criterion
4*n*	Total number of equations

n, number of slice.

Table 2.4 Summary of unknowns

Unknowns	*Description*
1	Safety factor
n	Normal force at the base of slice
n	Location of normal force at the base of slice
n	Shear force at the base of slice
n − 1	Interslice horizontal force
n − 1	Interslice tangential force
n − 1	Location of interslice force (line of thrust)
6*n* − 2	Total number of unknowns

the interslice normal and shear forces. That will further reduce the number of unknowns by $n-1$ (n slice has only $n-1$ interfaces), so the problem will become over-specified by 1 after the introduction of the assumptions in the individual method. There is no LEM method which can avoid this limitation and satisfy the global and local force/moment equilibrium automatically. The extreme principle by Cheng et al. (2010, 2011b), which is practically equivalent to the lower bound method, is an advancement over the classical LEM in that all the local and global force/moment equilibrium can be satisfied, but the amount of computations required for extremum determination far exceed that based on the classical LEMs. Based on different assumptions along the interfaces between slices, there are more than 10 existing methods of analysis at present.

The LEM can be broadly classified into two main categories: *simplified* methods and *rigorous* methods. For the simplified methods, either force or moment equilibrium can be satisfied, but not both at the same time. For the rigorous methods, both force and moment equilibrium can be satisfied, but usually the analysis is more tedious and may sometimes experience non-convergence problems. The authors have noticed that many engineers have a concept that methods which can satisfy both the force and moment equilibrium are accurate or even *exact*. This is a wrong concept as all methods of analysis require some assumptions to make the problem statically determinate. The authors have even come across many cases where very

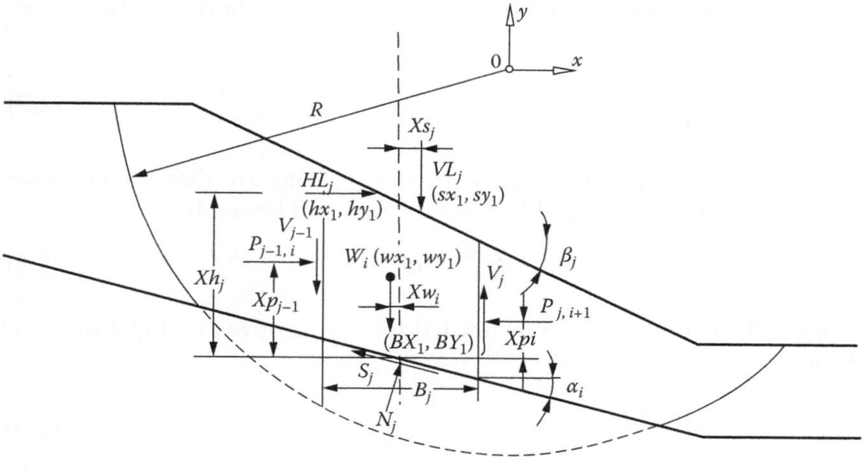

Figure 2.1 Internal forces in a failing mass.

strange results can come from *rigorous* methods (which should be eliminated because the internal forces are unacceptable), but the situation is usually better for *simplified* methods. In this respect, no method is particularly better than the others, though methods that take internal forces into consideration will usually be more effective than the others in most cases. Morgenstern (1992), Cheng, and many other researchers have found that most of the commonly used methods of analysis give similar results. In this respect, there is no urgent need to fine-tune the *rigorous* slope stability formulations except for isolated cases, as the interslice shear forces have only a small effect on the factor of safety in general.

To begin with the generalized formulation, consider the equilibrium of force and moment for a general case shown in Figure 2.1. The assumptions used in the present unified formulation are as follows:

1. The failure mass is a rigid body.
2. The base normal force acts in the middle of each slice base.
3. The Mohr–Coulomb failure criterion is used.

2.2.1.1 Force equilibrium

The horizontal and vertical force equilibrium conditions for slice i are given by

$$P_i \sin\alpha_i - T_i \cos\alpha_i + HL_i = E_{i,i+1} - E_{i-1,i} \tag{2.5}$$

$$W_i + VL_i - P_i \cos\alpha_i - T_i \sin\alpha_i = E_{i,i+1} \tan\phi_{i,i+1} - E_{i-1,i} \tan\phi_{i-1,i} \tag{2.6}$$

The Mohr–Coulomb relation is applied to the base normal force P_i and shear force T_i as

$$T_i = \frac{P_i \tan\phi_i + c_i l_i}{F} \tag{2.7}$$

The boundary conditions to these three equations are that the interslice normal and shear forces will be 0 for the first and last ends:

$$E_{0,1} = X_{0,1} = 0; \quad E_{n,n+1} = X_{n,n+1} = 0 \tag{2.8}$$

When $i=1$ (first slice), the base normal force P_1 as given by Equations 2.5 through 2.7 is

$$P_1 = \frac{A_1 \times F + C_i}{H_1 + AE_1 \times F}, \quad E_{1,2} = \frac{L_1 + K_1 \times F + M_1}{H_1 + AE_1 \times F} \tag{2.9}$$

where $E_{1,2}$ is a first-order function of the factor of safety F. For slice i, the base normal force is given by

$$P_i = \frac{(\tan\phi_{i-1,i} - \tan\phi_{i,i+1}) F \times E_{i-1,i} + A_i \times F + C_i}{H_i + AE_i \times F} \tag{2.10}$$

$$E_{i,i+1} = \frac{(J_i \times f_i + G_i \times F) E_{i-1,i} + L_i + K_i \times F + M_i}{H_i + AE_i \times F} \tag{2.11}$$

When $i=n$ (last slice), the base normal force P is given by

$$P_n = \frac{AA_n \times F + D_n}{J_n + G_n \times F}, \quad E_{n-1,n} = -\frac{L_n + K_n \times F + M_n}{J_n + G_n \times F} \tag{2.12}$$

Equations 2.11 and 2.12 relate the left and right interslice normal forces of a slice, and the subscript i, $i+1$ represents the internal force between slice i and $i+1$.

The symbols used in these equations are defined as follows:

$$A_i = W_i + VL_i - HL_i \tan\phi_{i,i+1}, \qquad AA_i = W_i + VL_i - HL_i \tan\phi_{i-1,i}$$

$$C_i = (\sin\alpha_i + \cos\alpha_i \tan\phi_{i,i+1}) c_i A_i, \quad D_i = (\sin\alpha_i + \cos\alpha_i \tan\phi_{i-1,i}) c_i A_i$$

$$AE_i = \cos\alpha_i + \tan\phi_{i,i+1} \sin\alpha_i, \qquad G_i = \cos\alpha_i + \tan\phi_{i-1,i} \sin\alpha_i$$

$$H_i = (-\sin\alpha_i - \tan\phi_{i,i+1} \cos\alpha_i) f_i, \quad J_i = (-\sin\alpha_i - \tan\phi_{i-1,i} \cos\alpha_i) f_i$$

$$K_i = (W_i + VL_i)\sin\alpha_i + HL_i\cos\alpha_i, \qquad X_i = E_{i,i+1}\tan\Phi_{i,i+1}$$

$$L_i = \left(-(W_i + VL_i)\cos\alpha_i - HL_i\sin\alpha_i\right)f_i, \quad M_i = \left(\sin^2\alpha_i - \cos^2\alpha_i\right)c_iA_i$$

$$B_i = W_i + VL_i - HL_i\tan\phi_{i-1,i}$$

where

α is the base inclination angle, clockwise is taken as positive
β is the ground slope angle, anticlockwise is taken as positive
W is the weight of the slice
VL is the external vertical surcharge
HL is the external horizontal load
E is the interslice normal force
X is the interslice shear force
P is the base normal force
T is the base shear force
F is the factor of safety
c, f is the base cohesion c' and $\tan\phi'$, where ϕ' is the friction angle
B is the width of the slice
l is the base length l of the slice $= B/\cos\alpha$
$\tan\Phi = \lambda f(x)$
$\{BX, BY\}$ are coordinates of the mid-point of the base of each slice
$\{wx, wy\}$ are coordinates for the centre of gravity of each slice
$\{sx, sy\}$ are coordinates for the point of application of the vertical load for each slice
$\{hx, hy\}$ are coordinates for the point of application of the horizontal load for each slice
Xw, Xs, Xh, h are the lever arm from the middle of the base for self-weight, vertical load, horizontal load and line of thrust, where $Xw = BX - wx$; $Xs = BX - sy$; $Xh = BY - hy$

2.2.1.2 Moment equilibrium equation

Taking the moment of about any given point O in Figure 2.1, the overall moment equilibrium is given by

$$\sum_{i=1}^{n}\begin{bmatrix} W_i wx_i + VL_i sx_i + HL_i hy_i + (P_i\sin\alpha_i - T_i\cos\alpha_i)BY_i \\ -(P_i\cos_i - T_i\sin\alpha_i)BX_i \end{bmatrix} = 0 \tag{2.13}$$

It should be noted that most of the *rigorous* methods adopt the overall moment equilibrium instead of the local moment equilibrium in the formulation, except for the Janbu rigorous method (1973) and the extremum method by Cheng et al. (2010), which will be introduced in Sections 3.12 and 3.13. The line of thrust can be back-computed from the internal forces after the stability analysis. Since the local moment equilibrium equation is not adopted explicitly, the line of thrust may fall outside the slice, which is clearly unacceptable; hence, the local moment equilibrium cannot be maintained under this case. The effects of the local moment equilibrium are, however, usually not critical toward the factor of safety, as the effect of the interslice shear force is usually small in most cases. Engineers should, however, check the location of the thrust line as a good practice after performing *rigorous* analyses. Sometimes, the local moment equilibrium can be maintained by fine-tuning the interslice force function $f(x)$, but there is no systematic way to achieve this except by manual trial and error or by using the lower bound method by Cheng et al. (2010) as discussed in Section 3.12.

2.2.2 Interslice force function

The interslice shear force X is assumed to be related to the interslice normal force E by the relation $X=\lambda f(x)E$. There is no theoretical basis to determine $f(x)$ for a general problem, as the slope stability problem is statically indeterminate by nature. More detailed discussion about $f(x)$ by the lower bound method will be provided in Section 3.12. There are seven types of $f(x)$ commonly in use:

Type 1: $f(x)=1.0$. This case is equivalent to the Spencer method and is commonly adopted by many engineers. Consider the case of a sandy soil with $c'=0$. If the Mohr–Coulomb relation is applied to the interslice force relation: $X=E\tan\phi'$, then $f(x)=1.0$ and $\lambda=\tan\phi$. Since there is no urgent requirement to apply the Mohr–Coulomb relation for the interslice forces, $f(x)$ should be different from 1.0 in general. It will be demonstrated in Section 3.12 that $f(x)=1.0$ is actually not a realistic or good choice for some cases.

Type 2: $f(x)=\sin(x)$. This is a relatively popular alternative to $f(x)=1.0$. This function is adopted purely because of its simplicity.

Type 3: $f(x)=$trapezoidal shape shown in Figure 2.2. Type 3 $f(x)$ will reduce to type 1 as a special case, but is seldom adopted in practice.

Type 4: $f(x)=$error function or the Fredlund–Wilson–Fan force function (1986), which is in the form of $f(x)=\Psi\ \exp(-0.5c^n\eta^n)$, where Ψ, c and n are defined by the user. η is a normalized dimensional factor which has a value of −0.5 at the left exit end and 0.5 at the right exit end of the failure surface. η varies linearly with the x ordinates of the failure surface. This error function is actually based on an elastic finite element stress analysis by Fan et al. (1986). Since the stress state in the limit equilibrium analysis is the ultimate condition and is different from the elastic stress analysis by Fan et al. (1986), the suitability of this interslice force function cannot be

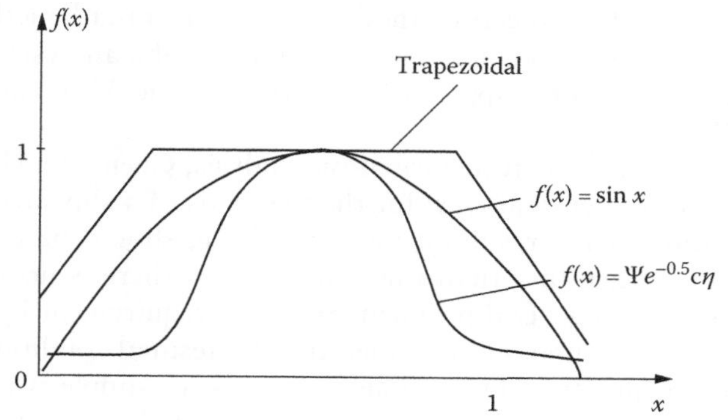

Figure 2.2 Shape of interslice shear force function.

justified simply by the elastic analysis. It is also difficult to define suitable parameters for a general problem with soil nails, water table and external loads. This function is also not applicable for complicated cases.

The first four types of functions shown earlier are commonly adopted in the Morgenstern–Price and GLE methods, and both the moment (globally) and force (locally and globally) equilibrium can be satisfied simultaneously. A completely arbitrary interslice force function is theoretically possible, but there is no simple way or theoretical background in defining this function except for the extremum principle introduced in Section 3.12, so the arbitrary interslice force function is seldom considered in practice.

Type 5: Corps of Engineers interslice force function. $\lambda f(x)$ is assumed to be constant and is equal to the slope angle defined by the two extreme ends of the failure surface. Moment equilibrium cannot be achieved by this function as $\lambda f(x)$ is prescribed to be known.

Type 6: Lowe–Karafiath interslice force function. $\lambda f(x)$ is assumed to be the average of the slope angle of the ground profile and the failure surface at the section under consideration. Similar to the Corps of Engineers method, moment equilibrium cannot be achieved by this function as $\lambda f(x)$ is prescribed to be known.

Type 7: $f(x)$ is defined as the tangent of the base slope angle at the section under consideration, and this assumption is used in the load factor method in China. Similar to the previous two cases, this method cannot satisfy moment equilibrium.

In type 5 to type 7 interslice force functions, only force equilibrium is enforced in the formulation. The factors of safety from these methods are, however, usually very close to those used in the *rigorous* methods, and are usually better than the results obtained by the Janbu simplified method (1957). In fact, the Janbu simplified method (1957) is given in the case of

$\lambda = 0$ for the Corps of Engineers method, the Lowe–Karafiath method (1960) and the load factor method, and results from the Janbu analysis (1957) can also be taken as the first approximation trial in the Morgenstern–Price analysis (1965).

Based on a Mohr circle transformation analysis, Chen and Morgenstern (1983) have established that $\lambda f(x)$ for the two ends of a slip surface, which is the inclination of the resultant interslice force, should be equal to the ground slope angle. Other than this requirement, there is no simple way to establish $f(x)$ for a general problem. Since the requirement by Chen and Morgenstern (1983) applies only under an infinitesimal condition, it is seldom adopted in practice. Even though there is no simple way to define $f(x)$, Morgenstern (1992), among others, has pointed out that for normal problems, F values from different methods of analyses are similar so that the assumptions on the internal force distributions are not major issues for practical use except in some particular cases. In view of the difficulty in prescribing a suitable $f(x)$ for a general problem, most engineers will choose $f(x) = 1.0$, which is satisfactory for most cases. Cheng et al. (2010) have, however, established the upper and lower bounds of the factor of safety and the corresponding $f(x)$ based on the extremum principle, which will be discussed in Section 3.12. Cheng et al. (2010) have also demonstrated that the extremum principle can satisfy the requirement of $\lambda f(x)$ for the two ends of a slip surface as stipulated by Chen and Morgenstern (1983).

2.2.3 Reduction to various methods and discussion

The present unified formulation by Cheng and Zhu (2005) can reduce to most of the commonly used methods of analysis as shown in Table 2.5. In Table 2.5, the angle of inclination of the interslice forces is prescribed for methods 2–9.

The classical Swedish method for undrained analysis (Fellenius analysis) considers only the global moment equilibrium and neglects all the internal forces between slices. For the Swedish method under drained analysis, the left and right interslice forces are assumed to be equal and opposite so that the base normal forces become known. The factor of safety can be obtained easily without the need of iteration analysis. The Swedish method is well known to be conservative, and sometimes its results can be 20%–30% less accurate than those from the *rigorous* methods; hence, the Swedish method is seldom adopted in practice. This method is, however, simple enough to be operated by hand or spreadsheet calculation, and the problem of non-convergence does not exist as iteration is not required.

The Bishop method is one of the most popular slope stability analysis methods and is used worldwide. This method satisfies only the moment equilibrium given by Equation 2.13 but not the horizontal force equilibrium given by Equation 2.5, and it applies only for a circular failure surface.

Table 2.5 Assumptions used in various methods of analysis

Method	*Assumptions*	*Force equilibrium* X	*Force equilibrium* Y	*Moment equilibrium*
1. Swedish	$E=X=0$	×	×	√
2. Bishop simplified	$X=0$ or $\Phi=0$	×	√	√
3. Janbu simplified	$X=0$ or $\Phi=0$	√	√	×
4. Lowe and Karafiath	$\Phi=(\alpha+\beta)/2$	√	√	×
5. Corps of Engineers	$\Phi=\beta$ or $\Phi_{i-1,i}=\dfrac{\alpha_{i-1}+\alpha_i}{2}$	√	√	×
6. Load transfer	$\Phi=\alpha$	√	√	×
7. Wedge	$\Phi=\phi$	√	√	×
8. Spencer	$\Phi=$ constant	√	√	√
9. Morgenstern–Price and GLE	$\Phi=\lambda f(x)$	√	√	√
10. Janbu rigorous	Line of thrust (Xp)	√	√	√
11. Leshchinsky	Magnitude and distribution of P	√	√	√

× means not satisfied; √ means satisfied.

The centre of the circle is taken as the moment point in the moment equilibrium equation given by Equation 2.13. The Bishop method has been used for some non-circular failure surfaces, but Fredlund et al. (1992) have demonstrated that the factor of safety will depend on the choice of the moment point because there is a net unbalanced horizontal force in the system. The use of the Bishop method in the non-circular failure surface is generally not recommended (though allowed by some commercial programs) because of the unbalanced horizontal force problem, and this can be important for problems with loads from earthquake or soil reinforcement. This method is simple for hand calculation and the convergence is fast. It is also virtually free from convergence problems, and its results are very close to those obtained by the *rigorous* methods. If a circular failure surface is sufficient for the design and analysis, this method can be a very good solution for engineers. When applied to an undrained problem with $\phi=0$, the Bishop method and the Swedish method show identical results.

For the Janbu simplified method (1957), force equilibrium is completely satisfied while moment equilibrium is not satisfied. This method is also popular worldwide as it is fast in computation with only a few convergence problems. This method can be used for non-circular failure surfaces, which are commonly observed in sandy-type soil. Janbu (1973) later proposed a *rigorous* formulation, which is more tedious in computation. Based on the ratio of the factors of safety from the *rigorous* and *simplified* analyses, Janbu (1973) proposed a correction factor f_0 given by Equation 2.14 for the Janbu simplified method (1957). When the factor of safety from the simplified method is multiplied with this correction factor, the result will be close to that from the *rigorous* analysis.

$$\begin{aligned}
&\text{For } c,\phi > 0, \quad f_0 = 1 + 0.5\left[\frac{D}{l} - 1.4\left(\frac{D}{l}\right)^2\right] \\
&\text{For } c = 0, \quad f_0 = 1 + 0.3\left[\frac{D}{l} - 1.4\left(\frac{D}{l}\right)^2\right] \qquad (2.14) \\
&\text{For } \phi = 0, \quad f_0 = 1 + 0.6\left[\frac{D}{l} - 1.4\left(\frac{D}{l}\right)^2\right]
\end{aligned}$$

For the correction factor shown in this equation and Figure 2.3, l is the length joining the left and right exit points, while D is the maximum thickness of the failure zone with reference to this line. Since the correction factors by Janbu (1973) are based on limited case studies, the use of these factors on to complicated non-homogeneous slopes is questioned by some engineers. Since the interslice shear force can sometimes generate a high factor of safety for some complicated cases, which may occur in dam and hydropower projects, the use of the Janbu method (1957) is preferred over other methods in these kinds of projects in China.

The Lowe and Karafiath method (1960) and the Corps of Engineers method are based on the interslice force functions type 5 and type 6. These two methods satisfy only force equilibrium but not moment equilibrium. In general, the Lowe and Karafiath method (1960) will give results close

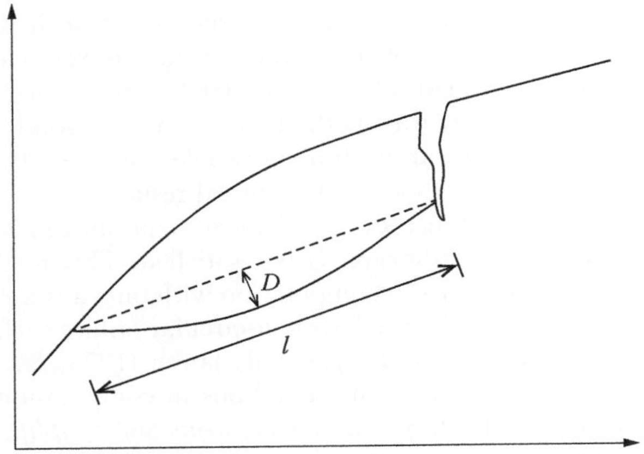

Figure 2.3 Definitions of *D* and *l* for the correction factor in the Janbu simplified method.

to those from the *rigorous* methods even though the moment equilibrium is not satisfied. For the Corps of Engineers method, it may lead to a high factor of safety in some cases, and some engineers actually adopt a lower interslice force angle to account for this problem (Duncan and Wright, 2005); this practice is also adopted by some engineers in China. The load transfer and the wedge methods in Table 2.5 satisfy only the force equilibrium. These methods are used in China only and are seldom adopted in other countries.

The Morgenstern–Price method is usually based on the interslice force function types 1 to 4, though the use of arbitrary function is possible and is occasionally used. If type 1 interslice force function is used, this method will reduce to the Spencer method. The Morgenstern–Price method satisfies the force and the global moment equilibrium. Since the local moment equilibrium equation is not used in the formulation, the internal forces of an individual slice may not be acceptable. For example, the line of thrust (centroid of the interslice normal force) may fall outside the soil mass from the Morgenstern–Price analysis. The GLE method is basically similar to the Morgenstern–Price method, except that the line of thrust is determined and is closed at the last slice. The acceptability of the line of thrust for any intermediate slice may still be unacceptable from the GLE analysis. In general, the results from these two methods of analysis are very close.

The Janbu rigorous method (1973) appears to be appealing in that the local moment equilibrium is used in the intermediate computation. The internal forces will hence be acceptable if the analysis can converge. As suggested by Janbu (1973), the line of thrust ratio is usually taken as one-third of the interslice height, which is basically compatible with the classical lateral earth pressure distribution. It should be noted that the equilibrium of the last slice is actually not checked in the Janbu rigorous method (1973), so the local moment equilibrium from the Janbu rigorous method (1973) is not strictly rigorous. A practical limitation of this method is the relatively poor convergence in the analysis, particularly when the failure surface is highly irregular or there are external loads. This is due to the fact that the line of thrust ratio is predetermined, with no flexibility in the analysis. The constraints in the Janbu rigorous method (1973) are more than those in the other methods, hence convergence is usually poorer. If the method is slightly modified by assuming $h_t/h = \lambda f(x)$, where h_t = height of line of thrust above slice base and h = length of the vertical interslice, the convergence of this method will be improved. There is, however, difficulty in defining $f(x)$ for the line of thrust, and hence this approach is seldom considered. Cheng has developed another version of the Janbu rigorous method, which is implemented in the program SLOPE 2000 and will be discussed in Chapter 8.

For the Janbu rigorous method (1973) and the Leshchinsky method (1985), Φ (or λ equivalently) is not known in advance. The relationship between the line of thrust Xp and the angle Φ in the Janbu rigorous method (1973) can be derived in the following ways:

a. Taking moment approximately in the middle of the slice base in the Janbu rigorous method (1973), the moment equilibrium condition is given by

$$W_i Xw_i + Vl_i Xs_i - HL_i Xh_i = E_{i,i+1} h_i - E_{i-1,i} h_{i-1} + \frac{1}{2}\left(E_{i,i+1} \tan \Phi_{i,i+1} + E_{i-1,i} \tan \Phi_{i-1,i}\right) B_i \tag{2.15}$$

From this, the interslice normal force is obtained as

$$E_{i,i+1} = \frac{Al_i}{2h_i + B_i \tan \Phi_{i,i+1}} \tag{2.16}$$

where

$$Al_i = 2W_i Xw_i + 2VL_i Xs_i - 2VL_i Xh_i + 2E_{i-1,i} h_{i-1} - B_i E_{i-1,i} \tan \Phi_{i-1,i} \tag{2.17}$$

From Equation 2.9, the interslice normal force is also obtained as

$$E_{i,i+1} = \frac{A2_i}{-f_i \sin \alpha_i - f_i \cos \alpha_i \tan \Phi_{i,i+1} + K \cos \alpha_i + K \sin \alpha_i \tan \Phi_{i,i+1}} \tag{2.18}$$

where

$$A2_i = \left(J_i + G_i \times F\right) E_{i-1,i} + M_i + L_i + K_i F \tag{2.19}$$

From Equations 2.15 and 2.17, the relation between the line of thrust h_i and the angle Φ is given by

$$\tan \Phi_{i,i+1} = -\frac{-2A2_i h_i - Al_i f_i \sin \alpha_i + Al_i F \cos \alpha_i}{-A2_i B_i - Al_i f_i \cos \alpha_i + Al_i F \sin \alpha_i} \tag{2.20}$$

b. For the Leshchinsky method (1985), where the distribution of the base normal force N is assumed to be known, Φ can then be determined as

$$\tan \Phi_{i,i+1} = \frac{-P_i f_i \sin \alpha_i + P_i F \cos \alpha_i - E_{i-1,i} F \tan \Phi_{i-1,i} - c_i A_i \sin \alpha_i - W_i F - VL_i F}{-P_i f_i \cos \alpha_i + P_i F \sin \alpha_i + E_{i-1,i} F - c_i A_i \sin \alpha_i + VL_i F} \tag{2.21}$$

Once Φ is obtained from Equation 2.19 or 2.20, the calculation can then proceed as described previously.

2.2.4 Solution of non-linear factor of safety equation

In Equation 2.11, the interslice normal force for slice i $E_{i,i+1}$ is controlled by the interslice normal $E_{i-1,i}$. If we put the equation for interslice normal force $E_{1,2}$ (Equation 2.9) from slice 1 into the equation for interslice normal force $E_{2,3}$ for slice 2 (Equation 2.11), we will get a second-order equation in the factor of safety F as

$$E_{2,3} = \frac{\left(J_2 \times f_2 + G_2 \times F\right) E_{1,2} + L_2 + K_2 \times F + M_2}{H_2 + E_2 \times F} \tag{2.22}$$

The term $(J_2 \times f_2 + G_2 \times F)\ E_{1,2}$ is a second order function in F. The numerator on the right-hand side of Equation 2.22 is hence a second-order function in F. Similarly, if we put the equation $E_{2,3}$ into the equation for $E_{3,4}$, a third-order equation in F will be achieved. If we continue this process to the last slice, we will arrive at a polynomial for F and the order of the polynomial is n for $E_{n,n+1}$, which is just 0! Sarma (1987) has also arrived at similar conclusion for a simplified slope model. The importance of this polynomial under the present formulation is that there are n possible factors of safety for any prescribed Φ. Most of the solutions will be physically unacceptable and are either imaginary numbers or negative solutions. Physically acceptable factors of safety are given by the positive real solutions from this polynomial.

In Equation 2.22, λ and F are the two unknowns and they can be determined by several different methods. In most of the commercial programs, the factor of safety is obtained by the use of iteration with an initial trial factor of safety (usually 1.0), which is efficient and effective for most cases. The use of the iteration method is actually equivalent to expressing the complicated factor of safety polynomial in a functional form as

$$F = f(F) \tag{2.23}$$

Chen and Morgenstern (1983) and Zhu et al. (2001) have proposed the use of the Newton–Rhapson technique in the evaluation of the factor of safety F and λ. The gradient-type methods are more complicated in the formulation but are fast in solution. Chen and Morgenstern (1983) suggested that the initial trial λ can be chosen as the tangent of the average base angle of the failure surface (which may, however, be unreasonable if soil nail or external loads are present), and these two values can be determined by the use of the Newton–Rhapson method. Chen and Morgenstern (1983) have also provided the expressions for the derivatives of the moment and shear terms required for the Newton–Rhapson analysis. Zhu et al. (2001) admitted that the initial trials of F and λ can greatly affect the efficiency of the computation. In some cases, poor initial trials can even lead to divergence in analysis. Zhu et al. (2001) proposed a technique to estimate the initial trial value, which appears to work fine for smooth failure surfaces, but this method can also break down and diverge, which is a fundamental problem in the LEM. The authors' experience is that for non-smooth or deep-seated failure surface, it is not easy to estimate a good initial trial value, and Zhu's proposal may not work for these cases. The fundamental problem of convergence will be further discussed in Sections 3.10 and 3.12.

As an alternative, Cheng and Zhu (2005) have proposed that the factor of safety based on the force equilibrium is determined directly from the polynomial as discussed earlier, and this can avoid the problems that may be encountered using the Newton–Rhapson method or iteration method. The present proposal can be effective with under difficult problems while Chen's (1983) and Zhu's (2001) methods are more efficient for general smooth failure surface. The additional advantage of the present proposal is that it can be applied to many slope stability analysis methods if the unified formulation is adopted. To solve for the factor of safety, the following steps can be used:

1. From slice 1 to n, based on an assumed value of λ and $f(x)$ and hence Φ for each interface, the factors of safety can be determined from the polynomial by the Gauss–Newton method with a line search step selection. The internal forces E, X, P and T can then be determined from Equations 2.5 through 2.11 *without* using any iteration analysis. The special feature of the present technique is that while determination of interslice forces is required for calculating the factor of safety in iterative analysis (for rigorous methods), the factor of safety is determined directly. Since the Bishop analysis does not satisfy horizontal force equilibrium, the present method cannot be applied to Bishop analysis. This is not important as the Bishop method can be solved easily by the classical iterative algorithm.
2. For rigorous methods, moment equilibrium has to be checked. Based on the internal forces as determined in step 1 for a specific physically acceptable factor of safety, the moment equilibrium in Equation 2.13 is checked. If moment equilibrium is not satisfied with that specific factor of safety based on the force equilibrium, the step should be repeated with the next factor of safety for checking the moment equilibrium.
3. If no acceptable factor of safety is found, try the next λ and repeat steps 1 and 2. In the actual implementation, the sign of the unbalanced moment from Equation 2.13 is monitored against λ and interpolation is used to accelerate the determination of λ, which satisfies the moment equilibrium.
4. For the Janbu rigorous method (1973) or the Leshchinsky method (1985), Equations 2.20 and 2.21 have to be used in the procedures during each step of analysis.

It will be demonstrated in Chapter 3 that there are many cases where iteration analysis may fail to converge but the factors of safety actually exist.

On the other hand, using the Gauss–Newton method and the polynomial form by Cheng and Zhu (2005) or the matrix form and the double QR method by Cheng (2003), it is possible to determine the factor of safety without iteration analysis. The root of the polynomial (factor of safety) close to the initial trial can be determined directly by the Gauss–Newton method. For the double QR method, the factor of safety and the internal forces are determined directly without the need of any initial trial at the expense of computer time for solving the matrix equation. Zheng et al. (2009) have also suggested procedures to solve the factor of safety, which are similar to the procedure suggested earlier.

Based on the fact that the interslice forces at any section are the same for the slices to the left and to the right of that section, an overall equation can be assembled in a way similar to that in the stiffness method, which will result in a matrix equation (Cheng, 2003). The factor of safety equation as given by Equation 2.22 can be cast into a matrix form instead of a polynomial (actually equivalent). The complete solution of all the real positive factors of safety from the matrix can be obtained by the double QR method suggested by Cheng (2003), which is a useful numerical method to calculate all the roots associated with the Hessenberg matrix arising from Equation 2.22. It should be noted that imaginary numbers may satisfy the factor of safety polynomial, so the double QR method instead of the classical QR method is necessary to determine the real positive factors of safety. If an F value from the double QR analysis cannot satisfy the above requirement, the next F value will be computed. Processes 1 to 4 will continue until all the possible F values are examined. If no factor of safety based on the force equilibrium can satisfy the moment equilibrium, λ will increase. The analysis is assumed to fail in convergence if λ takes a very large value, which is not realistic.

The advantage of the present method is that the factor of safety and the internal forces with respect to force equilibrium are obtained directly *without* any iteration analysis. Cheng (2003) has also demonstrated that there can be at most n possible factors of safety (including negative value and imaginary number) from the double QR analysis for a failure mass with n slices. The actual factor of safety can be obtained from the force and moment balance at a particular λ value. The time required for the double QR computation is not excessively long as interslice normal and shear forces are not required to be determined for obtaining a factor of safety. In general, if the number of slices used for the analysis is less than 20, the solution time for the double QR method is only 50%–100% longer than the iteration method. Cheng has also developed a faster version of double QR method suitable for problems with more slices.

Since all the possible factors of safety are examined, this method is the ultimate method in the determination of the factor of safety. If other methods of analysis fail to determine the factor of safety, this method may still work, which will be demonstrated in Section 3.10. On the other hand,

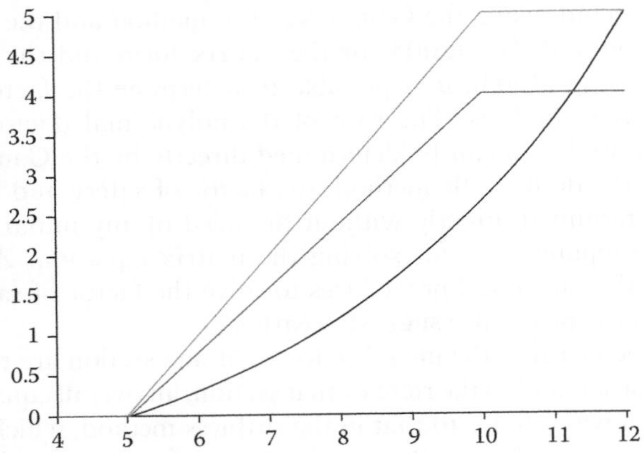

Figure 2.4 Numerical examples for a simple slope.

if no physically acceptable solution is found from the double QR method, the problem under consideration has no solution because of the nature of the specified $f(x)$. Failure to converge is thus equivalent to the specification of an unsuitable $f(x)$. More discussion about the use of the double QR method and convergence will be provided in Section 3.12.

2.2.5 Examples of slope stability analysis

Figure 2.4 is a simple slope given by coordinates (4,0), (5,0), (10,5) and (12,5) while the water table is given by (4,0), (5,0), (10,4) and (12,4). The soil parameters are unit weight = 19 kN/m^3, c' = 5 kPa and ϕ' = 36°. To define a circular failure surface, the coordinates of the centre of rotation and the radius should be defined. Alternatively, a better method is to define the x ordinates of the left and right exit ends and the radius of the circular arc. The latter approach is better as the left and right exit ends can usually be easily estimated by engineering judgement. In the present example, the x ordinates of the left and right exit ends are defined as 5.0 and 12.0 m while the radius is defined as 12 m. The soil mass is divided into 10 slices for analysis and the details are given in the following table.

Slice	Weight (kN)	Base angle (°)	Base length (m)	Base pore pressure (kPa)
1	2.50	16.09	0.650	1.57
2	7.29	19.22	0.662	4.52
3	11.65	22.41	0.676	7.09
4	15.54	25.69	0.694	9.26
5	18.93	29.05	0.715	10.99

6	21.76	32.52	0.741	12.23
7	23.99	36.14	0.774	12.94
8	25.51	39.94	0.815	13.04
9	32.64	45.28	1.421	7.98
10	11.77	52.61	1.647	0.36

The results of the analysis of the problem in Figure 2.4 are given in Table 2.6. For the Swedish method or the ordinary method of slices where only the moment equilibrium is considered while the interslice shear force is neglected, the factor of safety from the global moment equilibrium takes the simple form:

$$F_{\mathrm{m}} = \frac{\Sigma\left(c'l + (W\cos\alpha - ul)\tan\phi'\right)}{\Sigma W\sin\alpha} \tag{2.24}$$

A factor of safety of 0.991 is obtained directly from the Swedish method for this example without any iteration. For the Bishop method, which assumes the interslice shear force X to be zero, the factor of safety by the global moment equilibrium will reduce to

$$F_{\mathrm{m}} = \frac{\Sigma\left(c'b + (W - ul)\tan\phi'\right)\sec\alpha / m_{\alpha}}{\Sigma W\sin\alpha} \tag{2.25}$$

where

$$m_{\alpha} = \cos\alpha\left(1 + \tan\alpha\frac{\tan\phi'}{F}\right)$$

Based on an initial factor of safety 1.0, the successive factors of safety during the Bishop iteration analysis are 1.0150, 1.0201, 1.0219, 1.0225 and 1.0226. For the Janbu simplified method (1957), the factor of safety based on force equilibrium using the iteration analysis takes the form

$$F_{\mathrm{f}} = \frac{\Sigma\left[c'b + (W - ub)\tan\phi'\right]/n_{\alpha}}{\Sigma W\tan\alpha} \quad \text{and} \quad n_{\alpha} = \cos\alpha \cdot m_{\alpha} \tag{2.26}$$

Table 2.6 Factors of safety for the failure surface shown in Figure 2.4

F	*Bishop*	*Janbu simplified*	*Janbu rigorous*	*Swedish*	*Load factor*	*Sarma*	*Morgenstern price*
	1.023	1.037	1.024	0.991	1.027	1.026	1.028

Note: The correction factor is applied to the Janbu simplified method. The results for the Morgenstern–Price method using $f(x) = 1.0$ and $f(x) = \sin(x)$ are the same. Tolerance in iteration analysis is 0.0005.

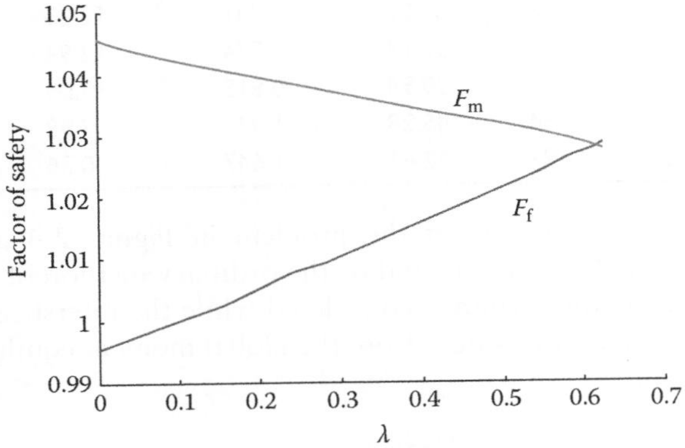

Figure 2.5 Variation of F_f and F_m with respect to λ for the example in Figure 2.4.

The successive factors of safety during the iteration analysis using the Janbu simplified method are 0.9980, 0.9974 and 0.9971. Based on a correction factor of 1.0402, the final factor of safety from the Janbu simplified analysis is 1.0372. If the double QR method is used for the Janbu simplified method (1957), a value of 0.9971 is obtained directly from the first positive solution of the Hessenberg matrix without using any iteration analysis. For the Janbu rigorous method (1973), the successive factors of safety based on the iteration analysis are 0.9980, 0.9974, 0.9971, 1.0102, 1.0148, 1.0164, 1.0170, 1.0213, 1.0228, 1.0233 and 1.0235. For the Morgenstern–Price method (1965), a factor of safety of 1.0282 and the internal forces are obtained directly from the double QR method without any iteration analysis. The variation of F_f and F_m with respect to λ using the iteration analysis for this example is shown in Figure 2.5. It should be noted that F_f is usually more sensitive to λ than F_m in general, and the two lines may not meet for some cases, in which case one may consider that there is no solution to the problem. There are cases where the lines are very close but actually do not intersect. If a large enough tolerance is defined, then the two lines can be considered to have an intersection and the solutions converge. This type of *false* convergence is experienced by many engineers in Hong Kong. These two lines may be affected by the choice of the moment point, and convergence can sometimes be achieved by adjusting the choice of the moment point. The results as shown in Figure 2.5 assume the interslice shear forces to be zero in the first solution step, and this solution procedure appears to be adopted in many commercial programs. Cheng et al. (2008a) have, however, found that the results shown in Figure 2.5 may not be the true results for some special cases, and this will be further discussed in Chapter 3.

From Table 2.6, it is clear that the Swedish method is very conservative as first suggested by Whitman and Bailey (1967). Besides, the Janbu simplified method (1957) will also give a smaller factor of safety if the correction factor is not used. After the application of the correction factor, Cheng found that the results from the Janbu simplified method (1957) are usually close to those of the *rigorous* methods. In general, the factors of safety from different methods of analysis are usually similar as pointed out by Morgenstern (1992).

2.3 MISCELLANEOUS CONSIDERATIONS ON SLOPE STABILITY ANALYSIS

2.3.1 Acceptability of the failure surfaces and results of the analysis

Based on an arbitrary interslice force function, the internal forces which satisfy both the force and moment equilibrium may not be kinematically acceptable. The acceptability conditions of the internal forces include the following:

1. Since the Mohr–Coulomb relation is not used along the vertical interfaces between different slices, it is possible, though not common, that the interslice shear forces and normal forces may violate the Mohr–Coulomb relation.
2. Except for the Janbu rigorous method and the extremum method as discussed in Section 3.12 under which the resultant interslice normal force must be acceptable, the line of thrust from other *rigorous* methods that are based on overall moment equilibrium may lie outside the failure mass and is unacceptable.
3. The interslice normal forces should not be in tension. For the interslice normal forces near the crest of the slope where the base inclination angles are usually high, if c' is high, it is highly likely that the interslice normal forces will be in tension in order to maintain the equilibrium. This situation can be eliminated by the use of tension crack. Alternatively, the factor of safety with tensile interslice normal forces for the last few slices may be accepted, as the factor of safety is usually not sensitive to these tensile forces. On the other hand, tensile interslice normal forces near the slope toe are usually associated with special shape failure surfaces with kinks, steep upward slope at the slope toe or an unreasonably high/low factor of safety. The factors of safety associated with these special failure surfaces need special care in the assessment and should be rejected if the internal forces are unacceptable. Such failure surfaces should also be eliminated during the location of the critical failure surfaces.
4. The base normal forces may be negative near the toe and crest of the slope. For negative base normal forces near the crest of the slope,

the situation is similar to the tensile interslice normal forces and may be tolerable. For negative base normal forces near the toe of the slope, which is physically unacceptable, it is usually associated with deep-seated failure with a high upward base inclination. Since a very steep exit angle is not likely to occur, it is possible to limit the exit angle during automatic location of the critical failure surface.

If these criteria are strictly enforced on all the slices of a failure surface, many slip surfaces will fail to converge. One of the reasons is the effect of the last slice when the base angle is large. Based on the force equilibrium, a tensile interslice normal force will be created easily if c' is high. This result can propagate so that the results for the last few slices will be in conflict with the criteria described earlier. If the last few slices are not strictly enforced, the factor of safety will be acceptable when compared with other methods of analysis. A suggested procedure is that if the number of slices is 20, only the first 15 slices are checked against the criteria.

2.3.2 Tension crack

As the condition of limiting equilibrium develops with the factor of safety close to 1, a tension crack shown in Figure 2.3 may form near the top of the slope through which no shear strength can be developed. If the tension crack is filled with water, a horizontal hydrostatic force P_w will generate additional driving moment and driving force, which will reduce the factor of safety. The depth of a tension crack z_c can be estimated as

$$z_c = \frac{2c\sqrt{K_a}}{\gamma} \tag{2.27}$$

where K_a is the Rankine active pressure coefficient. The presence of a tension crack will tend to reduce the factor of safety of a slope, but the precise location of a tension crack is difficult to be estimated for a general problem. It is suggested that if a tension crack is required to be considered, it should be specified at different locations and the critical results can then be determined. Sometimes, the critical failure surface with and without a tension crack can differ appreciably, and the location of the tension crack needs to be assessed carefully. In SLOPE 2000 by Cheng or in some other commercial programs, the location of the tension crack can be varied automatically during the location of the critical failure surface.

2.3.3 Earthquake

Earthquake loadings are commonly modelled as vertical and horizontal loads applied at the centroid of the sliding mass, and the values are given by the earthquake acceleration factors k_v/k_h (vertical and horizontal) multiplied with the weight of the soil mass. This quasi-static simulation of earthquake loading is simple in implementation but should be sufficient for most design purposes, unless the strength of the soil may be reduced by more than 15% due to the impact of the earthquake. Beyond that, a more rigorous dynamic analysis may be necessary, which will be more complicated, and more detailed information about the earthquake acceleration as well as the soil constitutive behaviour is required. Usually, a single earthquake coefficient may be sufficient for the design, but a more refined earth dam earthquake code is specified in DL5073-2000 in China. The design earthquake coefficients will vary according to the height under consideration, which will be different for different slices. Though this approach appears to be more reasonable, most of the design codes and existing commercial programs do not adopt this approach. The program SLOPE 2000 by Cheng can, however, accept this special earthquake code.

2.3.4 Water

Increase in pore water pressure is one of the main factors of slope failure. Pore water pressure can be defined in several ways. The classical pore pressure ratio r_u is defined as $u/\gamma h$, and an average pore pressure for the whole failure mass is usually specified for the analysis. Several different types of stability design charts are also designed using an average pore pressure definition. The use of a constant averaged pore pressure coefficient is obviously a highly simplified approximation. With the advancement in computer hardware and software, the use of these stability design charts is now limited to preliminary designs only. Pore pressure coefficient is also defined as a percentage of the vertical surcharge applied on the ground surface in some countries. This definition of the pore pressure coefficient is, however, not commonly used.

If pore pressure is controlled by the groundwater table, u is commonly taken as $\gamma_w h_w$, where h_w is the height of the water table above the base of the slice. This is the most commonly used method of defining the pore pressure, which assumes that there is no seepage and the pore pressure is hydrostatic. Alternatively, a seepage analysis can be conducted and the pore pressure can be determined from the flow-net or the finite element analysis. This approach is more reasonable but less commonly adopted in practice due to the extra effort of performing a seepage analysis. More importantly, it is not easy to construct a realistic and accurate hydrogeological model to perform the seepage analysis.

Pore pressure can also be generated from the presence of a perched water table. In a multilayered soil system, a perched water table may exist together with the presence of a water table if there are great differences in the permeability of the soil. This situation is rather common for the slopes

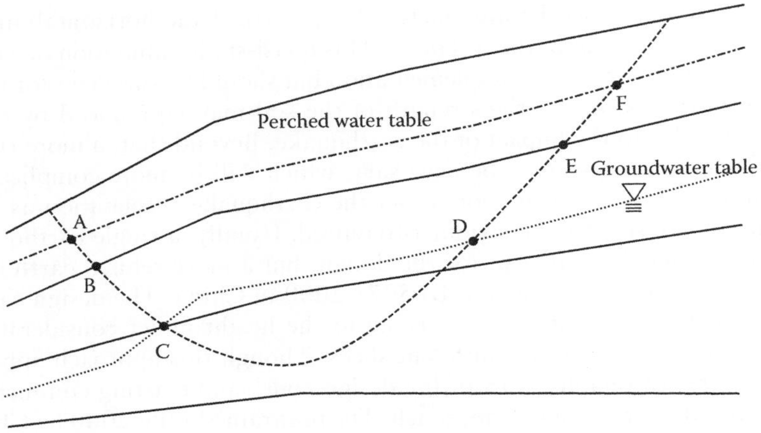

Figure 2.6 Perched water table in a slope.

in Hong Kong. For example, slopes at mid-levels in Hong Kong Island are commonly composed of fill at the top, which is underlain by colluvium and completely decomposed granite. Since the permeability of completely decomposed granite is 1–2 orders less than that for colluvium and fill, a perched water table can easily be established within the colluvium/fill zone during heavy rainfall while the permanent water table may be within the completely decomposed granite zone. Consider Figure 2.6, where a perched water table may be present in soil layer 1 with respect to the interface between soils 1 and 2 due to the permeability of soil 2 being 10 times less than that of soil 1. A slice base between A and B is subjected to the perched water table effect, and pore pressure should be included in the calculation. For a slice base between B and C, no water pressure is required in the calculation while the water pressure for a slice base between C and D is calculated using the ground water table only.

For the problem shown in Figure 2.7, if EFG, which is below ground surface, is defined as the groundwater table, the pore water pressure will be determined by EFG directly. If the groundwater table is above the ground surface and undrained analysis is adopted, the ground surface CDB is impermeable and the water pressure arising from AB will act as an external load on surface CDB. For drained analysis, the water table given by AB should be used, but vertical and horizontal pressures corresponding to the hydrostatic pressure should be applied on surfaces CD and DB. Thus, a trapezoidal horizontal and vertical pressure will be applied to surfaces CD and DB while the water table AB will be used to determine the pore pressure.

For the treatment of interslice forces, total stresses instead of effective stresses are generally used. This approach, though slightly less rigorous in formulation, can greatly simplify the analysis and is adopted in virtually all

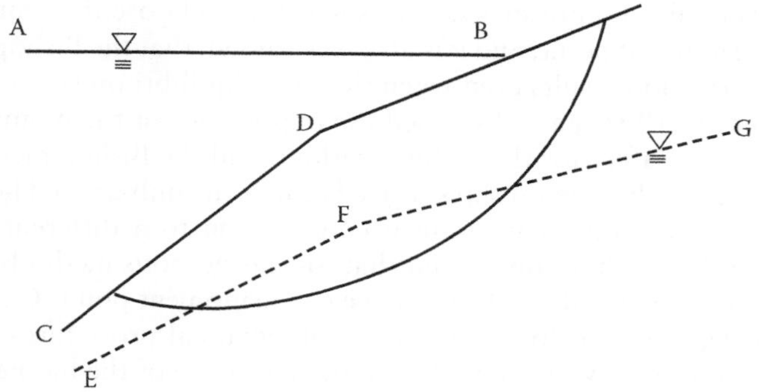

Figure 2.7 Modelling of ponded water.

commercial programs. Greenwood (1987) and Morrison and Greenwood (1989) have reported that this error is particularly significant where the slices have high base angles with a high water table. King (1989) and Morrison and Greenwood (1989) have also proposed revisions to the classical effective stress LEM. Duncan and Wright (2005) have also reported that some *simplified* methods can be sensitive to the assumption of the total or effective interslice normal forces in the analysis.

2.3.5 Saturated density of soil

The unit weights of the soil above and below the water table are not the same and may differ by 1–2 kN/m^3. For computer programs which cannot accept the input of saturated density, this can be modelled by the use of two different types of soil for a soil which is partly submerged. Alternatively, some engineers assume the two unit weights to be equal in view of the small differences between them.

2.3.6 Moment point

For simplified methods which satisfy only the force or moment equilibrium, the Janbu (1957) and Bishop methods (1955) are the most popular ones adopted by engineers. There is a perception among some engineers that the factor of safety from the moment equilibrium is more stable and is more important than the force equilibrium in stability formulation (Abramson et al., 2002). However, true moment equilibrium depends on the satisfaction of force equilibrium. Without force equilibrium, there is actually no moment equilibrium. Force equilibrium is, however, totally independent of moment equilibrium. For methods which satisfy only the moment equilibrium, the factor of safety actually depends on the choice of the moment point. For circular failure surfaces, it is natural to choose the centre of the circle as the moment point, and it is also well known that the Bishop method can yield very good results even when the force equilibrium is not satisfied. Fredlund et al. (1992) have discussed the importance of the moment point on the factor of safety for the Bishop method, and the Bishop method cannot be applied to the general slip surface because the unbalanced horizontal force will create a different moment contribution to a different moment point. Baker (1980) has pointed out that for the *rigorous* methods, the factor of safety is independent of the choice of the moment point. Cheng et al. (2008a) have, however, found that the mathematical procedures to evaluate the factor of safety may be affected by the choice of the moment point. Actually, many commercial programs allow the user to choose the moment point for analysis. The double QR method by Cheng (2003) is not affected by the choice of the moment point in the analysis and is a very stable solution algorithm.

2.3.7 Use of soil nail/reinforcement

Soil nailing is a slope stabilization method by introducing a series of thin elements called nails to resist tension, bending and shear forces in the slope. The reinforcing elements are usually made of round cross-sectional steel bars. Nails are installed sub-horizontally into the soil mass in a pre-bore hole, which is fully grouted. Occasionally, the initial portions of some nails are not grouted but this practice is not commonly adopted. Nails can also be driven into the slope, but this method of installation is uncommon in practice.

2.3.7.1 Advantages of soil nailing

Soil nailing presents the following advantages that have contributed to the widespread use of this technique:

- Economy: economical evaluation has led to the conclusion that soil nailing is a cost-effective technique compared to a tieback wall. Cost of soil nailing may be 50% of a tieback wall.
- Rate of construction: fast rates of construction can be achieved if adequate drilling equipment is employed. Shotcrete is also a rapid technique for placement of the facing.
- Facing inclination: there is virtually no limit to the inclination of the slope face.
- Deformation behaviour: observation of actual nailed structures demonstrated that horizontal deformation at the top of the wall ranges from 0.1% to 0.3% of the wall height for well-designed walls (Clouterre, 1991; Elias and Juran, 1991).

- Design flexibility: soil anchors can be added to limit the deformation in the vicinity of existing structures or foundations.
- Design reliability in saprolitic soils: saprolitic soils frequently present relict weak surfaces which can be undetected during site investigation. Hong Kong has encountered such a situation, which led to slope failures in such weak planes. Soil nailing across these surfaces may lead to an increased factor of safety and increased reliability, compared to other stabilization solutions.
- Robustness: deep-seated stability is maintained.

The fundamental principle of soil nail is the development of tensile force in the soil mass that renders the soil mass stable. Although only tensile force is considered in the analysis and design, soil nail function by a combination of tensile force, shear force and bending action is difficult to analyse. The use of finite element by Cheng et al. (2007a) has demonstrated that the bending and shear contribution to the factor of safety is generally not significant, and the current practice in soil nail design should be good enough for most cases. Nails are usually constructed at an angle of inclination from 10° to 20°. For ordinary steel bar soil nail, a thickness of 2 mm is assumed as the corrosion zone so that the design bar diameter is totally 4 mm less than the actual diameter of the bar according to the practice in Hong Kong. The nail is usually protected by galvanization, paint, epoxy and cement grout. For critical location, protection by expensive sleeving similar to that in rock anchor may be adopted. Alternatively, fibre-reinforced plastic (FRP), or carbon fibre–reinforced plastic (CFRP) may be used for soil nail, and these methods are currently under consideration.

The practical limitations of soil nails include the following:

1. Lateral and vertical movement may be induced from excavation and the passive action of the soil nail is not as effective as the active action of the anchor.
2. There may be difficulty in installation under some groundwater conditions.
3. Some engineers doubt the suitability of soil nail in loose fill as engineers – stress transfer between nail and soil is difficult to establish.
4. In some ground conditions, the drill hole may collapse before the nail is installed.
5. In the case of a very long nail hole, it is not easy to maintain the alignment of the drill hole.

There are several practices in the design of soil nails. One of the precautions in the adoption of soil nails is that the factor of safety of a slope without soil nail must be greater than 1.0 if the soil nail is going to be used. This is because the soil nail is a passive element, and the strength of the soil nail cannot be mobilized until soil tends to deform. The effective nail load is usually taken as the minimum of

a. The bond strength between cement grout and soil
b. The tensile strength of the nail, which is limited to 55% of the yield stress in Hong Kong, and 2 mm sacrificial thickness of the bar surface is allowed for corrosion protection
c. The bond stress between the grout and the nail

In general, only (a) and (b) are the controlling factors in design. The bond strength between cement grout and soil is usually based on one of the following criteria:

a. The effective overburden stress between grout and soil controls the unit bond stress on the soil nail, and it is estimated from the formula ($\pi c'D + 2D\sigma_v'\tan\phi'$) for practice in Hong Kong, while the Davis method (Shen et al., 1981) allows an inclusion of the angle of inclination, and D is the diameter of the grout hole. A safety factor of 2.0 is commonly applied to this bond strength in Hong Kong. During the calculation of the bond stress, only the portion behind the failure surface is taken into calculation.
b. Some laboratory tests suggest that the effective bond stress between nail and soil is relatively independent of the vertical overburden stress. This is based on the stress redistribution after the nail hole is drilled and the surface of the drill hole is stress free. The effective bond load will then be controlled by the dilation angle of the soil. Some of the laboratory tests in Hong Kong have shown that the effective overburden stress is not important for the bond strength. On the other hand, some field tests in Hong Kong have shown that the nail bond strength depends on the depth of embedment of the soil nail. It appears that the bond strength between cement grout and soil may be governed by the type of soil, method of installation and other factors, and the bond strength may depend on the overburden height in some cases, but this is not a universal behaviour.
c. If the bond load is independent of the depth of embedment, the effective nail load will then be determined in a proportional approach shown in Figure 2.8.

For a soil nail of length L, bonded length L_b and total bond load T_{sw}, L_e for each soil nail and T_{mob} for each soil nail are determined from the following formula:

For slip 1:

$$T_{mob} = T_{sw}$$

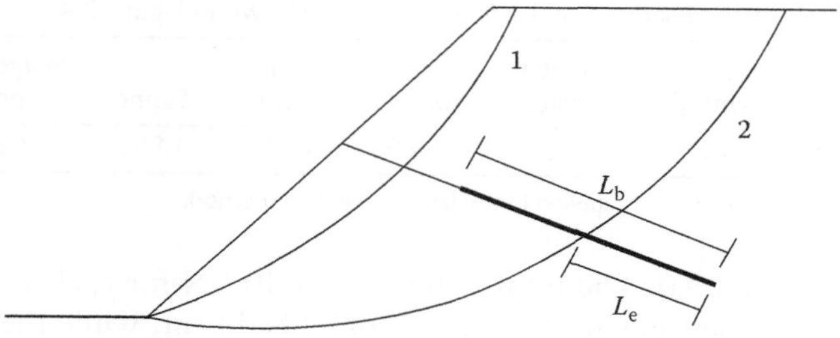

Figure 2.8 Definition of effective nail length in the bond load determination.

In this case, the slip passes in front of the bonded length and the full magnitude is mobilized to stabilize the slip.

For slip 2:

$$T_{mob} = T_{sw} \times \left(\frac{L_e}{L_b} \right)$$

In this case, the slip intersects the bonded length and only a proportion of the full magnitude provided by the nail length behind the slip is mobilized to stabilize the slip.

The effective nail load is usually applied as a point load on the failure surface in the analysis. Some engineers, however, model the soil nail load as a point load at the nail head or as a distributed load applied on the ground surface. In general, there is no major difference in the factors of safety from these minor variations in treating the soil nail forces.

The effectiveness of soil nail can be illustrated by adding 2 rows of 5 m length soil nails inclined at an angle of 15° to the problem shown in Figure 2.4, which is shown in Figure 2.9. The x ordinates of the nail

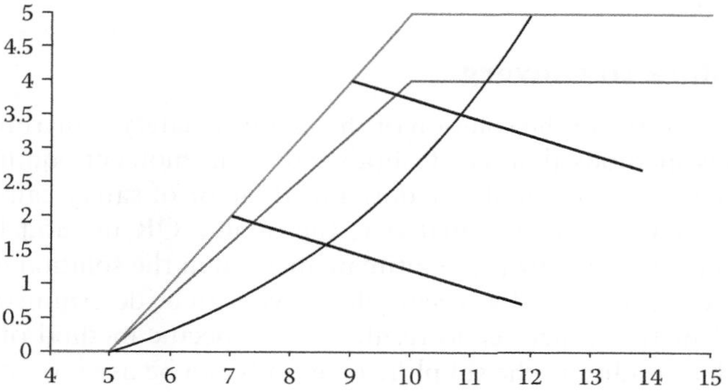

Figure 2.9 Two rows of soil nail are added to the problem in Figure 2.4.

Table 2.7 Factor of safety for the failure surface shown in Figure 2.4

F	Bishop	Janbu simplified	Janbu rigorous	Swedish	Load factor	Sarma	Morgenstern price
	1.807	1.882	Fail	1.489	1.841	1.851	1.810

Note: Correction factor is applied in the Janbu simplified method.

heads are 7.0 and 9.0. The total bond load is 40 kN for each nail, which is taken to be independent of the depth of embedment, while the effective nail loads are obtained as 27.1 and 24.9 kN considered by a simple proportion as given in Figure 2.8. The results of analysis shown in Table 2.7 have illustrated that (1) the Swedish method is a conservative method in most cases and (2) the Janbu rigorous method (1973) is more difficult to converge compared to other methods. It is also noticed that when external load is present, there are greater differences between the results from different methods of analysis.

During the computation of the factor of safety, this value can be defined as

$$F = \frac{\text{Shear strength}}{\text{Mobilized shear} - \text{Contribution from reinforcement}} \tag{2.28a}$$

or

$$F = \frac{\text{Shear strength} + \text{Contribution from reinforcement}}{\text{Mobilized shear}} \tag{2.28b}$$

The results shown in Table 2.7 are based on Equation 2.28a, which is the more popular definition of the factor of safety with soil reinforcement. Some commercial software also offers an option for Equation 2.28b, and engineers must be clear about the definition of the factor of safety. In general, the factor of safety using Equation 2.28a will be greater than that based on Equation 2.28b.

2.3.8 Failure to converge

Failure to converge in the solution of the factor of safety sometimes occurs for *rigorous* methods that satisfy both force and moment equilibrium. If this situation is encountered, the initial trial factor of safety can be varied to achieve convergence. Alternatively, the double QR method by Cheng (2003) can be used as this is the ultimate method in the solution of the factor of safety. If no physically acceptable answer can be determined from the double QR method, there is no result for the specific method of analysis. Under such a condition, the simplified methods can be used to estimate the factor of safety or the extremum principle in Section 3.12 may be adopted to determine the factor of safety. The convergence problem of the *rigorous* method will be studied in more detail in Sections 3.10 and 3.12, and further case studies are provided in the user guide of SLOPE 2000.

2.3.9 Location of the critical failure surface

The minimum factor of safety as well as the location of the critical failure surface are required for the proper design of a slope. For a homogeneous slope with a simple geometry and no external load, the log-spiral failure surface will be a good solution for the critical failure surface. In general, the critical failure surface for a sandy soil with a small c' value and high ϕ' will be close to the ground surface while the critical failure surface will be a deep-seated one for a soil with a high c' value and small ϕ'. With the presence of external vertical load or soil nail, the critical failure surface will generally drive the critical failure surface deeper into the soil mass. For a simple slope with a heavy vertical surcharge on top of the slope (typical abutment problem), the critical failure surface will be approximately a two-wedge failure from the non-circular search. This failure mode is also specified by the German code for abutment design. For a simple slope without any external load or soil nail, the critical failure surface will usually pass through the toe. Based on these characteristics of the critical failure surface, engineers can manually locate the critical failure surface easily for a simple problem. The use of the factor of safety from the critical circular or log-spiral failure surface (Frohlich, 1953; Chen, 1975), which will be slightly higher than that from the non-circular failure surface, is also adequate for simple problems.

For complicated problems, however, these guidelines may not be applicable, and it will be tedious to carry out the manual trial and error in locating the critical failure surface. An automatic search for the critical circular failure surface is available in nearly all commercial slope stability programs. A few commercial programs also offer an automatic search for a non-circular critical failure surface with some limitations. Since the automatic determination of the effective nail load (controlled by the overburden stress) appears to be unavailable in most commercial programs, engineers often have to perform the search for the critical failure surface by manual trial and error and the effective nail load is separately determined for each trial failure surface. To save time, only limited failure surfaces will be considered in the routine design. The authors have found that relying only on the manual trial and error in locating the critical failure surface may not be adequate, and the adoption of the modern optimization methods to overcome this problem will be discussed in Chapter 3.

2.3.10 Three-dimensional analysis

All failure mechanism is 3D in nature but 2D analysis is performed at present. The difficulties associated with true 3D analysis are (1) sliding direction,

(2) satisfaction of 3D force and moment equilibrium, (3) relating the factor of safety to the previous two factors and (4) extensive computational geometrical calculations that are required. At present, there are still many practical limitations in the adoption of 3D analysis, and there are only a few 3D slope stability programs that are suitable for ordinary use. Simplified 3D analysis for symmetric slope is available in SLOPE 2000 by Cheng, and true 3D analysis for general slope is under development in SLOPE3D. Three-dimensional slope stability analysis will be discussed in detail in Chapter 5.

2.4 LIMIT ANALYSIS METHOD

The limit analysis adopts the concept of an idealized stress–strain relationship, that is, the soil is assumed as a rigid perfectly plastic material with an associated flow rule. Without carrying out the step-by-step elasto-plastic analysis, the limit analysis can provide solutions to many problems. Limit analysis is based on the bound theorems of classical plasticity theory (Drucker et al., 1951; Drucker and Prager, 1952). The general procedure of limit analysis is to assume a kinematically admissible failure mechanism for an upper bound solution or a statically admissible stress field for a lower bound solution, and the objective function will be optimized with respect to the control variables. Early efforts of limit analysis focused mainly on using the direct algebraic method or analytical method to obtain the solutions for slope stability problems with simple geometry and soil profile (Chen, 1975). Since closed-form solutions for most practical problems are not available, attention has been shifted to employing the slice techniques in the traditional limit equilibrium to the upper bound limit analysis (Michalowski, 1995; Donald and Chen, 1997).

Limit analysis is based on two theorems: (1) the lower bound theorem, which states that any statically admissible stress field will provide a lower bound estimate of the true collapse load, and (2) the upper bound theorem, which states that when the power dissipated by any kinematically admissible velocity field is equated with the power dissipated by the external loads, then the external loads are upper bounds on the true collapse load (Drucker and Prager, 1952).

A statically admissible stress field is one that satisfies the equilibrium equations, stress boundary conditions and yield criterion. A kinematically admissible velocity field is one that satisfies strain and velocity compatibility equations, velocity boundary conditions and the flow rule. When combined, the two theorems provide a rigorous bound on the true collapse load. Application of the lower bound theorem usually proceeds as follows: (1) First, a statically admissible stress field is constructed. Often it will be a discontinuous field in the sense that we have a patchwork of regions of constant stress that together cover the whole soil mass. There will be one or more particular values of stress that are not fully specified by the conditions of equilibrium. (2) These unknown stresses are then adjusted so that the load on the soil is maximized but the soil remains unyielded. The resulting load becomes the lower bound estimate for the actual collapse load.

Stress fields used in lower bound approaches are often constructed without a clear relation to the real stress fields. Thus, lower bound solutions for practical geotechnical problems are often difficult to find. Collapse mechanisms used in the upper bound calculations, however, have a distinct physical interpretation associated with actual failure patterns and thus have been extensively used in practice.

2.4.1 Lower bound approach

The application of the conventional analytical limit analysis was usually limited to simple problems. Numerical methods, therefore, have been employed to compute the lower and upper bound solutions for the more complex problems. The first lower bound formulation based on FEM was proposed by Lysmer (1970) for plain strain problems. The approach used the concept of finite element discretization and linear programming. The soil mass was subdivided into simple three-node triangular elements where the nodal normal and shear stresses were taken as the unknown variables. The stresses were assumed to vary linearly within an element, while stress discontinuities were permitted to occur at the interface between adjacent triangles. The statically admissible stress field was defined by the constraints of the equilibrium equations, stress boundary conditions and linearized yield criterion. Each nonlinear yield criterion was approximated by a set of linear constraints on the stresses that lie inside the parent yield surface, thus ensuring that the solution was a strict lower bound. This led to an expression for the collapse load, which was maximized subject to a set of linear constraints on the stresses. The lower bound load could be solved by optimization, using the techniques of linear programming. Other investigations have worked on similar algorithms (Anderheggen and Knopfel, 1972; Bottero et al., 1980). The major disadvantage of these formulations was the linearization of the yield criterion, which generated a large system of linear equations, and required excessive computational times, especially if the traditional simplex or revised simplex algorithms were used (Sloan, 1988a). Therefore, the scope of the early investigations was mainly limited to small-scale problems.

Efficient analyses for solving numerical lower bounds by FEM and linear programming method have been developed recently (Bottero et al., 1980; Sloan, 1988a,b). The key concept of these analyses was the introduction of an active set algorithm (Sloan, 1988b) to solve the linear programming problem where the constraint matrix was sparse. Sloan (1988b) has shown that the active set algorithm was ideally suited to the numerical lower bound formulation and could solve a large-scale linear programming problem efficiently. A second problem associated with the numerical lower bound solutions occurred when dealing with statically admissible conditions for an infinite half space. Assdi and Sloan (1990) have solved this problem by adopting the concept of infinite elements, hence obtaining rigorous lower bound solutions for general problems.

Lyamin and Sloan (1997) proposed a new lower bound formulation which used linear stress finite elements, incorporating non-linear yield conditions, and exploiting the underlying convexity of the corresponding optimization problem. They showed that the lower bound solution could be obtained efficiently by solving the system of non-linear equations that define the Kuhn–Tucker optimality conditions directly.

Recently, Zhang (1999) presented a lower bound limit analysis in conjunction with another numerical method – the rigid finite element method (RFEM) to assess the stability of slopes. The formulation presented satisfies both static and kinematical admissibility of a discretized soil mass without requiring any assumption. The non-linear programming method is employed to search for the critical slip surface.

2.4.2 Upper bound approach

Implementation of the upper bound theorem is generally carried out as follows: (1) First, a kinematically admissible velocity field is constructed. No separations or overlaps should occur anywhere in the soil mass. (2) Second, two rates are calculated: the rate of internal energy dissipation along the slip surface and discontinuities that separate the various velocity regions, and the rate of work done by all the external forces, including gravity forces, surface tractions and pore water pressures. (3) Third, these two rates are set to be equal. The resulting equation, called energy–work balance equation, is solved for the applied load on the soil mass. This load would be equal to or greater than the true collapse load.

The first application of the upper bound limit analysis to the slope stability problem was by Drucker and Prager (1952) to find the critical height of a slope. A failure plane was assumed, and analyses were performed for isotropic and homogeneous slopes with various angles. In the case of a vertical slope, it was found that the critical height obtained by the upper bound theorem was identical with that obtained by the LEM. Similar studies have been conducted by Chen and Giger (1971) and Chen (1975). However, their attention was mainly limited to a rigid body sliding along a circular or log-spiral slip surface passing both through the toe and below the toe in cohesive materials. The stability of slopes was evaluated by the stability factor, which could be minimized using an analytical technique.

Karel (1977a,b) presented an energy method for soil stability analysis. The failure mechanisms used in the method included (1) a rigid zone with a planar or a log-spiral transition layer, (2) a soft zone confined by plane or log-spiral surfaces and (3) a composed failure mechanism consisting of rigid and soft zones. The internal dissipation of energy occurred along the transition layer for the rigid zone, within the zone and along the transition layer for the soft zone. However, no numerical technique was proposed to determine the least upper bound of the factor of safety.

Izbicki (1981) presented an upper bound approach to the slope stability analysis. A translational failure mechanism confined by a circular slip surface in the form of rigid blocks similar to the traditional slice method was used. The factor of safety was determined by an energy balance equation and the equilibrium conditions of the field of force associated with the assumed kinematically admissible failure mechanism. However, no numerical technique was provided to search for the least upper bound of the factor of safety in the approach.

Michałowski (1995) presented an upper bound (kinematical) approach of limit analysis in which the factor of safety for slopes is derived that is associated with a failure mechanism in the form of rigid blocks analogous to the vertical slices used in traditional LEMs. A convenient way to include pore water pressure has also been presented and implemented in the analysis of both translational and rotational slope collapse. The strength of the soil between blocks was assumed explicitly to be zero or its maximum value set by the Mohr–Coulomb yield criterion.

Donald and Chen (1997) proposed another upper bound approach to evaluate the stability of slopes based on a multi-wedge failure mechanism. The sliding mass was divided into a small number of discrete blocks, with linear interfaces between the blocks and with either linear or curved bases to individual blocks. The factor of safety was iteratively calculated by equating the work done by external loads and body forces to the energy dissipated along the bases and interfaces of the blocks. Powerful optimization routines were used to search for the lowest factor of safety and the corresponding critical failure mechanism.

Other efforts have been made in solving the limit analysis problems by FEM, which represents an attempt to obtain the upper bound solution by numerical methods on a theoretically rigorous foundation of plasticity. Anderheggen and Knopfel (1972) appear to have developed the first formulation based on the upper bound theorem, which used constant-strain triangular finite elements and linear programming for plate problems. Bottero et al. (1980) later presented the formulation for plain-strain problems. In the formulation, the soil mass was discretized into three-node triangular elements whose nodal velocities were the unknown variables. Each element was associated with a specific number of unknown plastic multiplier rates. Velocity discontinuities were permitted along prespecified interfaces of adjacent triangles. Plastic deformation could occur within the triangular element and at the velocity discontinuities. Kinematically admissible velocity fields were defined by the constraints of compatibility equations, the flow rule of the yield criterion and the velocity boundary conditions. The yield criterion was linearized using a polygonal approximation. Thus, the finite element formulation of the upper bound theorem led to a linear programming problem whose objective function was the minimization of the collapse load and was expressed in terms of the unknown velocities and plastic multipliers. The upper bound loads were obtained using the revised simplex algorithm. Sloan (1988b, 1989) adopted the same basic formulation as Bottero et al. (1980) but solved the linear programming problem using an active set algorithm. The major problem encountered by Bottero et al. (1980) and Sloan (1988b, 1989) was caused by the incompressibility condition of perfectly plastic deformation. The discretization using linear triangular elements must be arranged such that four triangles form a quadrilateral with the central nodes lying at its centroid. Yu et al. (1994) have shown that this constraint can be removed using higher-order (quadratic) interpolation of the nodal velocities.

Another problem of the formulation used by Bottero et al. (1980) and Sloan (1988b, 1989) was that it could only handle a limited number of velocity discontinuities with prespecified directions of shearing. Sloan and Kleeman (1995) have made significant progress in developing a more general numerical upper bound formulation in which the direction of shearing was solved automatically during the optimization solution. Yu et al. (1998) compares rigorous lower and upper bound solutions with conventional limit equilibrium results for the stability of simple earth slopes.

Many researchers (Mroz and Drescher, 1969; Collins, 1974; Chen, 1975; Michalowski, 1989; Drescher and Detournay, 1993; Donald and Chen, 1997; Yu et al., 1998) pointed out that an upper bound limit analysis solution may be regarded as a special limit equilibrium solution but not vice versa. The equivalence of the two approaches plays a key role in the derivations of the limit load or factor of safety for materials following the non-associated flow rule.

Classically, algebraic expressions for the upper bound method are determined for simple problems. Assuming a log-spiral failure mechanism for failure surface A shown in Figure 2.10, the work done by the weight of the soil is dissipated along the failure surface based on the upper bound approach by Chen (1975) using an associated flow rule, and the height of the slope can be expressed as

$$H = \frac{c'}{\gamma} f(\phi', \alpha, \beta, \theta_h, \theta_0) \tag{2.29}$$

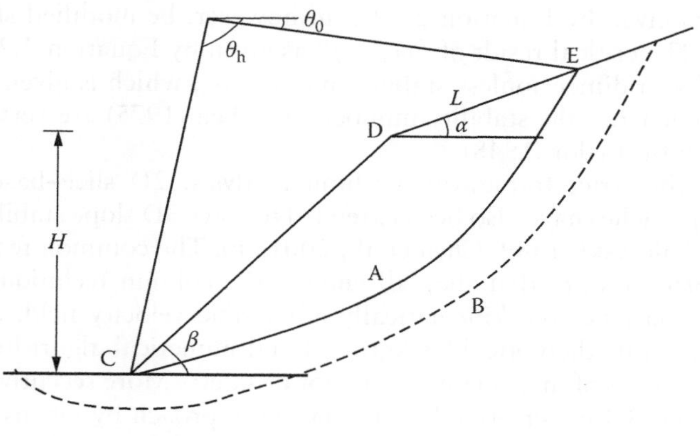

Figure 2.10 Critical log-spiral failure surface by limit analysis for a simple homogeneous slope.

where

$$f = \frac{\sin\beta\left\{\exp\left[2(\theta_h - \theta_0)\tan\phi\right] - 1\right\}}{2\sin(\beta - \alpha)\tan\phi(f_1 - f_2 - f_3)}$$

$$\times\left\{\sin(\theta_h + \alpha)\exp\left[(\theta_h - \theta_0)\tan\phi\right] - \sin(\theta_0 + \alpha)\right\}$$

$$f_1 = \frac{1}{3\left(1 + 9\tan^2\phi\right)}\left\{(3\tan\phi\cos\theta_h + \sin\theta_h)\right.$$

$$\left.\times\exp\left[3(\theta_h - \theta_0)\tan\phi\right] - (3\tan\phi\cos\theta_0 + \sin\theta_0)\right\}$$

$$f_2(\theta_h, \theta_0) = \frac{1}{6}\frac{L}{r_0}\left(2\cos\theta_0 - \frac{L}{r_0}\cos\alpha\right)\sin(\theta_0 + \alpha)$$

$$f_3(\theta_h, \theta_0) = \frac{1}{6}\exp\left[(\theta_h - \theta_0)\tan\phi\right]\left[\sin(\theta_h - \theta_0) - \frac{L}{r_0}\sin(\theta_h + \alpha)\right] *$$

$$\left\{\cos\theta_0 - \frac{L}{r_0}\cos\alpha + \cos\theta_h\exp\left[(\theta_h - \theta_0)\tan\phi\right]\right\}$$

The critical height of the slope is obtained by minimizing Equation 2.29 with respect to θ_0 and θ_h, which has been obtained by Chen (1975). Chen has also found that failure surface A is the most critical log-spiral failure surface unless β is small. When β and ϕ' are small, a deep-seated failure shown by failure surface B in Figure 2.10 may be more critical. The basic solution as given by Equation 2.29 can, however, be modified slightly in this case. The critical result of $f(\phi', \alpha, \beta)$ as given by Equation 2.29 can be expressed as a dimensionless stability number N_s, which is given by Chen (1975). In general, the stability numbers by Chen (1975) are very close to those given by Taylor (1948).

Within the strict framework of limit analysis, 2D slice–based upper bound approaches have also been extended to solve 3D slope stability problems (Michalowski, 1989; Chen et al., 2001a,b). The common features for these approaches are that they all employ the column techniques in 3D LEMs to construct the kinematically admissible velocity field, and have exactly the same theoretical background and numerical algorithm, which involve a process of minimizing the factor of safety. More recently, a promising 2D and 3D upper bound limit analysis approach by means of linear finite elements and non-linear programming (Lyamin and Sloan, 2002b) has emerged. The approach obviates the need to linearize the yield surface as adopted in the 2D approach using linear programming (Sloan, 1989; Sloan and Kleeman, 1995). However, the approach, nonetheless, involves stress while performing upper bound calculations.

2.5 RIGID ELEMENT METHOD

The REM originated from the rigid-body spring model (RBSM) proposed by Kawai (1977). More recently, Zhang and Qian (1993) used the RBSM to evaluate the static and dynamic stability of slopes or dam foundations within the framework of stress deformation analysis. Qian and Zhang (1995) and Zhang and Qian (1993) expanded the research field of REM to the stability analysis. Zhang (1999) performed a lower bound limit analysis in conjunction with the rigid elements to assess the stability of slopes. Recently, Zhuo and Zhang (2000) conducted a systematic study on the theory, methodologies and algorithms of REM, and demonstrated its application to a wide range of discontinuous mechanics problems with linear and non-linear material behaviour, beam and plate bending, as well as to the static and dynamic problems. It should be noted that this approach is known by different names such as RBSM, RFEM and interface element method; however, the general form *rigid element method* (REM) is adopted here. The REM provides an effective approach to the numerical analysis of the stability of soils, rocks or discontinuous media. Further studies and applications of REM are still being carried out, attracting the interest of many researchers.

The pre-processing and solution procedure in the REM are quite similar to that in the conventional FEM, except that the two main components in REM are elements and interfaces while they are nodes and elements in FEM.

In REM, each element is assumed to be rigid. The media under study is partitioned into a proper number of rigid elements mutually connected at the interfaces. Displacement of any point in a rigid element can be described as a function of the translation and rotation of the element centroid. The deformation energy of the system is stored only at the interfaces between rigid elements. The concept of contact *overlap*, though physically inadmissible because elements should not interpenetrate each other, may be accepted as a mathematical means to represent the deformability of the contact interfaces. In such a discrete model, though the relative displacements between adjacent elements show a discontinuous feature of deformation, the studied media can still be considered to be a continuum as a whole mass body.

In REM, the element centroid displacements are the primary variables, while in FEM the nodal displacements are selected. In the case of stress deformation analysis, the forces at the element interfaces are calculated in the REM, which are different from the Gauss point stress tensor calculated in FEM. Thus, while using the Mohr–Coulomb failure (yield) condition, the normal and shear stresses on each interface can be directly incorporated into the failure function for checking. This treatment in fact assumes that interfaces between the adjacent rigid elements may be the failure surfaces, and makes the calculation results quite sensitive to the mesh partition.

2.5.1 Displacements of the rigid elements

For the sake of convenience, a local reference coordinate system of *n-d-s* axes for the REM calculations is introduced. Consider the face in Figure 2.11, where the *n*-axis is pointing along the outward normal of the face; the *d*-axis is the dip direction (the steepest descent on the face); and the

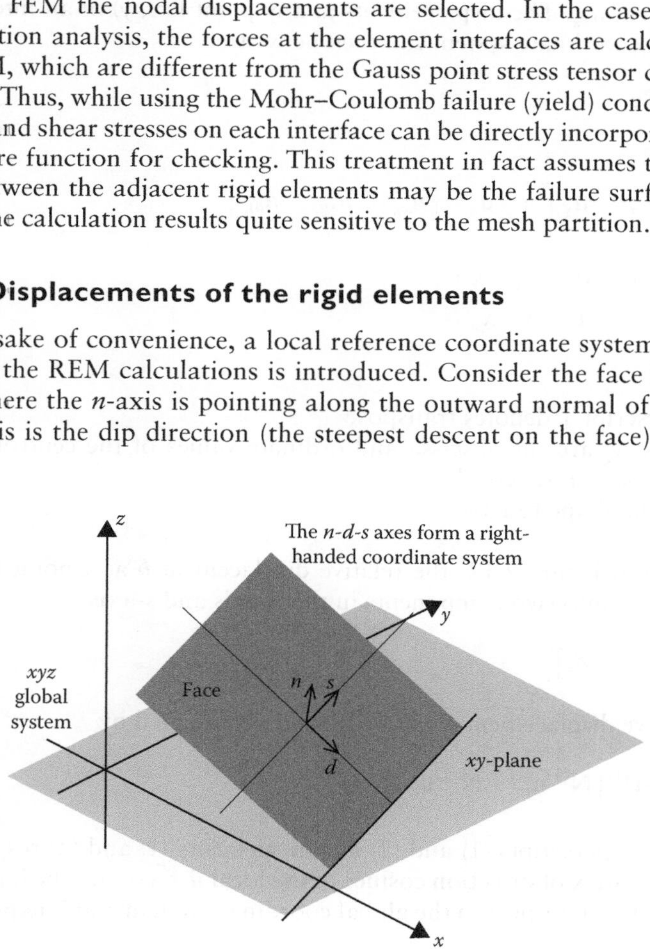

Figure 2.11 Local coordinate system defined by *n* (normal direction), *d* (dip direction) and *s* (strike direction).

s-axis is the strike direction (parallel to the projected intersection between the xy-plane and the face). The n-d-s axes form a right-handed coordinate system.

In order to illustrate the key features of rigid element analysis in a simple way, we restrict our attention to 2D computation in this chapter. In the 2D case, any point has two degrees of freedom, the x and y displacements, denoted as u_x and u_y. Each rigid element is associated with a 3D vector $\mathbf{u}_g$ of displacement variable at its centroid (similar to the DDA by Shi, 1996, Cheng et al., 1998 and Cheng and Zhang, 2000, 2002), that is, the rigid element has both translational displacements u_{xg} and u_{yg}, and rotational displacement $u_{\theta g}$. The displacements at any point P(x, y) of an interface in the global coordinate system can then be written as

$$\mathbf{u} = \mathbf{N}\mathbf{u}_g \tag{2.30}$$

where

$$\mathbf{u} = \begin{bmatrix} u_x & u_y \end{bmatrix}^T; \quad \mathbf{u}_g = \begin{bmatrix} u_{xg} & u_{yg} & u_{\theta g} \end{bmatrix}^T \tag{2.31}$$

$$\mathbf{N} = \begin{bmatrix} 1 & 0 & y_g - y \\ 0 & 1 & x - x_g \end{bmatrix} \tag{2.32}$$

where

Superscript T denotes transpose

x_g and y_g are the abscissa and ordinate values of the centroid of the element, respectively

N is the shape function

As shown in Figure 2.12, the relative displacement $\boldsymbol{\delta}$ at a point P can be decomposed into two components in the n-axis and s-axis:

$$\delta = \begin{bmatrix} \delta_n & \delta_s \end{bmatrix}^T \tag{2.33}$$

The relative displacement $\boldsymbol{\delta}$ can be further represented by

$$\delta = -\mathbf{L}^{(1)}\left(\mathbf{N}^{(1)}\mathbf{u}_g^{(1)} - \mathbf{N}^{(2)}\mathbf{u}_g^{(2)}\right) \tag{2.34}$$

where the superscripts (1) and (2) denote elements (1) and (2) respectively; $\mathbf{L}^{(1)}$ is the matrix of direction cosines of the local n-s axes on the interface of element (1) with respect to the global coordinate system and is expressed by

$$\mathbf{L}^{(1)} = \begin{bmatrix} \cos(\mathbf{n},\mathbf{x}) & \cos(\mathbf{n},\mathbf{y}) \\ \cos(\mathbf{s},\mathbf{x}) & \cos(\mathbf{s},\mathbf{y}) \end{bmatrix} \tag{2.35}$$

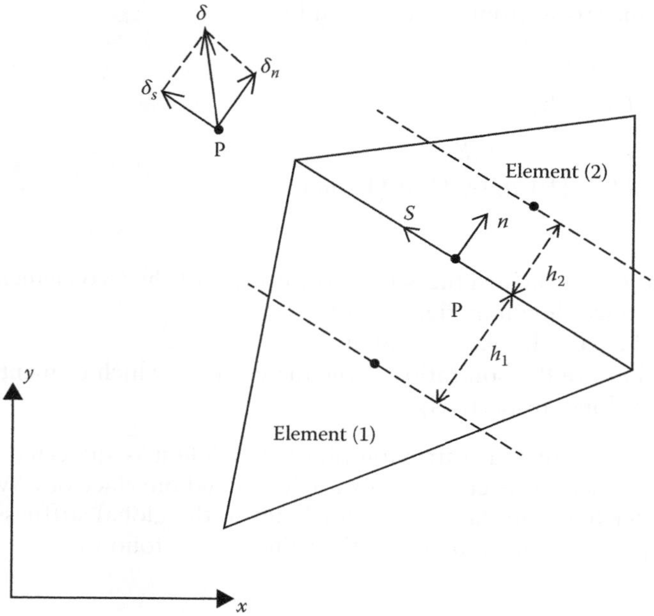

Figure 2.12 Two adjacent rigid elements.

2.5.2 Contact stresses between rigid elements

From elasticity theory, the relation of the contact stress and displacement in REM is expressed as

$$\sigma = \mathbf{D}\delta \tag{2.36}$$

$$\sigma = \begin{bmatrix} \sigma_n & \tau_s \end{bmatrix}^{\mathrm{T}} \tag{2.37}$$

$$\mathbf{D} = \begin{bmatrix} d_n & 0 \\ 0 & d_s \end{bmatrix} \tag{2.38}$$

D is termed the elasticity matrix, and for the plane strain problem, it is given by

$$\begin{cases} d_n = \dfrac{E_1E_2}{E_1h_2\left(1-\mu_2^2\right)+E_2h_1\left(1-\mu_1^2\right)} \\ d_s = \dfrac{E_1E_2}{2E_1h_2\left(1+\mu_2\right)+2E_2h_1\left(1+\mu_1\right)} \end{cases} \tag{2.39}$$

For the plane stress problem, it is given by

$$\begin{cases} d_n = \dfrac{E_1 E_2}{E_1 h_2 + E_2 h_1} \\ d_s = \dfrac{E_1 E_2}{2E_1 h_2 (1+\mu_2) + 2E_2 h_1 (1+\mu_1)} \end{cases} \tag{2.40}$$

where

h_1 and h_2 are the distances from the centres of the two elements to the interface shown in Figure 2.12
E_1 and E_2 are the elastic moduli
μ_1 and μ_2 are Poisson ratios of the materials to which elements (1) and (2) belong, respectively

An interface is called a restriction interface while it is subjected to a certain displacement restriction, for example, a fixed interface or a symmetric interface. Such an interface also contributes to the global stiffness matrix. For example, for a fixed interface, the values are as follows:

Plane stress:

$$d_n = \frac{E_1}{h_1(1-\mu_1^2)}; \quad d_s = \frac{E_1}{2h_1(1+\mu_1)} \tag{2.41}$$

Plane strain:

$$d_n = \frac{E_1}{h_1}; \quad d_s = \frac{E_1}{2h_1(1+\mu_1)} \tag{2.42}$$

For a symmetric plane stress interface:

$$d_n = \frac{E_1}{h_1(1-\mu_1^2)}; \quad d_s = 0 \tag{2.43}$$

For a symmetric plane strain interface:

$$d_n = \frac{E_1}{h_1}; \quad d_s = 0 \tag{2.44}$$

2.5.3 Principle of virtual work

The previous section describes how all the important quantities can be expressed in terms of the displacements of the element centroid. These relationships can be used to derive the rigid element stiffness matrix. The principle of virtual work states that when a structure is in equilibrium the external work done by any virtual displacement equals the internal energy dissipation. For REM, the deformation energy of the system is stored only at the interfaces between the rigid elements. The rigid element itself has no strain and thus there is no internal energy dissipation within the element. The virtual work done by the traction force at the interface can be viewed as an external work for the observed element. The total virtual external work done is the sum of the work done of the individual elements. The virtual work equation can be written as

$$\sum_e \left[\iiint_{\Omega_e} \delta \mathbf{u}^{\mathrm{T}} \mathbf{F} \mathrm{d}\Omega + \iint_{S_\sigma^e} \delta \mathbf{u}^{\mathrm{T}} \mathbf{X} \,\mathrm{d}S \right] + \sum_e \iint_{S_0^e} \delta \mathbf{u}_{\mathrm{l}}^{\mathrm{T}} \sigma \mathrm{d}S = 0 \tag{2.45}$$

where

F and *X* are body forces and boundary loadings
$\mathbf{u}_{\mathrm{l}}$ is the interface displacement represented in the local reference coordinate system
S, Ω are the surface and volume of the structure body, respectively

Using Equations 2.34 and 2.36 in Equation 2.45 gives

$$\sum_e \delta \mathbf{u}_{\mathrm{g}}^{(1)\mathrm{T}} \iint_{S_0^e} \mathbf{N}^{(1)\mathrm{T}} \mathbf{L}^{(1)\mathrm{T}} \mathbf{D} \mathbf{L}^{(1)} \left(\mathbf{N}^{(1)} \mathbf{u}_{\mathrm{g}}^{(1)} - \mathbf{N}^{(2)} \mathbf{u}_{\mathrm{g}}^{(2)} \right) \mathrm{d}S$$

$$= \sum_e \delta \mathbf{u}_{\mathrm{g}}^{\mathrm{T}} \left[\iiint_{\Omega_e} \mathbf{N}^{\mathrm{T}} \mathbf{F} \mathrm{d}\Omega + \iint_{S_\sigma^e} \mathbf{N}^{\mathrm{T}} \mathbf{X} \mathrm{d}S \right] \tag{2.46}$$

In REM formulations, we introduce a selection matrix $\mathbf{C}_e$ for each element, which is defined by

$$\mathbf{u}_{\mathrm{g}} = \mathbf{C}_e \mathbf{U} \tag{2.47}$$

and for element *i*, $\mathbf{C}_{ie}$ is given by

$$\mathbf{C}_{ie} = \begin{bmatrix} 0 & \cdots & \overset{3i-2}{\overbrace{1}} & \overset{3i-1}{\overbrace{0}} & \overset{3i}{\overbrace{0}} & \cdots & 0 \\ 0 & \cdots & 0 & 1 & 0 & \cdots & 0 \\ 0 & \cdots & 0 & 0 & 1 & \cdots & 0 \end{bmatrix} \tag{2.48}$$

where **U** is the global displacement matrix

$$\mathbf{U} = \left[\mathbf{u}_{\mathrm{g}}^{(1)}, \mathbf{u}_{\mathrm{g}}^{(2)}, \ldots \right]^{\mathrm{T}} \tag{2.49}$$

Using the notations given by Equations 2.50 and 2.51a, Equation 2.46 can be written as

$$\mathbf{C}_e^* = \begin{bmatrix} \mathbf{C}_e^{(1)} & -\mathbf{C}_e^{(2)} \end{bmatrix}^{\mathrm{T}} \tag{2.50}$$

$$\mathbf{N}^* = \begin{bmatrix} \mathbf{N}^{(1)} & \mathbf{N}^{(2)} \end{bmatrix} \tag{2.51a}$$

$$\delta\mathbf{U}^{\mathrm{T}}\left[\sum_e \mathbf{C}_e^{(1)\mathrm{T}} \iint_{s_0^e} \mathbf{N}^{(1)\mathrm{T}}\mathbf{L}^{(1)\mathrm{T}}\mathbf{D}\mathbf{L}^{(1)}\mathbf{N}^* \mathrm{d}S \cdot \mathbf{C}_e^*\right]\mathbf{U} = \delta\mathbf{U}^{\mathrm{T}}\left[\sum_e \mathbf{C}_e^{\mathrm{T}}\left[\iiint_{\Omega_e} \mathbf{N}^{\mathrm{T}}\mathbf{F}\mathrm{d}\Omega + \iint_{s_\sigma^e} \mathbf{N}^{\mathrm{T}}\mathbf{X}\,\mathrm{d}S\right]\right] \tag{2.51b}$$

2.5.4 Governing equations

Considering the arbitrary feature of a virtual displacement $\delta\mathbf{U}$ in Equation 2.52a, the governing equation can be given in the form

$$\mathbf{KU} = \mathbf{R} \tag{2.52a}$$

$$\mathbf{K} = \sum \mathbf{C}_e^{*\mathrm{T}}\mathbf{k}_i\mathbf{C}_e^* \tag{2.52b}$$

$$\mathbf{k}_i = \iint_{s_i} \mathbf{N}^{*\mathrm{T}}\mathbf{L}^{(1)\mathrm{T}}\mathbf{D}\mathbf{L}^{(1)}\mathbf{N}^* \mathrm{d}S \tag{2.53}$$

$$\mathbf{R} = \sum_e \mathbf{C}_e^{\mathrm{T}}\mathbf{R}^e \tag{2.54}$$

$$\mathbf{R}^e = \iiint_{\Omega_e} \mathbf{N}^{\mathrm{T}}\mathbf{F}\mathrm{d}\Omega + \iint_{s_\sigma^e} \mathbf{N}^{\mathrm{T}}\mathbf{X}\mathrm{d}S \tag{2.55}$$

where

- $\mathbf{K}$ and $\mathbf{R}$ are the global stiffness matrix and global force matrix, respectively
- $\mathbf{k}_i$ is the stiffness matrix of each interface
- $\mathbf{R}^e$ is the force matrix at the centroid of the rigid element

2.5.5 General procedure of REM computation

The REM is a numerical procedure for solving engineering problems. Linear elastic behaviour is assumed here. The six steps of REM analysis are summarized as follows:

1. Discretize the domain – this step involves subdividing the domain into elements and nodes. As one of the main components of REM is interface, it is necessary to set up the topological relations of nodes, elements and interfaces.
2. Select the element centroid displacements as primary variables – the shape function and elastic matrix need to be set up.
3. Calculate the global loading matrix – this will be done according to Equations 2.54 and 2.55.
4. Assemble the global stiffness matrix – this will be done according to Equations 2.52b and 2.53 after calculating the stiffness matrix for each interface.
5. Apply the boundary conditions – add supports and applied loads and displacements.
6. Solve the global equations – to obtain the displacement of each element centroid. The relative displacement and stress of each interface can then be obtained according to Equations 2.34 and 2.36, respectively.

2.6 RELATION BETWEEN THE REM AND THE SLICE-BASED APPROACH

This section demonstrates that the present formulation based on REM can be easily reduced to the formulations of other upper bound limit analysis approaches proposed by Michalowski (1995) and Donald and Chen (1997), respectively, where slice techniques and translational failure mechanics are used.

We herein purposely divide the failing mass of the soil into rigid elements in the same way as in the case of inclined slices (or 2D wedges) considered in the upper bound limit analysis approach by Donald and Chen (1997). As shown in Figure 2.13, the rigid elements below the assumed failure surface ABCDE are fixed with zero velocities and thus called base elements. The index k denotes the element number, ϕ_k is the internal friction angle on the base interface (the interface between element k and the base element below) and $\bar{\phi}_k$ denotes the internal friction angle at the left interface (the interface between elements k and k – 1) of the kth element, respectively. α_k is the angle of inclination of the kth element base from the horizontal direction (anticlockwise positive), and β_k is the inclination angle of the kth element's left interface from the vertical direction (anticlockwise positive). Suppose the kth element has a velocity $\mathbf{V}_k$ (magnitude denoted as V_k, with v_{xk} and

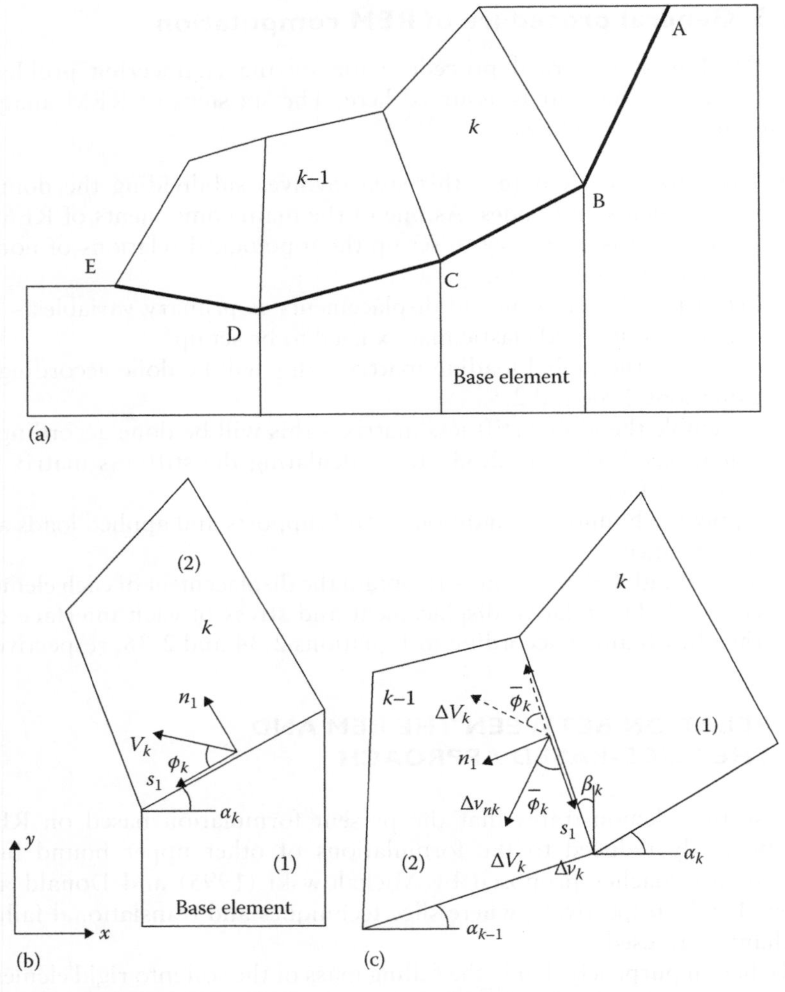

Figure 2.13 **Failure mechanism similar to traditional slice techniques: (a) rigid elements and base element division, (b) interface between rigid element and base element and (c) interface between two adjacent rigid elements.**

v_{yk} in x and y directions, respectively) in the global coordinate system. Note here that due to the assumption of a translational collapse mechanism, the rotation velocity of the kth element equals zero.

As shown in Figure 2.13b, the direction cosine matrix of the base interface of the kth element with respect to the base element can be written as

$$\mathbf{L}^{(1)} = \begin{bmatrix} -\sin\alpha_k & \cos\alpha_k \\ -\cos\alpha_k & -\sin\alpha_k \end{bmatrix} \tag{2.56}$$

The relative velocity of the base interface, $\mathbf{V}'_k$, can be expressed as follows:

$$\mathbf{V}'_k = \begin{bmatrix} v_{nk} \\ v_{sk} \end{bmatrix} = \begin{bmatrix} -v_{xk}\sin\alpha_k + v_{yk}\cos\alpha_k \\ -v_{xk}\cos\alpha_k - v_{yk}\sin\alpha_k \end{bmatrix} \tag{2.57}$$

As shown in Figure 2.13b, the element k has the tendency to move leftward with respect to the base element. According to the Mohr–Coulomb failure criterion (or yield criterion for perfect plasticity material) and the associated flow rule, the relationship between the normal velocity magnitude (Δv_n) and tangential velocity magnitude (Δv_s) jumps across the discontinuity can be written as

$$\Delta v_n = -\left|\Delta v_s\right|\tan\phi' \tag{2.58}$$

Using Equation 2.58, we have

$$\frac{v_{nk}}{v_{sk}} = \tan\phi_k \tag{2.59}$$

Thus, the following relationships can be obtained:

$$\begin{aligned} v_{xk} &= V_k\cos(\alpha_k - \phi_k) \\ v_{yk} &= V_k\sin(\alpha_k - \phi_k) \end{aligned} \tag{2.60}$$

Similarly, we can get

$$\begin{aligned} v_{x,k-1} &= V_{k-1}\cos\left(\alpha_{k-1} - \phi_{k-1}\right) \\ v_{y,k-1} &= V_{k-1}\sin\left(\alpha_{k-1} - \phi_{k-1}\right) \end{aligned} \tag{2.61}$$

From Figure 2.13c, the direction cosine matrix of the left interface of the kth element with respect to the $(k-1)$th element can be written as

$$\mathbf{L}^{(1)} = \begin{bmatrix} -\cos\beta_k & -\sin\beta_k \\ \sin\beta_k & -\cos\beta_k \end{bmatrix} \tag{2.62}$$

Similarly, the relative velocity of the left interface of the kth element, $\Delta\mathbf{V}_k$, can be given in the form

$$\Delta\mathbf{V}_k = \begin{bmatrix} \Delta v_{nk} \\ \Delta v_{sk} \end{bmatrix} = \begin{bmatrix} \cos\beta_k\left(v_{xk} - v_{x,k-1}\right) + \sin\beta_k\left(v_{yk} - v_{y,k-1}\right) \\ -\sin\beta_k\left(v_{xk} - v_{x,k-1}\right) + \cos\beta_k\left(v_{yk} - v_{y,k-1}\right) \end{bmatrix} \tag{2.63}$$

From Equation 2.58, we can get

$$\frac{\Delta v_{nk}}{\Delta v_{sk}} = \pm \tan \bar{\phi}_k \tag{2.64}$$

where the case with a negative sign in Equation 2.64 coincides with the case where the $(k - 1)$th element has a tendency to move upward with respect to the kth element shown in Figure 2.13c with the dashed lines. It should be noted that this case is identical to Case 1 defined in the method proposed by Donald and Chen (1997), and, similarly, the case with a positive mark in Equation 2.64 corresponds to Case 2 discussed in Donald and Chen's method (1997).

Putting Equations 2.60, 2.61 and 2.63 into Equation 2.64, we obtain the following relationship:

$$V_k = V_{k-1} \frac{\cos\left[\left(\alpha_{k-1} - \phi_{k-1}\right) - \left(\beta_k \mp \bar{\phi}_k\right)\right]}{\cos\left[\left(\alpha_k - \phi_k\right) - \left(\beta_k \mp \bar{\phi}_k\right)\right]} \tag{2.65}$$

With Equation 2.65, and according to Equation 2.60, we can express v_{xk} and v_{yk} in terms of V_{k-1} thus:

$$\begin{aligned} v_{xk} &= V_{k-1} \frac{\cos\left[\left(\alpha_{k-1} - \phi_{k-1}\right) - \left(\beta_k \mp \bar{\phi}_k\right)\right]}{\cos\left[\left(\alpha_k - \phi_k\right) - \left(\beta_k \mp \bar{\phi}_k\right)\right]} \cos\left(\alpha_k - \phi_k\right) \\ v_{yk} &= V_{k-1} \frac{\cos\left[\left(\alpha_{k-1} - \phi_{k-1}\right) - \left(\beta_k \mp \bar{\phi}_k\right)\right]}{\cos\left[\left(\alpha_k - \phi_k\right) - \left(\beta_k \mp \bar{\phi}_k\right)\right]} \sin\left(\alpha_k - \phi_k\right) \end{aligned} \tag{2.66}$$

Together with Equation 2.61, we put Equation 2.66 into 2.63 and obtain

$$\Delta V_k = V_{k-1} \frac{\sin\left(\alpha_k - \phi_k - \alpha_{k-1} + \phi_{k-1}\right)}{\cos\left[\left(\alpha_k - \phi_k\right) - \left(\beta_k \mp \bar{\phi}_k\right)\right]} \tag{2.67}$$

In the method proposed by Donald and Chen (1997), the velocities of 2D wedges can be determined by a hodograph:

$$\begin{aligned} V_r &= V_l \frac{\sin\left(\theta_l - \theta_j\right)}{\sin\left(\theta_r - \theta_j\right)} \\ V_j &= V_l \frac{\sin\left(\theta_r - \theta_l\right)}{\sin\left(\theta_r - \theta_j\right)} \end{aligned} \tag{2.68}$$

Using the definitions

$$\begin{aligned} \theta_l &= \pi + \alpha_l - \phi_{el} \\ \theta_r &= \pi + \alpha_r - \phi_{er} \end{aligned} \tag{2.69}$$

and

$$\begin{aligned} \theta_j &= \frac{\pi}{2} - \delta + \phi_{ej} \quad \text{for case 1} \\ \theta_j &= \frac{3\pi}{2} - \delta - \phi_{ej} \quad \text{for case 2} \end{aligned} \tag{2.70}$$

Variables V_l, V_r, V_j, α_l, ϕ_{el}, α_r, ϕ_{er} and ϕ_{ej} in Donald and Chen's approach are identical to those V_{k-1}, V_k, ΔV_k, α_{k-1}, ϕ_{k-1}, α_k, ϕ_k and $\bar{\phi}_k$ defined in the present method respectively. It should be noted that δ in their formulations is equal to $-\beta_k$ in the present formulation, since the direction definition of δ (clockwise positive) is opposite to that of β_k used in the present method (anticlockwise positive).

Substituting $V_{k-1}, V_k, \Delta V_k, \alpha_{k-1}, \phi_{k-1}, \alpha_k, \phi_k, \bar{\phi}_k, \beta_k$ into Equation 2.68, and keeping the consistency between corresponding cases in the two approaches, Equation 2.68 arrives at exactly the same form of Equations 2.65 and 2.67 in the proposed method.

In the method proposed by Michalowski (1995), vertical slices were employed. For vertical slices, β_k equals zero, and Equations 2.65 and 2.67 can be reduced to the following two equations:

$$V_k = V_{k-1} \frac{\cos\left(\alpha_{k-1} - \phi_{k-1} - \bar{\phi}_k\right)}{\cos\left(\phi_k + \bar{\phi}_k - \alpha_k\right)} \tag{2.71}$$

$$\Delta V_k = V_k \frac{\sin\left(\phi_k - \phi_{k-1} - \alpha_k + \alpha_{k-1}\right)}{\cos\left(\alpha_{k-1} - \phi_{k-1} - \bar{\phi}_k\right)} \tag{2.72}$$

These equations correspond to the case where the $(k-1)$th element moves downward with respect to the kth element, that is, $\Delta v_{nk} / \Delta v_{sk} = \tan \bar{\phi}_k$. In such a case, the velocity relationships in the present method are identical to those under the translational failure mechanism in the method proposed by Michalowski (1995).

It has thus been proved that the present formulations in REM reduce to exactly the same formulations of the methods proposed by Donald and Chen (1997) and Michalowski (1995) if the same slices with the same translational failure mechanism are used. In other words, the upper bound

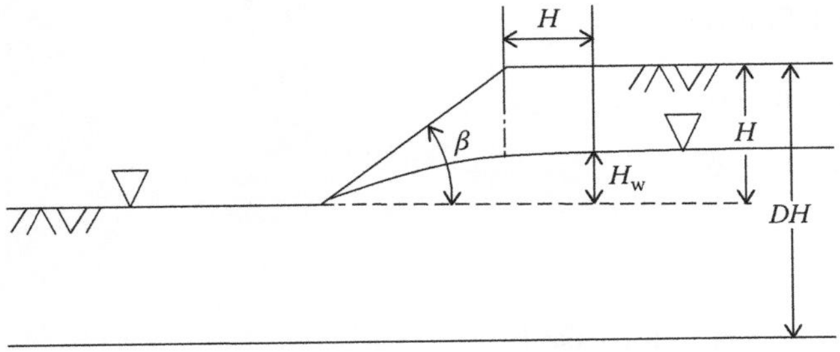

Figure 2.14 Simple homogeneous slope with pore water pressure.

Table 2.8 Comparisons of factors of safety for various conditions of water table

	Study by Kim et al. (1999)				*Present method (upper bound)*		
H_w *(m)*	*Lower bound*	*Bishop method (Bishop 1955)*	*Upper bound*	*Janbu chart (Janbu 1968)*	*Coarse mesh*	*Medium mesh*	*Fine mesh*
4	1.036	1.101	1.166	1.030	1.403	1.276	1.202
6	0.971	1.036	1.068	0.973	1.284	1.162	1.096

limit analyses using slices (or 2D wedges) may be viewed as a special and simple case of the formulation of the present method.

As shown in Figure 2.14, Kim et al. (1999) have studied the slope in nine cases with different depth factors D and slope inclinations β. In this study, we only take one case to investigate the feasibility of the present method; for example, consider the slope with a depth factor $D=2$, $H=10$ m and $\beta=45°$, and with soil properties $\gamma=18$ kN/m^3, $c'=20$ kN/m^3 and $\phi'=15°$. To assess the effects of pore water pressure, two locations of the water table with $H_w=4$ and 6 m are considered in this study. Figure 5.11 shows three rigid finite element meshes (coarse, medium and fine meshes) used in the analysis, in the case of a water table with $H_w=6$ m. The relations between the number of rigid elements used in the mesh and calculated factor of safety, in the case of a water table with $H_w=4$ m and $H_w=6$ m, are shown in Table 2.8 and Figures 2.15 and 2.16.

2.7 USES OF DESIGN FIGURES AND TABLES FOR SIMPLE PROBLEMS

For a simple homogeneous slope with geometry shown in Figure 2.17, the critical factor of safety can be determined by the use of a stability table instead of using a computer program. Stability tables and figures have been prepared by Taylor (friction circle), Morgenstern (Spencer method),

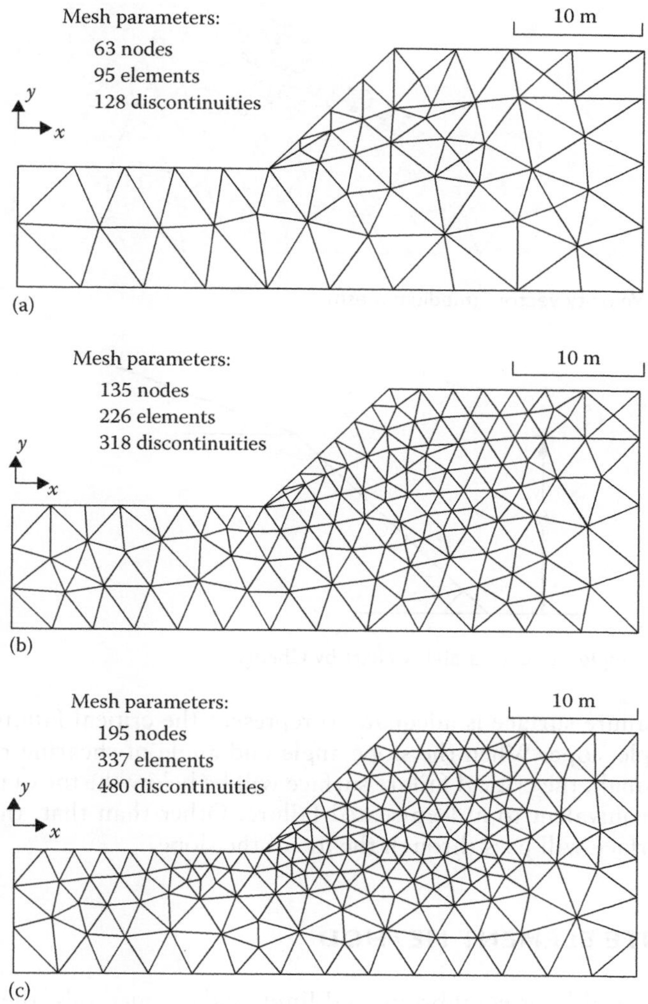

Figure 2.15 REM meshes with $H_w = 6$ m: (a) coarse mesh, (b) medium mesh and (c) fine mesh.

Chen (limit analysis) and Cheng. In general, most of the results from these stability tables are similar. All the previous stability tables/figures are, however, designed for 2D problems. Cheng has prepared stability tables using both the 2D and 3D Bishop methods based on SLOPE 2000, which are given in Tables 2.9 and 2.10.

For the 2D stability table by Cheng, the results are very close to those by Chen (1975) using a log-spiral failure surface. This also indicates that a

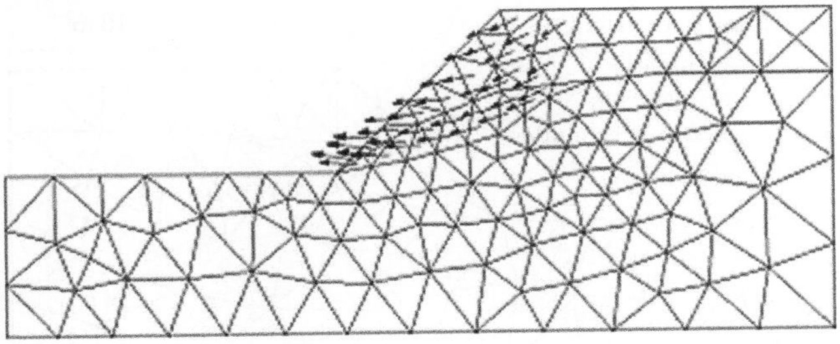

Figure 2.16 Velocity vectors (medium mesh).

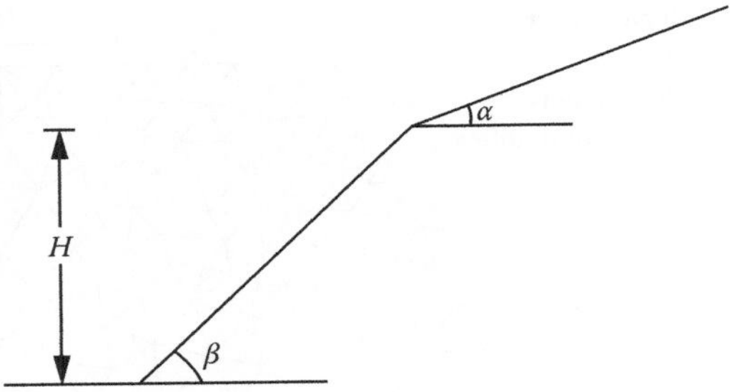

Figure 2.17 Simple slope for stability chart by Cheng.

circular failure surface is adequate to represent the critical failure surface for a simple slope. When the slope angle and angle of shearing resistance are both small, the critical failure surface will be below the toe of the slope, which is equivalent to a deep-seated failure. Other than that, the critical failure surface will pass through the toe of the slope.

2.8 FINITE ELEMENT METHOD

In the classical limit equilibrium and limit analysis methods, the progressive failure phenomenon cannot be estimated except in the method by Pan (1980). Some researchers propose to use the FEM to overcome some of the basic limitations in the traditional methods of analysis. At present, there are two major applications of the finite element in the slope stability analysis.

The first approach is to perform an elastic (or elastoplastic) stress analysis by applying the body force (weight) due to soil to the slope system. Once the stresses are determined, the local factors of safety can be determined easily from the stresses and the Mohr–Coulomb criterion. The global factor of safety can also be defined in a similar way by determining the ultimate shear force and the actual driving force along the failure surface. Pham and Fredlund (2003) have adopted the dynamic programming method to

Table 2.9 Stability chart using 2D Bishop simplified analysis

ϕ(°)	β/α(°)	*70*	*65*	*60*	*55*	*50*	*45*	*40*	*35*	*30*	*25*	*20*	*15*
0	0	4.80	5.03	5.25	5.46	5.67	5.87	5.41[a]	5.43[a]	5.45[a]	5.43[a]	5.45[a]	5.46[a]
5	0	5.41	5.73	6.09	6.46	6.85	7.29	7.79	8.37	9.08[a]	9.97[a]	11.43[a]	14.38[a]
	5	5.30	5.63	5.96	6.32	6.70	7.11	7.59	8.14	8.77[a]	9.60[a]	10.96[a]	13.69[a]
10	0	6.05	6.52	7.09	7.71	8.40	9.21	10.22	11.54	13.45	16.62	23.14	45.57
	5	5.95	6.44	6.99	7.58	8.26	9.05	10.06	11.36	13.24	16.33	22.78	45.00
	10	5.84	6.33	6.86	7.41	8.07	8.83	9.78	11.11	12.84	15.79	21.90	42.86
15	0	6.94	7.58	8.37	9.36	10.50	11.94	13.90	16.79	21.69	32.14	69.23	—
	5	6.77	7.50	8.30	9.25	10.36	11.80	13.74	16.59	21.48	31.86	68.97	
	10	6.67	7.38	8.18	9.09	10.20	11.61	13.51	16.33	21.13	31.36	68.18	
	15	6.53	7.22	8.01	8.89	9.96	11.25	13.14	15.85	20.45	30.25	68.18	
20	0	7.97	9.01	10.14	11.61	13.51	16.07	20.00	26.67	41.38	94.74	—	
	5	8.04	8.91	10.04	11.50	13.38	15.93	19.82	26.55	41.10	94.74		
	10	7.69	8.82	9.92	11.35	13.22	15.76	19.61	26.28	40.72	93.75		
	15	7.60	8.66	9.75	11.16	12.97	15.49	19.25	25.79	40.18	92.78		
	20	7.59	8.44	9.55	10.91	12.66	15.06	18.71	25.00	38.54	88.24		
25	0	9.42	11.01	12.57	14.80	18.04	22.93	31.47	50.28	120.00	—		
	5	9.28	10.91	12.46	14.69	17.91	22.78	31.03	50.00	120.00			
	10	9.50	10.84	12.33	14.57	17.73	22.56	31.03	49.72	119.21			
	15	9.00	10.60	12.16	14.35	17.51	22.28	30.72	49.32	118.42			
	20	8.97	10.51	12.00	14.12	17.22	21.95	30.25	48.65	116.88			
	25	8.81	10.17	11.73	13.74	16.74	21.25	29.27	46.75	111.80			

(continued)

Table 2.9 (continued) Stability chart using 2D Bishop simplified analysis

$\phi(°)$	$\beta/\alpha(°)$	*70*	*65*	*60*	*55*	*50*	*45*	*40*	*35*	*30*	*25*	*20*	*15*
30	0	11.89	13.79	16.07	19.65	25.53	35.64	58.63	144.00	—			
	5	11.54	13.74	16.00	19.52	25.35	35.64	58.44	144.00				
	10	11.43	13.75	15.83	19.35	25.17	35.43	58.25	144.00				
	15	11.04	13.53	15.65	19.19	25.00	35.16	57.32	142.86				
	20	10.71	13.31	15.49	18.95	24.66	34.75	57.32	142.29				
	25	10.81	12.93	15.23	18.60	24.19	34.16	56.60	140.63				
	30	10.43	12.11	14.86	18.09	23.44	32.97	54.22	134.33				
35	0	14.83	18.00	21.25	27.48	39.30	65.93	166.67	—				
	5	14.94	17.82	21.18	27.40	39.30	65.69	166.67					
	10	14.25	17.65	21.08	27.27	39.13	65.45	165.90					
	15	14.04	17.54	20.93	27.03	38.79	65.45	165.14					
	20	13.85	17.65	20.69	26.87	38.63	64.98	165.14					
	25	13.19	16.32	20.55	26.55	38.30	64.29	165.14					
	30	12.82	15.76	20.18	26.01	37.50	63.16	163.64					
	35	12.54	15.67	19.35	25.14	36.14	61.02	155.17					
40	0	20.07	24.03	30.10	42.35	72.29	185.57	—					
	5	19.13	23.68	29.41	42.06	72.00	185.57						
	10	19.82	23.72	29.27	42.35	71.43	185.57						
	15	18.95	23.53	30.28	41.86	71.43	183.67						
	20	17.61	23.38	29.32	41.47	71.43	183.67						
	25	16.93	23.23	28.85	41.10	70.87	183.67						
	30	16.36	22.70	28.57	40.72	70.04	181.82						
	35	16.04	21.05	28.13	40.00	68.97	180.00						
	40	15.72	20.00	27.69	38.46	65.93	171.43						

[a] Means below toe failure.

Table 2.10 Stability chart using 3D Bishop simplified analysis by Cheng

ϕ(°)	β/α(°)	*70*	*65*	*60*	*55*	*50*	*45*	*40*	*35*	*30*	*25*	*20*	*15*
0	0	6.16	6.43[a]	6.55[a]	6.67[a]	6.79[a]	6.92[a]	6.27[a]	6.25[a]	6.25[a]	6.25[a]	6.21[a]	6.15[a]
5	0	6.81	7.20	7.66	8.11[a]	8.57[a]	9.04[a]	9.52[a]	10.11[a]	10.84[a]	12.00[a]	13.74[a]	17.48[a]
	5	6.72	7.09	7.53	7.96[a]	8.41[a]	8.82[a]	9.23[a]	9.73[a]	10.47[a]	11.46[a]	13.14[a]	16.45[a]
10	0	7.74	8.20	8.96	9.78	10.71	11.69	12.86	14.52	16.82	20.69	29.03	58.06
	5	7.77	8.09	8.82	9.68	10.53	11.52	12.68	14.17	16.44	20.36	28.48	56.25
	10	7.63	7.99	8.70	9.50	10.29	11.26	12.33	13.74	15.93	19.57	27.27	53.25
15	0	9.23	9.78	10.65	11.92	13.33	15.25	17.65	21.43	27.69	41.10	90.00	—
	5	9.23	9.68	10.53	11.76	13.24	15.06	17.48	21.18	27.27	40.72	90.00	
	10	9.08	9.57	10.40	11.61	12.99	14.75	17.14	20.69	26.87	40.18	86.54	
	15	8.96	9.40	10.23	11.39	12.77	14.40	16.82	20.22	26.09	38.30	87.38	
20	0	11.39	11.84	13.28	15.13	17.22	20.45	25.35	33.96	52.94	124.14	—	
	5	11.46	11.84	12.90	15.03	17.14	20.22	25.35	33.96	52.94	122.45		
	10	11.39	11.69	12.77	14.52	16.98	20.00	25.00	33.33	51.72	121.62		
	15	11.07	11.54	12.59	14.42	16.67	19.78	24.66	32.73	51.43	118.42		
	20	10.17	11.39	12.41	13.95	16.32	19.19	23.97	31.86	49.18	112.50		
25	0	14.46	14.81	16.81	19.62	23.68	29.03	40.00	64.29	155.17	—		
	5	14.63	14.88	16.67	20.00	23.62	29.03	40.00	64.29	155.17			
	10	13.93	14.75	17.82	19.21	22.78	28.57	39.30	63.38	153.85			
	15	14.62	14.81	15.93	18.71	22.50	28.21	38.30	63.38	152.54			
	20	12.68	14.52	15.79	18.09	21.95	27.69	38.22	61.64	151.26			
	25	12.00	14.40	15.57	17.65	21.18	26.87	37.11	59.02	142.86			

(*continued*)

Table 2.10 (continued) Stability chart using 3D Bishop simplified analysis by Cheng

ϕ(°)	β/α(°)	*70*	*65*	*60*	*55*	*50*	*45*	*40*	*35*	*30*	*25*	*20*	*15*
30	0	18.56	18.95	21.95	28.13	33.33	45.23	75.00	187.50	—			
	5	18.37	18.91	21.63	28.13	33.33	45.23	75.00	183.67				
	10	18.56	18.93	21.18	28.13	32.85	45.00	73.77	183.67				
	15	17.79	18.87	20.93	25.64	32.61	44.33	72.00	183.67				
	20	16.29	18.95	20.69	24.39	32.73	44.12	72.58	183.67				
	25	15.25	18.65	20.22	24.00	31.03	43.27	72.00	176.47				
	30	14.63	17.14	20.16	23.38	29.51	41.67	69.23	168.22				
35	0	23.38	25.00	30.00	40.00	50.56	82.57	209.30	—				
	5	24.03	25.00	30.03	38.22	51.43	82.57	209.30					
	10	23.50	24.49	29.80	38.30	51.28	81.82	206.90					
	15	23.47	24.26	28.57	38.30	51.58	82.57	206.90					
	20	21.69	24.39	28.13	35.86	50.56	82.57	206.90					
	25	20.00	24.32	27.69	34.62	48.91	80.36	209.30					
	30	19.15	24.23	26.87	33.33	48.65	78.95	204.55					
	35	20.11	21.69	26.09	32.14	45.00	76.92	195.65					
40	0	29.95	34.09	45.23	60.40	90.91	236.84	—					
	5	30.00	33.64	43.90	58.06	90.00	236.84						
	10	30.15	33.33	43.90	60.00	90.00	233.77						
	15	30.86	32.73	41.47	59.21	90.00	233.77						
	20	30.82	32.26	40.91	59.41	90.00	230.77						
	25	28.13	32.14	39.13	59.02	88.24	230.77						
	30	26.47	32.49	38.30	54.55	85.71	227.85						
	35	25.55	33.09	37.50	51.43	86.54	227.85						
	40	22.73	28.57	36.79	48.65	84.11	214.29						

[a] Means below toe failure.

perform this optimization search, and suggested that this approach can overcome the limitations of the classical LEM. The authors, however, realize that the elastic stress analysis is not a realistic picture of the slope at the ultimate limit state. In view of these limitations, the authors do not think that this approach is really better than the classical approach. It is also interesting to note that both the factor of safety and the location of the critical failure surface from such analysis are usually similar to those obtained by the LEM. To adopt the elastoplastic finite element slope stability analysis, one precaution must be taken. If the deformation is too large so that the finite element mesh is greatly modified, the geometric non-linear effect may induce a major effect on the results. The authors have come across a case where the geometric non-linear effect has induced more than 10% change in the factor of safety. An illustration of this approach will be given in Chapter 4.

The second finite element slope stability approach is the strength reduction method (SRM). In the SRM, the gravity load vector for a material with unit weight γ_s is determined from the following equation as

$$\{f\} = \gamma_s \int [N]^{\mathrm{T}} \mathrm{d}v \tag{2.73}$$

where

$\{f\}$ is the equivalent body force vector
$[N]$ is the shape factor matrix

The constitutive model adopted in the non-linear element is usually the Mohr–Coulomb criterion, but other constitutive models are also possible, though seldom adopted in practice. The material parameters c' and ϕ' are reduced according to

$$c_{\mathrm{f}} = \frac{c'}{F}; \quad \phi_{\mathrm{f}} = \tan^{-1}\left\{\tan\left(\frac{\phi'}{F}\right)\right\} \tag{2.74}$$

The factor of safety F keeps on changing until the ultimate state of the system is attained, and the corresponding factor of safety will be the factor of safety of the slope. The termination criterion is usually based on one of the following:

1. The non-linear equation solver cannot achieve convergence after a pre-set maximum number of iterations.
2. There is a sudden increase in the rate of change of displacement in the system.
3. A failure mechanism has developed.

The location of the critical failure surface is usually determined from the contour of the maximum shear strain or the maximum shear strain rate.

The main advantages of the SRM are as follows: (1) the critical failure surface is found automatically from the localized shear strain arising from the application of gravity loads and the reduction of shear strength; (2) it requires no assumption on the interslice shear force distribution; (3) it is applicable to many complex conditions and can give information such as stresses, movements, and pore pressures, which is not possible with the LEM. Griffiths and Lane (1999) pointed out that the widespread use of the SRM should be seriously considered by geotechnical practitioners as a powerful alternative to the traditional LEMs. One of the important critics on SRM is the relative poor performance of the FEM in capturing the localized shear band formation. While the determination of the factor of safety is relatively easy and consistent, many engineers find that it is not easy to determine the critical failure surfaces in some cases as the yield zone is spread over a wide domain instead of localizing within a soft band. Other limitations of the SRM include the choice of an appropriate constitutive model and parameters, boundary conditions and the definition of the failure condition/failure surface; the detailed comparison between the SRM and LEM will be given in Chapter 4.

2.9 DISTINCT ELEMENT METHOD

One may consider two versions of the distinct element method (DEM) that can be used for slope stability analysis. The first approach is proposed by Chang (1992) where each slice is connected with the others through normal and tangential springs along the interfaces. The relative displacements between adjacent slices can be related to the interslice normal and shear force and moment by the normal and tangential springs. In this formulation, elastoplastic behaviour can be incorporated into the springs, and the concept of residual strength and stress redistribution can also be implemented. The equilibrium of the whole system is then assembled, and this will result in a 3N simultaneous equation that will need to be solved. Cheng has programmed this method (Chan, 1999) and has found that the factor from this method is very similar to that by the other methods. If fact, if the normal and shear stiffnesses are kept constant throughout the system, the factor of safety is practically independent of the values of the stiffnesses. This method appears not to be used for practical or research purposes, and the selection of the normal and shear stiffnesses for general non-homogeneous conditions is actually difficult as no guideline or theoretical background can be provided for this approach. Although Cheng has found that this approach will give results similar to the classical LEMs for normal problems, the suitability of this method for highly irregular problems is, however, unknown.

When a soil slope reaches its critical failure state, the critical failure surface is developed and FOS of the slope is 1.0. The continuum methods like FEM or LEM will fail to characterize the post-failure mechanisms in the subsequent failure process. For a rock slope comprising multiple joint sets that control the mechanism of failure, a discontinuum modelling approach is actually more appropriate. Discontinuum methods treat the problem domain as an assemblage of distinct, interacting bodies or blocks, which are subjected to external loads and are expected to undergo significant motion with time. This methodology is collectively named the DEM. DEM will be a more suitable tool for the study of progressive failure and the flow of the soil mass after initiation of slope failure, though this method is not efficient for the analysis of a stable slope. The development of DEM represents an important step in the modelling and understanding of the mechanical behaviour of joint rock masses or soil slopes with FOS < 1.0. Although continuum codes can be modified to accommodate discontinuities, this procedure is often difficult and time-consuming. In addition, any modelled inelastic displacements are further limited to elastic orders of magnitude by the analytical principles exploited in developing the solution procedures. In contrast, the discontinuum analysis permits sliding along and opening/closure between particles or blocks. Nowadays, discontinuum modelling constitutes the commonly applied numerical approach to rock slope analysis.

DEM was initially developed by Cundall (1971, 1974) for the study of rock mechanics problems. The method was later enhanced to suit various applications (Cundall and Strack, 1979). DEM has been used extensively to study physical and geotechnical phenomena such as mechanisms of deformation, constitutive relations of soil, stability of rock masses, flow of granular media, ground collapse and other types of geotechnical phenomena. The distinct element approach describes the mechanical behaviour of both the discontinuities and the solid material. This method is based on a force displacement law, which is specifies the interaction between the deformable rock blocks and a law of motion which determines displacements caused in the blocks by out-of-balance forces. Joints are treated as boundary conditions. Deformable blocks are discretized into internal constant-strain elements (Eberhardt, 2003).

In DEM, the medium under study is divided into discrete elements with arbitrary shapes. The interaction of these elements is viewed as a transient problem with the states of equilibrium developing whenever the internal forces are balanced (Cundall and Strack, 1979). The calculation cycle alternates between the sum of forces acting on an element resulting from a force displacement law at the contacts, and Newton's law is used to find the incremental displacements, velocities and accelerations of the element. The time step is chosen to be small enough so that disturbances do not propagate to more than the adjacent particles during each time step, and the accelerations can be assumed to be constant during that time step (Cundall and Strack, 1979).

There are a series of computer codes available in the literature for the application of DEM. Most of these codes follow the pioneering work by Cundall (1974, 1979) with variations either in the modelling of the contact forces or in the solution algorithm. The first DEM code was developed for the study of rock mass behaviour by Cundall (1971, 1974). Cundall also developed the universal discrete element code (UDEC) to model jointed rock mass and BALL and TRUEBALL for the study of granular media. UDEC is suitable for highly jointed rock slopes subjected to static or dynamic external loading. Two-dimensional analysis of the translational failure mechanism allows for simulating large displacements, modelling deformation or material yielding (Itasca, 2000). Three-dimensional DEMs were developed in the early 1990s. Three-dimensional discontinuum code 3DEC contains modelling of multiple intersecting discontinuities and is therefore suitable for the analysis of wedge instabilities or the influence of rock support (e.g. rock bolts, cables) (Eberhardt, 2003).

Discontinuous soil mass can be modelled with the help of the DEM in the form of a particle flow code, for example program PFC2D/3D (Itasca, 1999) or other codes. Open-source particle flow codes are available that are useful for some relatively simple problems. Spherical particles interact through frictional sliding contacts. Simulation of joint bounded blocks may be realized through specified bond strengths. The law of motion is repeatedly applied to each particle and the force displacement law to each contact. The particle flow method enables modelling of granular flow, fracture of intact rock, transitional block movements, dynamic response to blasting or seismicity, and deformation between particles caused by shear or tensile forces. These codes also allow modelling subsequent failure processes of rock slopes, for example the simulation of rock fracture (Eberhardt, 2003). The method of particle flow code employs a time-stepping, explicit calculation scheme (Cundall and Strack, 1979), which has advantages over traditional implicit calculation schemes in that it can handle a large number of particles with modest memory requirements because there are no matrices to be inverted as because the solution scheme solves the full dynamic equations of motion by dynamic propagation of waves through the material, and the velocities of the waves are dependent on the stiffness, density and packing of particles (Hazzard et al., 1998). This also allows physical instabilities such as shear-band formation to be modelled without numerical difficulty because kinetic energy that accompanies shear-band formation is released and dissipated in a physically realistic way (Itasca, 1999). On the other hand, the use of dynamic relaxation to solve the system equation in DEM requires extensive cycle time to reach a state of equilibrium. If the time step chosen is excessive, the results obtained from the analysis can be meaningless. In practical application, greater care, knowledge and judgement is required for DEM as compared with FEM, and a significant problem in using DEM is that it may require days or even weeks to obtain a solution.

A more recent development in the discontinuum modelling techniques is the application of DEM in the form of particle flow codes, such as PFC2D/3D (Itasca, 1999). This code allows the soil to be represented as a circular (2D) or spherical (3D) particle, or the rock mass as a series of circular or spherical particles, which interact through frictional sliding contacts. Clusters of particles may also be bonded together through specified bond strength in order to simulate joint bounded blocks. The calculation cycle then involves the repeated application of the law of motion to each particle and a force displacement law to each contact.

The particle flow code is employed by Cheng for slope stability studies, and it employs the DEM with the following assumptions: (1) particles are circular in PFC2D/spherical in PFC3D and elastic; (2) contacts between particles occur over a vanishingly small area (i.e. at a point); (3) the magnitude of elastic particle *overlap* of nominal radii is related to the contact force by a force displacement law similar to linear springs that mimic the physics of Hertz elastic spheres (Cundall, 1988); and (4) tensional and shear bonds can exist between particles. With these codes, it is possible to model granular flow, translational movement of blocks, fracture of intact rock and dynamic response to blasting or seismicity. The breaking of bonds between circular particles roughly simulates intact rock fracture and failure (although not fracture propagation). Deformation between particles due to shear or tensile forces can also be included, where slip between adjacent particles is prescribed in terms of the frictional coefficients that limit the contact shear force. Particle flow codes are thus able to simulate material from the macro-level of fault- or joint-bounded blocks to the micro-scale of grain-to-grain contact, with the main limiting factors being the computing time and memory requirements. In this sense, it becomes possible to model a number of soil/rock slope failure processes and, subsequently, the runout of the failed material down the slope. At present, these codes are predominantly a research tool, but their potential is being widely recognized in mining, petroleum industries and civil engineering.

PFC runs according to a time difference scheme in which calculations include the repeated application of the law of motion to each particle, a force displacement law to each contact, and a contact updating a wall position. Generally, there are two types of contacts that exist in the program which are ball-to-wall contact and ball-to-ball contact. In each cycle, the set of contacts is updated from the known particle and known wall position. The force displacement law is first applied on each contact. New contact force is calculated and this replaces the old contact force. The force calculations are based on preset parameters like normal stiffness, density and friction. Next, the law of motion is applied to each particle to update its velocity, direction of travel based on the resultant force, moment and contact acting on the particle. The force displacement law is then applied to continue the circulation.

2.9.1 Force displacement law and the law of motion

The force displacement law is described for both ball-to-ball and ball-to-wall contacts. The contact occurs at a point. For the ball-to-ball contact, the normal vector is directed along the line between the ball centres. For the ball-to-wall contact, the normal vector is directed along the line defining the shortest distance between the ball centre and the wall. The contact force vector F_i is composed of normal and shear components in a single plane surface as

$$F_i = F_{ij}^n(t) + F_{ij}^s(t+\Delta t) \tag{2.75}$$

The force acting on particle i in contact with particle j at time t is given by

$$F_{ij}^n(t) = k_n\left(r_i + r_j - l_{ij}(t)\right) \tag{2.76}$$

where
r_j and r_i stand for particle i and particle j radii
$l_{ij}(t)$ is the vector joining both centres of the particles
k_n represent the normal stiffness at the point of contact

The shear force acting on particle i during contact with particle j is determined by

$$F_{ij}^s(t+\Delta t) = \pm\min\left(F_{ij}^s(t) + k_s\Delta s_{ij},\ f\left|F_{ij}^n(t+\Delta t)\right|\right) \tag{2.77}$$

where
f is the particle friction coefficient
k_s represents the tangent shear stiffness at the contact point

The new shear contact force is found by summing the old shear force (min $F_{ij}(t)$) with the shear elastic force. Δs_{ij} stands for shear contact displacement increment occurring over a time step Δt.

$$\Delta s_{ij} = V_{ij}^S \Delta t \tag{2.78}$$

where V_{ij}^s is the shear component of the relative velocity at the contact point between particles i and j over the time step Δt.

The motion of the particle is determined by the resultant force and moment acting on it. The motion induced by the resultant force is called translational motion. The motion induced by the resulting moment is the rotational motion. The equations of motion are written in vector form as follows:

- Translational motion:

$$\sum_j F_{ij} + m_i g + F_i^{\mathrm{d}} = m_i x_i'' \tag{2.79}$$

- Rotational motion:

$$\sum_j r_i F_{ij} + M_i^{\mathrm{d}} = I_r \theta_i'' \tag{2.80}$$

where

x_i'' and θ_i'' stand for the translational acceleration and rotational acceleration of particles i

I_r stands for the moment of inertia

F_i^{d} and M_i^{d} stand for damping force and damping moment

2.9.2 Limitations of the distinct element method

The constitutive model acting at a particular contact point consists of three parts: a stiffness model, a slip model and a bonding model. The stiffness model provides an elastic relation between the contact force and relative displacement. The slip model enforces a relation between the shear and normal contact forces such that the two contacting balls may slip relative to one another. The bonding model serves to limit the total normal and shear forces that the contact can carry by enforcing the bond strength limits.

A number of quantities in a PFC model are defined with respect to a specified measurement circle. These quantities include coordinate number, porosity, sliding fraction, stress and strain rate. The coordination number and stress are defined as the average number of contacts per particle. Only particles with centroids that are contained within the measurement circle are considered in computation. In order to account for the additional area of particles that is being neglected, a corrector factor based on the porosity is applied to the computed value of stress.

Since a measurement circle is used, stress in the particle is described as the two in-plane forces acting on each particle per volume of particle. Average stress is defined as the total stress in the particle divided by the volume of measurement circle. Thus, the shape of the particle is independent of the average stress measurement because the reported stress is easily scaled by volume unity. The reported stress is interpreted as the stress per volume of measurement circle.

While DEMs are powerful tools for the numerical analysis of discontinuous material in geotechnical problems and geological mechanics, there are also some fundamental limitations to be considered. Granular materials are considered to be packed assemblies of particles in DEM simulation, and the mechanical interaction between particles is assumed to be simple. However, in reality, the contact between particles is highly complex and is hard to detect explicitly. Moreover, it is difficult to generate complicated model geometry. In addition, Thornton (1997) mentioned that the development of accurate contact constitutive models for use in discrete element analysis is non-trivial. The effective stress in a soil will govern its response, and including pore water pressures in a discrete element framework is non-trivial. The relation between the micro-parameters and the macro-parameters appears to be only partially considered, and most engineers determine the micro-parameters by a back analysis with some basic test results. Empirical relations between the micro- and macro-parameters are available for some problems, but the applicability of such empirical relations to general conditions is questionable.

2.9.3 Case studies for slope stability analysis using PFC

To assess the complete failure mechanism, the DEM (in terms of particle flow approach) can provide a qualitative assessment. Two problem generation approaches have been used by Cheng. A slope can be formed by an assembly of particles or triangular rigid blocks. To avoid the use of excessive number of particles or rigid blocks, which requires extensive computation time for analysis, a limited number in the range of 10,000–100,000 is used by Cheng. Initially, the initial stress state of the system is generated from the known soil mechanism principle. Model generation is quite complicated and crucial in PFC. Circle particles as balls in PFC2D are used to simulate the soil particles. The boundaries of the model are defined by a series of walls that contain the particles. In general, the usual objective of creating an irregular packing is to fill some given space with particles at a given porosity and to ensure that the assembly is in equilibrium. In this study, particles fall under self-weight gravity at random lying on the bottom of the model layer by layer, which represents the slope generation in nature. This kind of specimen generation technique is called *rainy method*. Radius expansion, which is called *expand method*, is also used in particle generation to achieve a solution in the trial-and-error manner to satisfy the requirement of porosity ratio. Thus, two methods of specimen generation are combined and incorporated in this simulation model. Once the initial state is established, the change of the water table/pore pressure or the application of an external load will take place in the system. The complete displacement history of the system from initial movement to a complete collapse can be qualitatively assessed. While it is difficult to provide a factor of safety for the design using the DEM, the collapse mechanism can be assessed, which is not possible with all the classical methods as discussed earlier.

For the parameter determination, the friction coefficient μ at the contact point for each ball in PFC is taken as the value of tanϕ (ϕ is the macro-friction angle of soil). For the bond strength, the definition in PFC is relatively complicated. Particle flow code is based on contact constitutive law in theory. To simulate a medium with cohesion, optional bonding is defined in PFC. Unlike the friction coefficient, the bond cohesion in PFC cannot be compared with the macro-cohesive strength of soil. It is specified that bonded contacts may carry tension, and the bonds have finite strengths in tension and shear. The contact bonds which reproduce the effect of adhesion over the vanishingly small area of the contact point are chosen to be used in this study, and the contact bonds constitute the tensile normal bond strength (n_bond) and shear bond strength (s_bond). Both bonds can be envisioned as a kind of glue joining the two particles. The contact bond glue is of a vanishingly small size and acts only at the contact point. For a particle flow analysis, it is vital to choose the appropriate material parameters for the model in the analysis. Because of the discrete, particle-based nature of the model, specification of material properties and boundary conditions is more difficult than with available continuum methods. The material parameters for the particle flow analysis are, however, different from the macro-material properties as they are controlled by the micro-properties such as the particle size distribution, packing, particle and bond stiffnesses, particle friction coefficients and bond strengths. The macro-behaviour reflects the average behaviour and is controlled by the micro-material parameters, grading of the materials and state of packing. It is extremely difficult to determine the micro-material parameters directly, so they will vary over a wide range until macro-behaviour is reproduced.

The distinct element approach by Cheng (1998) can reproduce the results obtained by the classical analytical/numerical method. When the applied load is large enough, failure is initiated, which can be captured easily by DEM but not by the classical method. The limitations of the DEM in slope stability analysis include the following:

1. Very long computation time is required.
2. The contact material parameters for the contact cannot be assessed easily.
3. The classical soil parameters cannot be introduced directly in the particle form distinct element analysis.
4. Sensitivity of the method to the various parameters and modelling method is tedious to be assessed.

As an illustration, a 5 m high 45° slope is modelled with DEM by imposing the initial condition in the first step. The vertical stress is basically equal to the overburden stress while an at-rest pressure coefficient 0.5 is employed in the present example. The unit weight of the particle is 17 kN/m^3 while the friction factor is 0.5. Due to rain, a 4 m water table is established, which is equivalent to a body force of −9.81 kN/m^3 applied to the particle system. The slope finally collapses, which is shown in Figure 2.18r. The results

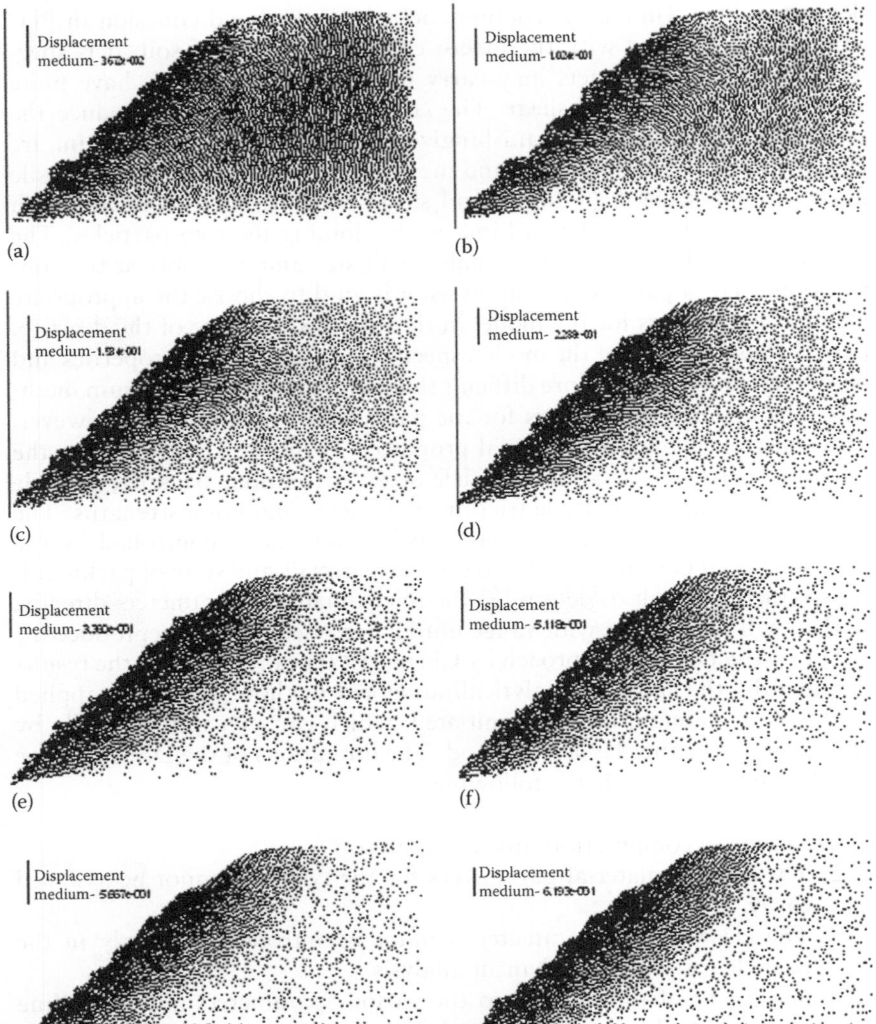

Figure 2.18 Displacement of the slope at different time steps when a 4 m water level is imposed: (a) 500 time steps, (b) 1,500 time steps, (c) 3,000 time steps, (d) 5,000 time steps, (e) 7,000 time steps, (f) 9,000 time steps, (g) 1,000 time steps, (h) 13,000 time steps.

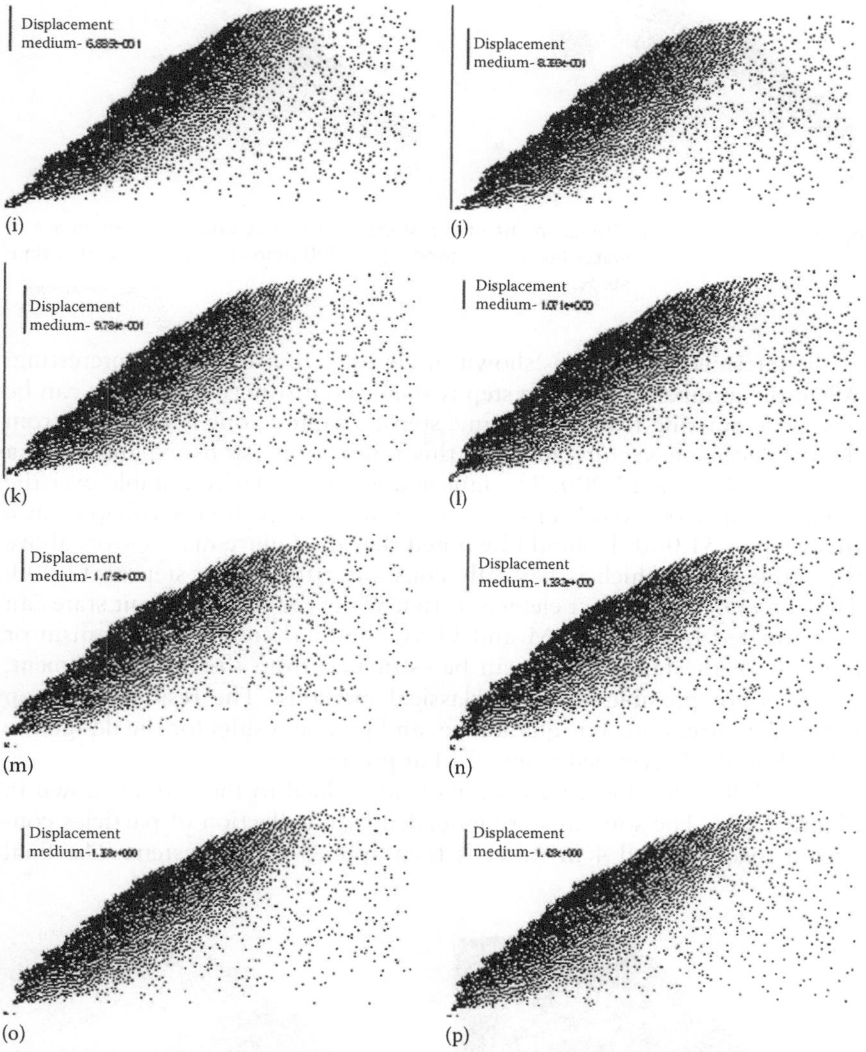

Figure 2.18 (continued) Displacement of the slope at different time steps when a 4 m water level is imposed: (i) 15,000 time steps, (j) 19,000 time steps, (k) 23,000 time steps, (l) 27,000 time steps, (m) 31,000 time steps, (n) 35,000 time steps, (o) 39,000 time steps, (p) 43,000 time steps.

(continued)

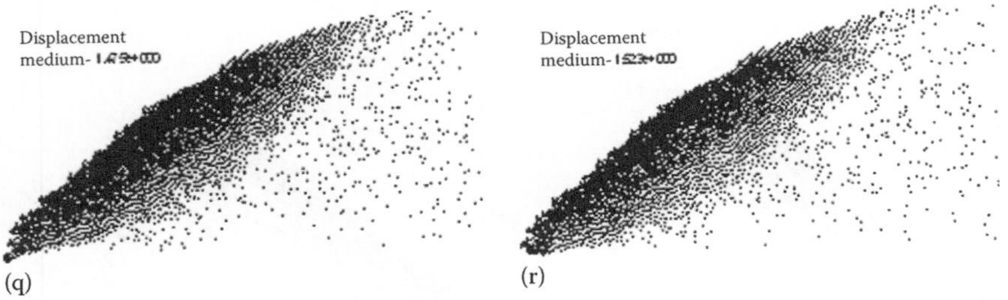

Figure 2.18 (continued) Displacement of the slope at different time steps when a 4 m water level is imposed: (q) 47,000 time steps and (r) 51,000 time steps.

of the intermediate analysis shown in Figure 2.18 are actually interesting. When the number of the time step is small, no distinct failure zone can be observed. Starting from 3,000 time steps, a failure zone is observed from the displacement vector plot, and this failure zone stops expanding at a time step of about 13,000. The failure domain is relatively stable over the remaining analysis and keeps moving until the slope finally collapses at a time step of 51,000. It should be noted that the failure mass moves above the stable zone, which is basically constant after a time step of 13,000. The power of the distinct element is that while the ultimate limit state can be estimated from the LEM and FEM, the final collapse mechanism or the flow of the failure mass can be estimated from the distinct element, which is not possible with the classical methods. The results shown in Figure 2.18 are, however, qualitative, and precise results for the design are difficult to be determined from DEM at present.

To stabilize the slope, two soil nails are added to the system shown in Figure 2.19a. The soil nails are modelled by a collection of particles connected together, and 4 m water is then applied to the system. The final

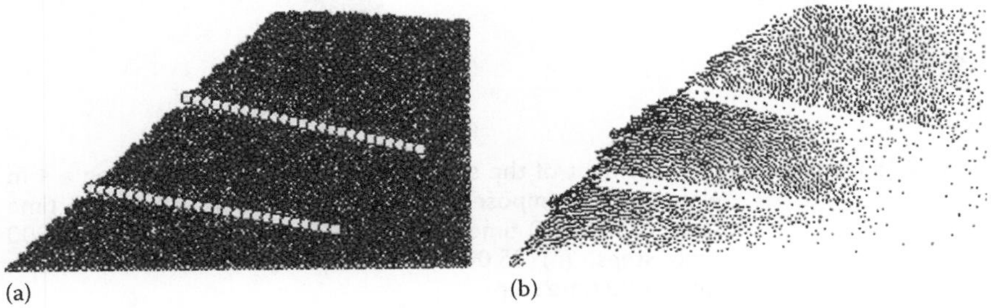

Figure 2.19 Effect of soil nail installation: (a) two soil nails inclined at 10° installed and (b) displacement field after 4 m water is imposed.

displacement field is shown in Figure 2.19b, which indicates that the soil nails have effectively inhibited the collapse of the slope.

2.9.4 Three-dimensional model and distinct element studies of a slope under a patch load

Cheng has carried out a series of studies on the use of DEM on slope stability analysis (Li, 2013; Sun, 2013), and some of the results will be discussed in this section.

Bridge foundation is often built on a slope, especially in a hilly terrain like Hong Kong and many other countries. The slope stability analysis and foundation design are crucial topics for engineers, and French engineers actually define the factor of safety of a slope in terms of the ultimate bridge abutment load against the applied load. In order to study the failure mechanism of a slope under surcharge loading, both macroscopic and microscopic aspects should be taken into account and considered in detail with comparisons. A qualitative study of the failure mechanism of 2D slope stability problems has already been carried out in the previous chapter using DEM. It gives decent results to illustrate the phenomenon and demonstrate the problems. On the other hand, a quantitative study is also required to testify the accuracy of the discrete element analysis. In general, it is difficult to carry out a quantitative study on DEM analysis, and therefore a qualitative study is the main application of DEM analysis at present.

Many previous studies using the DEM to analyse the discrete nature of soils provide insight into the constitutive behaviour of the soil mass. DEM starting with behaviour at the scale of a grain can help understand the constitutive relation without the need for contractive/dilative assumption, flow rule or hardening rule, which are merely curve-fitted by nature. Moreover, it is also widely used to simulate the behaviour of a large range of geomaterials mainly for qualitative analysis. It should be noted that there are only limited applications of DEM for slope stability analysis, especially regarding the progressive failure mechanism of a slope. DEM is also more suitable for qualitative rather than quantitative assessment of the stability of a slope, as it is difficult to define precisely the microscopic parameters, and most of the DEM models cannot completely reflect the grain distribution, initial stress or drainage condition.

There have also been many advances in DEM over the last 20 years, and complex mechanical interactions of a discontinuous system can now be analysed in three dimensions (though still very tedious at present). However, there is very limited literature on particle flow modelling in three dimensions due to the various constraints and difficulties such as image capturing, model generation in three dimensions and large volume of data for manipulation. Traditional approaches in DEM have modelled soil samples as an assembly of 2D discs or 3D spheres. PFC3D is an exclusively commercial 3D DEM program considering the assembly as spheres. It simulates the movement and interaction of spherical particles using DEM, described by Cundall and Strack (1979). PFC3D is designed to be an efficient tool to model complicated problems in solid mechanics and granular flow, but it still takes great effort to develop the desired model and a very long computation time in terms of days or weeks are necessary for a typical problem.

In this section, a laboratory slope model test is carried out with the corresponding 3D numerical simulation using DEM. The progressive failure mechanism of slope under local surcharge is determined both from the model test and numerical analysis for comparison. Furthermore, both qualitative and quantitative studies have been carried out in the present study, which is difficult to find in the literature.

2.9.4.1 Laboratory test on a model slope

Physical slope models were constructed to investigate the failure process of a slope under external local loading. A physical soil tank was built with a layout as shown in Figure 2.20, with the dimensions of the tank being approximately 1.5 m deep × 1.84 m wide × 1.2 m high. The model slope is 0.7 m in height with an angle of 45°. A sectional view of the soil slope model

Figure 2.20 Outlook of soil physical slope.

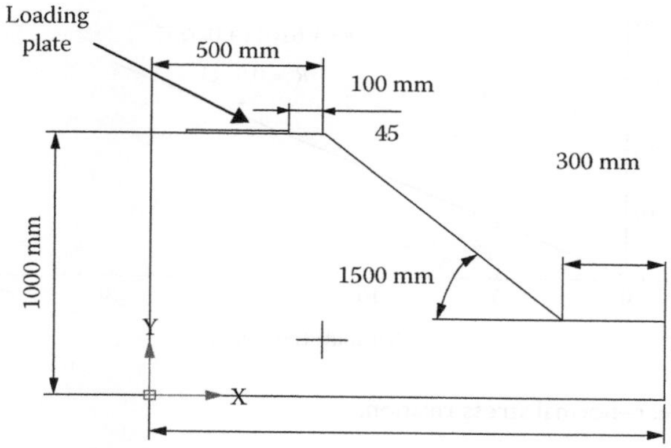

Figure 2.21 Sectional view of the model slope in Figure 2.20.

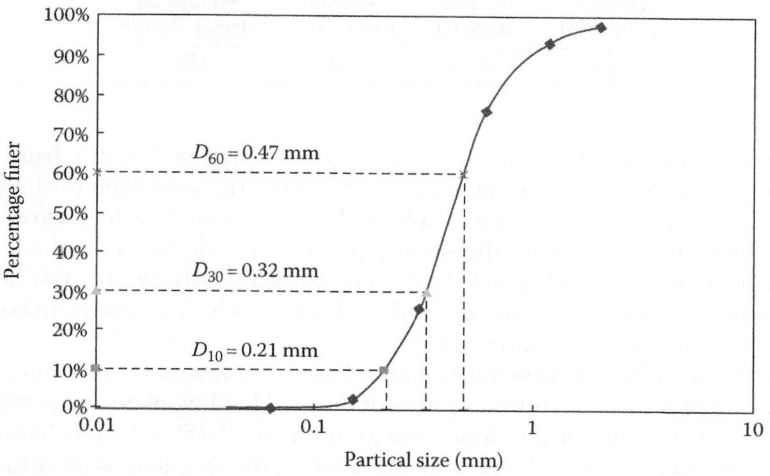

Figure 2.22 Particle size distribution for sand in the model test.

is illustrated in Figure 2.21. Sandy soil with 5% moisture content and particle size distribution shown in Figure 2.22 was compacted to form a slope model with an electric vibrator in seven layers. The average bulk density of the soil model is 1672 kg/m^3. Direct shear test is conducted to determine the shear strength parameters of the river sand used in the laboratory test, and the shear–normal stress relation is illustrated in Figure 2.23, which shows the friction angle of compacted sand to be 58.6° whereas the cohesion is 0.6 kPa. The parameters of sand used for the physical model are listed in Table 2.11.

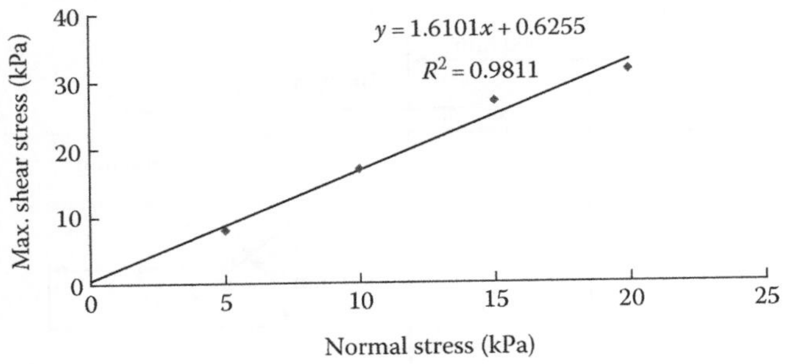

Figure 2.23 Shear–normal stress relation.

Table 2.11 Shear strength parameters of the river sand

Cohesion value (kPa)	*Friction angle (°)*	*Moisture content (%)*	*Average dry density (kg/m³)*
0.6255	58.61	5.2	1877

In this test, surcharge loading was applied on the *I* beam from the hydraulic jack and the loading was transferred to the steel plate to simulate the local distributed load. The loading plate was placed at the centre with 100 mm offset away from the slope crest and the side board of the frame to reduce the end effect. The loading was applied slowly and the test lasted for about 4 h. The maximum applied load was 35 kN. The complete failure processes are given in Figure 2.24.

From the failure process illustrated in Figure 2.24, we can see that cracks occurred first in the surrounding area of the steel loading plate and extended toward each corner of the slope crest at an angle of 45°, which is basically in accordance with the classical theory. During the test, flags with different colours were used to locate the time of appearance and location of cracks during loading, which can also be seen in Figure 2.24. The whole soil mass actually failed gradually instead of undergoing sudden failure. As the steel plate kept going down, the slope inclined surface was covered with vertical cracks and diagonal cracks toward the steel plate, which can be noticed in the intermediate process state of the modelling slope in Figure 2.24a. The soil mass at the slope surface was drawn down layer by layer to the slope toe by the action of the loading steel plate. After 4 h of loading, the sand at the middle top of the slope is pressed to form a big hollow, which can be noticed after moving the steel plate, while the largest cracks were generated in the critical failure surface within the slope body. The global soil mass was pushed down to the toe of the slope gradually, and the physical slope model collapsed eventually, as shown in Figure 2.24b.

(a)

(b)

Figure 2.24 Failure process of soil slope under increasing loading: (a) Intermediate process state and (b) eventual failure state.

2.9.4.2 Three-dimensional distinct numerical modelling of slope under local surcharge

For the present laboratory test, Cheng has adopted DEM in the analysis because of the development of cracks, face failure and the final collapse at the later stage of the test. In DEM, there are several methods of model generation. For the present problem, which is relatively simple in geometry and layout, the desired porosity is obtained by the radius expansion method. By using numerical biaxial tests, the micro-mechanical properties of the assembled material in the numerical models are calibrated in order to match with the macroscopic response of the real material in the physical test. Numerical simulations to reproduce the stress–strain and the normal stress–shear stress relations (Figure 2.23) similar to that by Cheng et al. (2009, 2010) are carried out under the same conditions as the physical experiments such as porosity, boundary conditions and loading. The micro-properties of the river sand as shown in Table 2.12 are determined by varying the micro-properties until the macro-properties obtained numerically match with the experimental results (angle of repose

Table 2.12 Microscopic parameters of the sands for particle flow analysis

Sand	*Diameter (mm)*	*Density of sphere (kg/m³)*	*Friction coefficient*	*Normal and shear stiffness (N/m²)*	*Normal and shear bond strength (N)*	*Friction coefficient of the wall*
Tested sand	0.2–0.5	2650	1.638	1e6	4	0

and stress–strain relation). The diameters of the particles in the DEM model are maintained as the same as those given in Figure 2.20. The frictional coefficient of sand was set to 1.638 (corresponding to a contact friction angle of 58.61° in Table 2.12). The bond strength is fixed at a value of 6N as referred to by Cheng (2003). The particle density of the sandy soil is 2650 kg/m³, while the bulk density for the sand soil should be 1650 kg/m³.

For the 3D DFM numerical simulation shown in Figure 2.25, the dimensions of the numerical model in the particle flow simulation are exactly the same as the physical model. In this study, two different loading patterns are modelled in order to compare their performance: (1) applying the force on the raft footing and (2) adding velocity on the loading wall, and these are illustrated in Figure 2.25. Actually, the results of analyses from these two approaches are very similar, so only the results for approach 1 will be given in this section. The failure process development under loading is illustrated in Figures 2.26 and 2.27. It may be noted that on top of the slope crest, the region below the loaded surface has deformed to form a depression zone from the DEM modelling, which is also observed

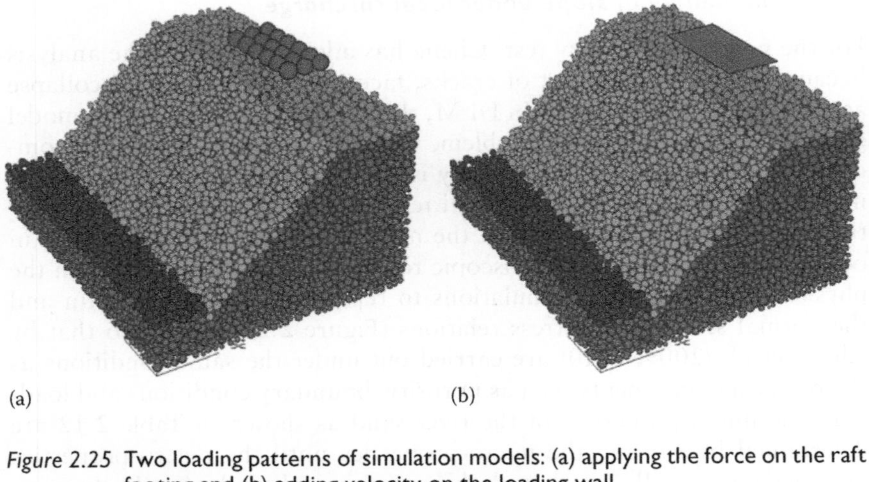

(a) (b)

Figure 2.25 Two loading patterns of simulation models: (a) applying the force on the raft footing and (b) adding velocity on the loading wall.

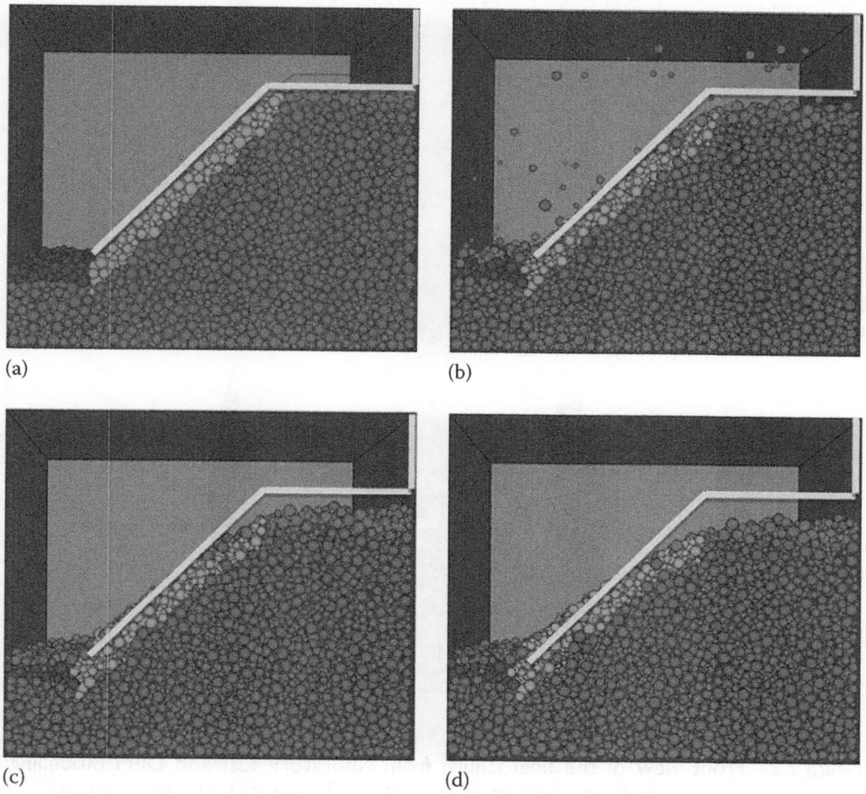

Figure 2.26 Failure process of slope under local loading wall in *XY* direction: (a) initial state, (b) 5×10^4 step, (c) 12×10^4 step and (d) 28×10^4 step.

from the test. Sand particles are triggered to move down the slope, dragging more and more sand downward. The depression zone develops larger and deeper, accompanied with considerable settlement at the top of the slope, and the inclined slope face moves forward with an upheaval at the slope toe as shown in the *XY* direction view in Figure 2.26d. When the ultimate state is reached, a distinct failure mechanism is formed, which is shown in Figures 2.27 and 2.28.

A slope with soil particles having higher bond strength is also simulated, where the normal and shear contact bond strengths of particles are increased to 60 N, which is 10 times larger than that in the former case. The results of the comparison are shown in Figure 2.29. It is observed that a noticeable collapse has taken place in Case 1, as well as a forward movement of the slope body and the extruding slope toe. On the contrary, the soil movement in Case 2 is not major and the slope is actually stable under

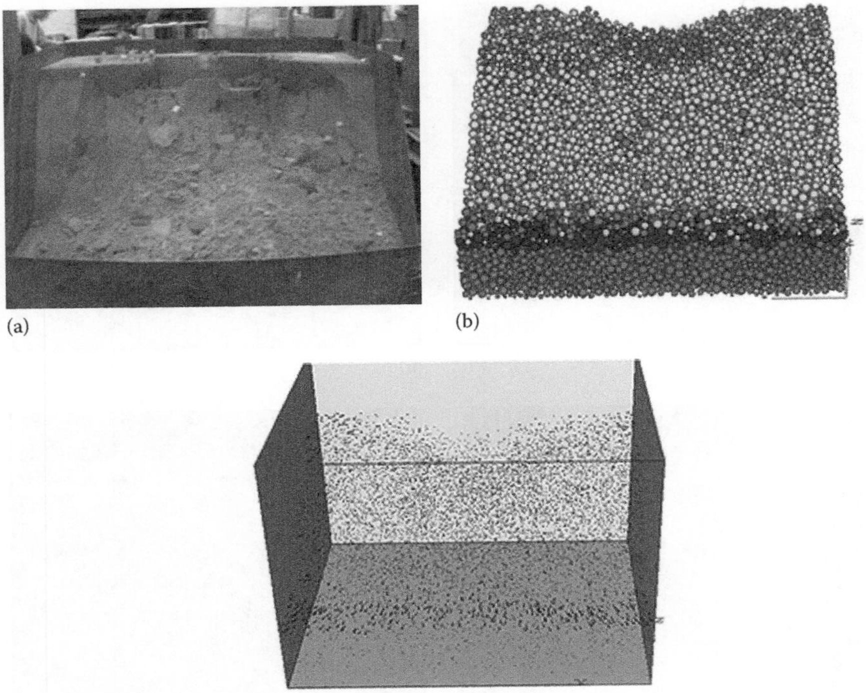

(a) (b) (c)

Figure 2.27 Front view of the final failure from laboratory test and DEM modelling: (a) test result, (b) DEM bond strength = 6 N and (c) DEM displacement.

(a) (b)

Figure 2.28 Side view of the final failure from laboratory test and DEM modelling: (a) lab test of slope without soil nail and (b) DEM displacement vector.

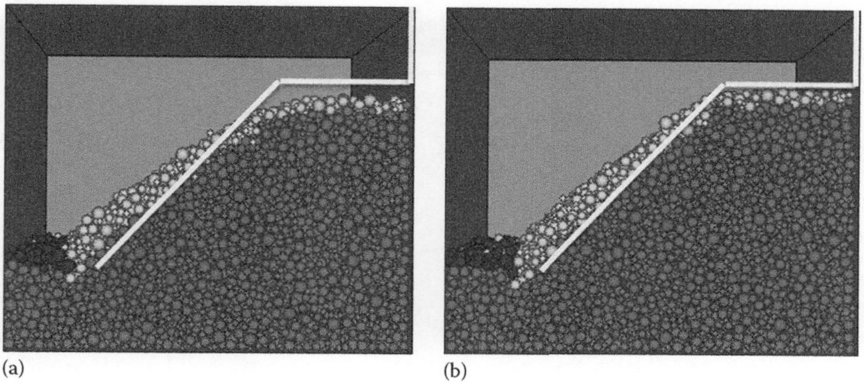

Figure 2.29 Eventual failure of two modelling cases under loading raft in XY direction: (a) Case 1: bond strength = 6 N and (b) Case 2: bond strength = 60 N.

the loading as shown in Figure 2.29b. It has been demonstrated that larger bond strength between soils is followed by slower and smaller failure, or that the soil is more stable under the application of the external load. In fact, the importance of cohesive strength is well recognized by engineers in Hong Kong, and there is a tendency to limit the amount of cohesive strength that can be used for design even if the laboratory tests give high values for the cohesive strength. On the contrary, the friction angle that can be used for the slope design is practically equal to that from the laboratory test in Hong Kong. The rationale behind this design practice in Hong Kong is that cohesive strength is more important than friction angle (for normal problem) in design, and the long-term cohesive strength of soil can be impaired by soil erosion or other processes. It is also interesting to note that the critical slip surface from the laboratory test is close to that from the PFC analysis; further discussion about this laboratory test with 3D limit analysis will be provided in Chapter 6.

Chapter 3

Location of critical failure surface, convergence and advanced formulations

The various methods for the analysis of two-dimensional slope stability problems have been discussed in Chapter 2. There are other issues in slope stability analysis that have not been well addressed in the past, and some of these important issues will be addressed here. In this chapter, only two-dimensional limit equilibrium analysis will be considered, while other methods will be discussed in the later chapters.

3.1 DIFFICULTIES IN LOCATING THE CRITICAL FAILURE SURFACE

As given in Tables 2.1 and 2.2, the critical/minimum factor of safety (FOS) of a slope must be determined before slope stabilization works can be designed. According to the upper bound theory, any prescribed failure surface will be an upper bound to the true solution. For the critical failure surface that corresponds to the global minimum, some of the difficulties and interesting phenomena in locating the critical failure surface will be discussed. Consider a one-dimensional function $y=f(x)$ defined over a solution domain AB as shown in Figure 3.1. The local minima where the gradients of the function are equal to 0 ($f'(x)=0$) are given by points C and D, while the global minimum is defined by point E. If the y-ordinate of B is lower than the y-ordinate of E, point B will then be the global minimum, but the gradient of the function is not equal to 0 at B. Cheng (2003) has demonstrated that this situation can occur for slope stability problem using an example from the ACADS study (1989). Cheng et al. (2012) have constructed another problem for which the global minimum is extremely difficult to be determined using all kinds of methods, and there are practical cases, which are basically similar to the artificial constructed problem as mentioned by Cheng et al. (2012). For multi-variable optimization analysis required by the slope stability problem, the FOS objective function is highly complicated, and the problem will be complicated N-P hard type, which has attracted the attention of many researchers.

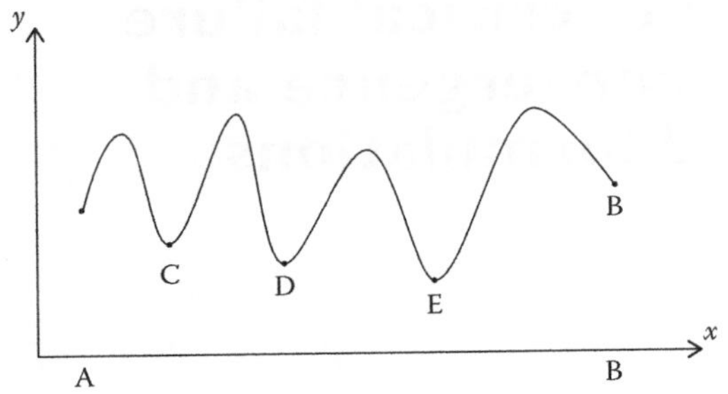

Figure 3.1 A simple one-dimensional function illustrating the local minima and the global minimum.

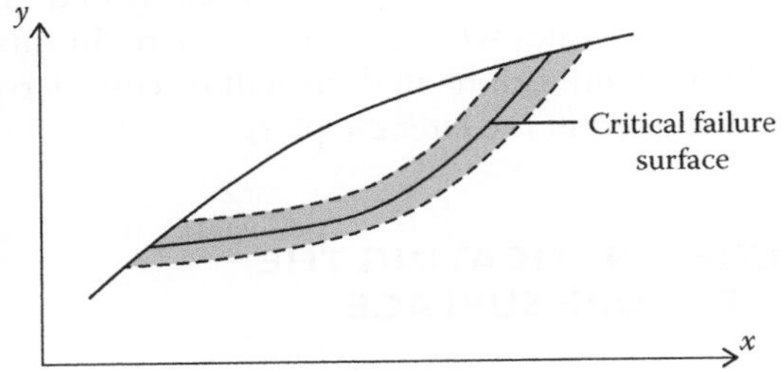

Figure 3.2 Region where FOSs are nearly stationary around the critical failure surface.

Another special feature about the critical failure surface for a simple slope is shown in Figure 3.2. There are only minor changes in the FOSs if the trial failure surfaces fall within the shaded region shown in Figure 3.2. In this respect, there is not a strong need to determine the precise location of the critical failure surface if the geometry and ground conditions for a slope are simple. For complicated slopes or slopes with soft bands, which will be illustrated in this chapter, it is however possible that a minor change in the location of the failure surface can induce a major change in the FOS. Under this condition, the robustness of the optimization algorithm will be important for success in locating the critical solution.

A failure surface can be divided into circular and non-circular failure surfaces. A circular failure surface is actually a subset of the non-circular failure surface, but it is useful because (1) some stability formulations apply to circular failure surface only and (2) critical circular failure surface is a good approximation to the critical solution for some simple problem and is simple to be evaluated. For circular failure surface, the location of the critical failure surface is usually determined by the method of grids shown in Figure 3.3. There are three control variables in this case: the

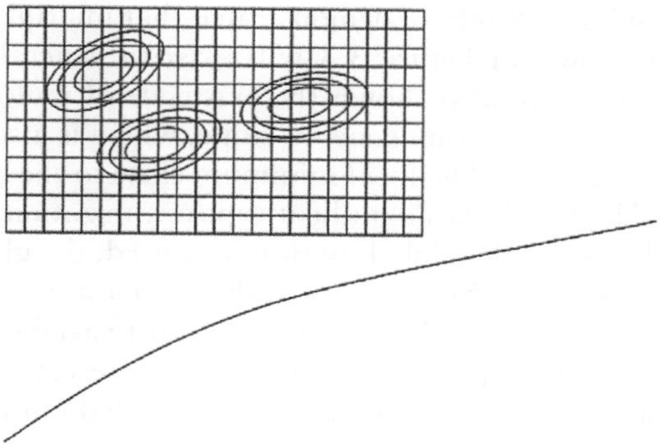

Figure 3.3 Grid method and presence of multiple local minima.

x- and y-ordinates of the centre of rotation and the radius of the failure surface. Each grid point is used as the centre of rotation, while different radii are considered for the circular failure surface, and the minimum FOS from different radii is assigned to this grid point. Different FOSs are hence assigned to different grid points, and the trend of the global minimum can be assessed by drawing the FOS contours from the FOSs associated with the gird points. This method is robust and is simple to operate, but accuracy would depend on the spacing between the grid points. The specified grid must also be large enough to embrace all the possible local minima and the global minimum in order to obtain a clear picture about the distribution of the FOS. The grid method is simple to implement and is available in most of the commercial slope stability programmes.

For a general non-circular failure surface, the number of control variables, which is controlled by the number of points for the failure surface, is usually much more than 3. To locate the critical failure surface, the geometric method similar to that for the circular failure surface will be very inefficient in application and requires effort from lots of engineers in defining the solution domain for each control variable (though adopted by some commercial programmes). Special features of the objective function of the safety factor F for this case include the following:

1. The objective function of the safety factor F is usually non-smooth, non-convex and discontinuous over the solution domain. Discontinuity of the objective function can be generated by: generation of an unacceptable failure surface; 'failure to converge' of the objective function; and presence of obstructions in the form of sheet piles, retaining walls, large boulders, tension cracks or others. Gradient-type optimization methods are applicable only to continuous functions and will break down if there are discontinuities in the objective function.

2. Chen and Shao (1983) have demonstrated that multiple minima similar to that shown in Figure 3.3 will exist in general. Duncan and Wright (2005) have also shown the existence of multiple local minima even for a simple homogeneous slope, which is also illustrated by Cheng et al. (2007b). The local minimum close to the initial trial will be obtained by the classical gradient-type optimization methods. If an initial trial close to the global minimum is used, the global minimum can usually be found by classical methods, but a good initial trial is difficult to be established for a general multi-variable problem. The success of a global optimization algorithm to escape from the local minima for an initial solution far from the global minimum is crucial in the slope analysis problem.
3. A good optimization algorithm should be effective and efficient over different topography, soil parameters and loadings. The analysis should also be insensitive to the optimization parameters as well.

Various classical optimization methods for the non-circular failure surface have been proposed and used in the past. Baker and Gaber (1978) have proposed the use of variational principle, but this method is complicated even for a simple slope and is not adopted for practical problems. Moreover, if the gradient of the global minimum is not 0, the variational principle will miss the critical solution. Chen and Shao (1983) and Nguyen (1985) have suggested the use of the simplex method for this problem, which is actually suitable only for linear problems. The simplex method has been adopted by the programme EMU developed by Chen and it works fairly well for simple problems. Cheng has however come across many complicated cases in China where manual interaction is required with the simplex method before a good solution can be found. The simplex method also fails to work automatically for cases where the local minimum and global minimum differ by a very small value, but differ significantly in location. Celestino and Duncan (1981) have adopted the alternating variable method, whereas Arai and Tagyo (1985) and Yamagami and Jiang (1997) have adopted the conjugate-gradient method and dynamic programming, respectively. These classical methods are applicable mainly to continuous functions, but they are limited by the presence of the local minimum, as the local minimum close to the initial trial will be obtained in the analysis. There is also a possibility that the global minimum within the solution domain is not given by the condition that the gradient of the objective function $\nabla f' = 0$, and a good example has been illustrated by Cheng (2003). The presence of the other local minima or the global minimum will not be obtained by the classical methods unless a good initial trial is adopted, but a good initial trial is difficult to be established for a general problem.

In view of the limitations of the classical optimization methods, the current approach to locate the critical failure surface is the adoption of the heuristic global optimization methods. The term heuristic is used for algorithms that find solutions among all possible ones, but they do not guarantee that the best would be found, and therefore, they may be considered as approximate and not accurate algorithms. These algorithms usually find a solution close to the best one, and they find it fast and easily. Another important feature is that the requirement of human judgment or interaction can be minimized or even eliminated, if possible, and Cheng has come across some hydropower projects in China where there are several weak zones (strong local minima) for which nearly all existing methods fail to work well.

Greco (1996) and Malkawi et al. (2001) have adopted the Monte Carlo technique for locating the critical slip surface, with success in some cases, but there is no precision control on the accuracy of the global minimum. Zolfaghari et al. (2005) adopted the genetic algorithm (GA), while Bolton et al. (2003) used the leap-frog optimization technique to evaluate the minimum FOS. All of the methods mentioned earlier are based on the use of static bounds to the control variables, which means that the solution domain for each control variable is fixed and pre-determined by engineering experience. Cheng (2003) has developed a procedure that transformed the various constraints and the requirement of a kinematically acceptable failure mechanism to the evaluation of the upper and lower bounds of the control variables, and the simulated annealing algorithm (SA) is used to determine the critical slip surface. The control variables are defined with dynamic domains, which are changing during the solution, and the bounds are controlled by the requirement of a kinematically acceptable failure mechanism. Through such an approach, there is no need to define the pre-determined static solution domain to each control variable based on engineering experience, and precision control during the search for the critical solution will be possible.

There are two major aspects in the location of the critical failure surface, which will be discussed in the following sections: the generation of the trial failure surface and the global optimization algorithms for the search of the critical failure surface.

3.2 GENERATION OF TRIAL FAILURE SURFACE

For the classical gradient-type optimization method, once an initial trial is defined, the refinement of the critical failure surface will be given by the gradient of the objective function (which can be obtained by a simple finite-difference operation). On the other hand, for the heuristic global optimization methods, trial failure surfaces are required to be generated, which are controlled by the bounds for each control variable. Different methods in generating the failure surfaces have been proposed by Greco (1996),

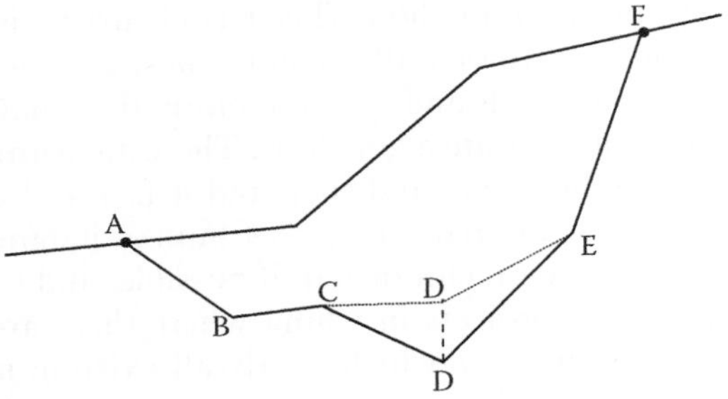

Figure 3.4 A failure surface with a kink or non-convex portion.

Malkawi et al. (2001), Cheng (2003), Cheng et al. (2007b,d, 2008c, 2012), Bolton et al. (2003), Li et al. (2005) and Zolfaghari et al. (2005). In general, these methods are very similar in the basic operations. The coordinates of the points defining the failure surface are taken as the control variables, and lower and upper bounds are assigned to each control variable. Consider the failure defined by ABCDEF shown in Figure 3.4. If each control variable is defined over static lower and upper bounds, point D, which is unlikely to be acceptable for a normal problem, can be generated by the random number generator. Since segment CD will be a kink that hinders the development of the failure, D is highly unlikely to be acceptable except for some special cases, which will be discussed later.

To generate a convex surface by the method proposed by Cheng (2003), consider a typical failure surface ACDEFB as shown in Figure 3.5. The x-ordinates of the two exit ends A and B are taken as the control

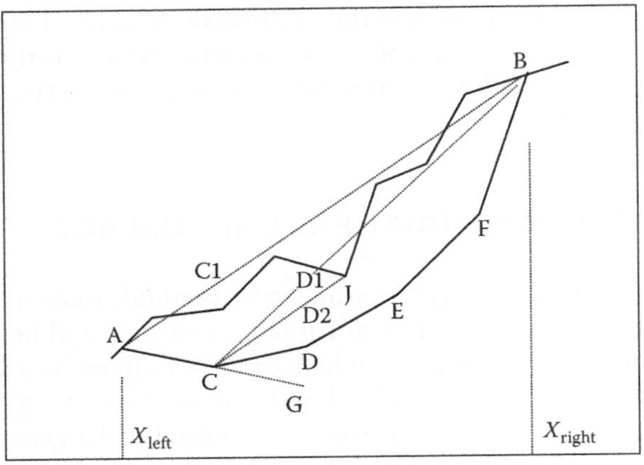

Figure 3.5 Generation of dynamic bounds for a non-circular surface.

variables of the objective function and the upper and lower bounds of these two variables are specified by the engineer (bounds for first two control variables are fixed). The static bounds for the first two control variables can be defined easily for the present problem with engineers' experience. Once the two exit ends A and B of the failure surface are defined, the requirements on a kinematically acceptable mechanism can then be implemented as follows:

1. The *x*-ordinates of the interior points C, D, E and F of the failure surface can be obtained by uniform division of the horizontal distance between A and B, which is $X_{right}-X_{left}$. The *x*-ordinates of C, D, E and F are hence not control variables. Alternatively, the division can be made to follow the slope profile and the *x*-ordinates of the interior points are also not control variables.
2. Points A and B are connected and C1 is determined as a point located vertically above C. The *y*-ordinate of C1 is the lower value of either (1) the *y*-ordinate of the ground profile as determined by the *x*-ordinate of C or (2) the *y*-ordinate of the point lying along the line joining points A and B and determined by the *x*-ordinate of C. C1 is the upper bound to the *y*-ordinate of the first inter-slice. The lower bound of the *y*-ordinate of C (third control variable) is set by Cheng (2003) as C1–AB/4. In fact, such a lower bound can allow a deep-seated failure surface and is adequate for all cases that Cheng has encountered. The lower bound of the *y*-ordinate of C can be set to C1–AB/5 (instead of C1–AB/4, which is a conservative estimation of the lower bound) in most situations without affecting the solution. The *y*-ordinate of point C is a control variable of the objective function and it is confined within the upper and lower bounds as determined in Step 2.
3. Once a *y*-ordinate of C is chosen in the simulated annealing analysis, connect A and C and extrapolate the line to G, which is defined by the *x*-ordinate of point D. The lower bound of the *y*-ordinate of point D will be point G in order to maintain a concave failure shape. The upper bound of D, which is D1, is determined in the same way as for point C1. If part of the ground profile lies below the line joining B and C and affects the determination of D1 (e.g. point J in Figure 3.5), connect C and J instead of B and C and determine the upper bound as D2 instead of D1.
4. Perform Step 3 for the remaining points until all upper and lower bounds of the control variables have been defined.
5. To allow a non-concave failure surface, which is unlikely to occur in reality, an option is allowed where the lower bound of point E will be set to the lower bound of the value as determined in Step 3 or the *y*-ordinate of point D. The *y*-ordinate of point E cannot be lower than

that of D or else there will be a kink in the failure surface, which prevents the failure to occur. The lower bound to the y-ordinate is sometimes totally eliminated, which is required for problem with a soft band. A non-convex failure surface can hence be generated from the present proposal by removing the lower bound requirement as required in the present method.

In Figure 3.5, the control variables are the x-ordinates of A and B and the y-ordinates of points C, D, E and F. A control variable vector X is used to store these control variables and the order of the control variables must be (X_A, X_B, Y_C, Y_D, Y_E, Y_F). For the location of the global minimum of the objective function, engineers need to define only the upper and lower bounds of the first two control variables. An initial trial will be determined in a way similar to the approaches in the steps mentioned. The upper and lower bounds of the other control variables will then be calculated according to Steps 2 and 3 mentioned earlier. If the number of slices is n, then the number of control variables will be $n+1$. If a rock is encountered in the problem, the lower bound determination shown earlier has to be modified slightly. In Steps 2 and 3, the lower bound will either be the y-ordinate of point G or the y-ordinate of the rock profile as determined by the x-ordinate of D.

For a circular failure surface, there are only three control variables, which are the x and y coordinates of the centre of rotation and the radius of the failure surface. Cheng (2003), however, adopts the x-ordinates of the two exit ends and the radius of the failure surface as the three control variables in the analysis as it is easier to define the upper and lower bounds for the two exit ends (see Figure 3.6). This approach is also used by many commercial programmes. The control variable vector X will be (X_A, X_B, r). For the lower and upper bounds of the radius, the lower bound is set to half of the length of line AB, which is the minimum possible radius. The upper

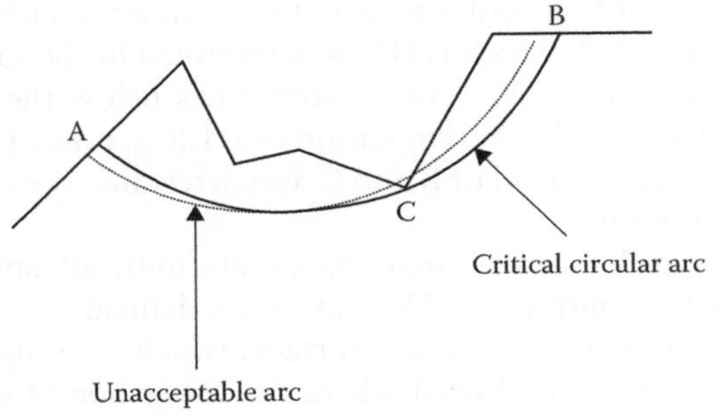

Figure 3.6 Dynamic bounds to an acceptable circular surface.

bound of the radius is set to 50 times AB (any value that is not too small would be acceptable). Unacceptable failure surface will not be generated in the analysis and the constraints will control the lower and upper bounds of the radius when the two exit ends are defined. The constraints include the following:

1. The failure surface cannot cut the ground profile at more than two points within the two exit ends. As shown in Figure 3.6, point C will control the upper bound of the radius.
2. The failure surface cannot cut into a rock stratum, which will control the lower bound of the radius.
3. The y-ordinate of the centre of rotation is higher than the y-ordinate of the right exit end. For this case, the last slice cannot be defined. This constraint will also control the lower bound of the radius.

In the present method, the first two variables, which are the x-ordinates for the left and right ends, are varied within the user-defined lower and upper bounds, which are constant during the analysis. Besides these two variables, the bounds for the remaining variables (y-ordinates of failure surface) are computed sequentially according to the guidelines shown earlier for circular and non-circular failure surfaces. The bounds from the present method are dynamic and are different from classical simulated annealing methods or other global optimization methods where the bounds remain unchanged during analysis. The generation of trial failure surfaces and the search direction will then proceed in accordance with the normal simulated annealing procedure and the global minimum can be located easily with very high accuracy under the present proposal. The minimization process in the present formulation will depend on the lower and upper bounds of the left and right exit ends shown in Figure 3.7, which can be decided easily with experience and engineering principle. For inexperienced engineers, wider ranges can be defined for the lower and upper bounds and the number of trials required for analysis will increase only slightly with increase in the left and

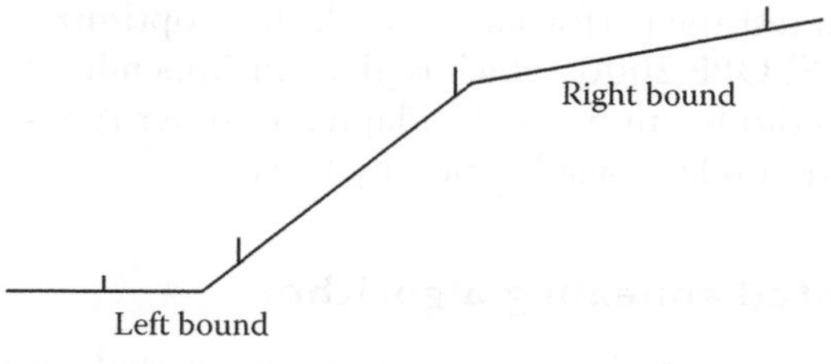

Figure 3.7 Domains for the left and right ends decided by engineers to define a search for the global minimum.

right ranges, which is another major advantage of the approach of Cheng (2003). For example, Cheng (2003) finds that when the ranges for the left and right exit ends are increased by two times, the number of trials required will remain unchanged in many cases and may increase by less than 15% in some rare cases.

In the present algorithm, the *x*-ordinates are not considered as the control variables in order to reduce the number of control variables. This is usually satisfactory as Cheng (2003) finds that the *y*-ordinates are more important than the *x*-ordinate in the FOS. Cheng et al. (2007d) have also proposed that the *x*-ordinates can be adopted as the control variables. This approach will approximately double the number of control variables, and is considered to be useful only for problems controlled by soft bands where the FOS is highly sensitive to the *x*-ordinates as well.

3.3 GLOBAL OPTIMIZATION METHODS

Global optimization problems are typically difficult to be solved, and in the context of combinatorial problems, they are often N–P hard type. The difficulties in performing the global optimization analysis and the requirement for a robust optimization algorithm have been discussed in Section 3.1. With developments in computer software and hardware, many artificial intelligence–based algorithms based on natural selection and mechanisms of population genetics have been developed. These algorithms are commonly applied in pattern recognition, electronics, production/control engineering or signal-processing systems. These new heuristic optimization algorithms have been applied successfully to many different disciplines for both continuous and discrete optimization problems, but there are only limited uses of these methods in slope stability problems.

Since most of the heuristic algorithms, which are artificial intelligence-based methods, are relatively new and are not familiar to geotechnical practitioners, a brief review on several simple but effective methods (with various improvements by Cheng et al., 2007b) will be given in this section. The readers can test the performance of all these optimization methods by using the demo SLOPE 2000, which is given in Appendix A. These modern optimization methods can be easily adapted to other types of geotechnical problems that are under consideration by Cheng.

3.3.1 Simulated annealing algorithm

The SA (Kirkpatrick et al., 1983) is a combinatorial optimization technique based on the simulation of the very slow cooling process of heated metal, called annealing. The concept of this algorithm is similar to heating a solid to a high temperature, and then cooling the molten material slowly in a controlled manner until it crystallizes, which is the minimum energy level of the system. The solution starts with a high temperature t_0, and a sequence of trial vectors is generated until the inner thermal equilibrium is reached. Once the thermal equilibrium is reached at a particular temperature, the temperature is reduced by using the coefficient λ and a new sequence of moves would start. This process is continued until a sufficiently low temperature t_e is reached, at which no further improvement in the objective function can be achieved.

The flowchart of the SA is shown in Figure 3.8, where t_0, t_e and λ are the initial temperature, the stopping temperature and the cooling temperature coefficient, respectively. Usually, the higher the value of t_0, the lower the value of t_e, and the smaller the value of λ, the more trials would be required in the optimization analysis. The parameter N identifies the number of iterations for a given temperature to reach its inner thermal equilibrium, and the array *ft*(***neps***) restores the objective function values obtained at the consecutive ***neps*** inner thermal equilibriums and terminates the optimization algorithm. $\mathbf{V}_g$, f_g are the best solution found so far and its associated objective function value. *Nit* is the number of iterations for the current temperature. r_s is a random number in the range [0,1], after N iterations are performed. If the termination criterion is not satisfied, $\mathbf{V}_g$, f_g are given to $\mathbf{V}_0$, f_0, and the procedure by Cheng (2003) is different from the classical SA in that the best solution found so far is used instead of the randomly adjusted solution to generate the next solution.

3.3.2 Genetic algorithms

The GA is developed by Holland (1975) and has received great attention in various disciplines. It is an optimization approach based on the concepts of genetics and natural reproduction, and evolution of living creatures, in which an optimum solution evolves through a series of generations. Each generation consists of a number of possible solutions (individuals) to the problem, defined by an encoding. The fitness of an each individual within the generation is evaluated, and it influences the reproduction of the next generation. The algorithm starts with an initial population of M individuals. An individual is composed of real coordinates associated with the variables of the objective function. The current generation is called parent generation, by which offspring generations are created using operators such as crossover and mutation. Another M individuals are re-chosen from the parent and offspring generations according to their fitness value. The flowchart for the GA is given in Figure 3.9, where ρ_c and ρ_m are the probability of the crossover and mutation of the algorithm, respectively.

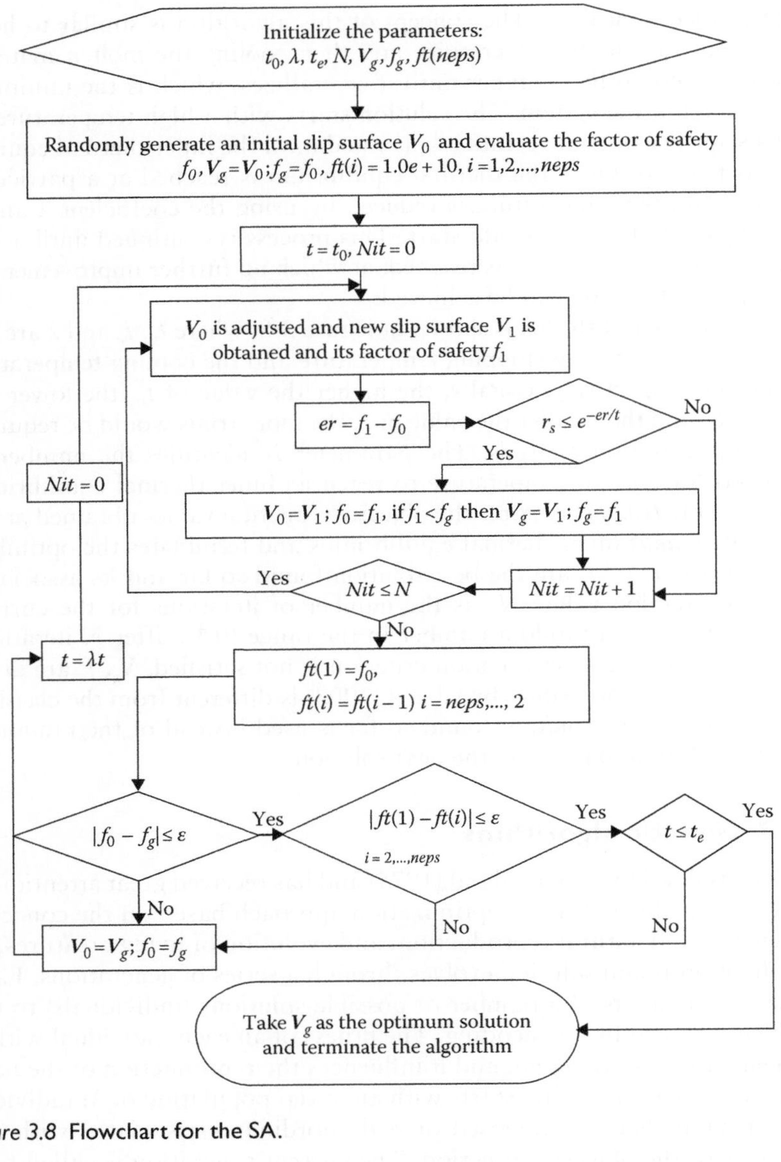

Figure 3.8 Flowchart for the SA.

Usually, the value of ρ_c varies from 0.8 to 0.9, whilst ρ_m falls in the range of 0.001–0.1. N_1 represents the number of iterations in the first stage, while N_2 represents the time interval by which the termination criterion is defined. If the best individual with the fitness value f_g remains unchanged after N_2 iterations, the algorithm will stop. *Niter* is the variable restoring

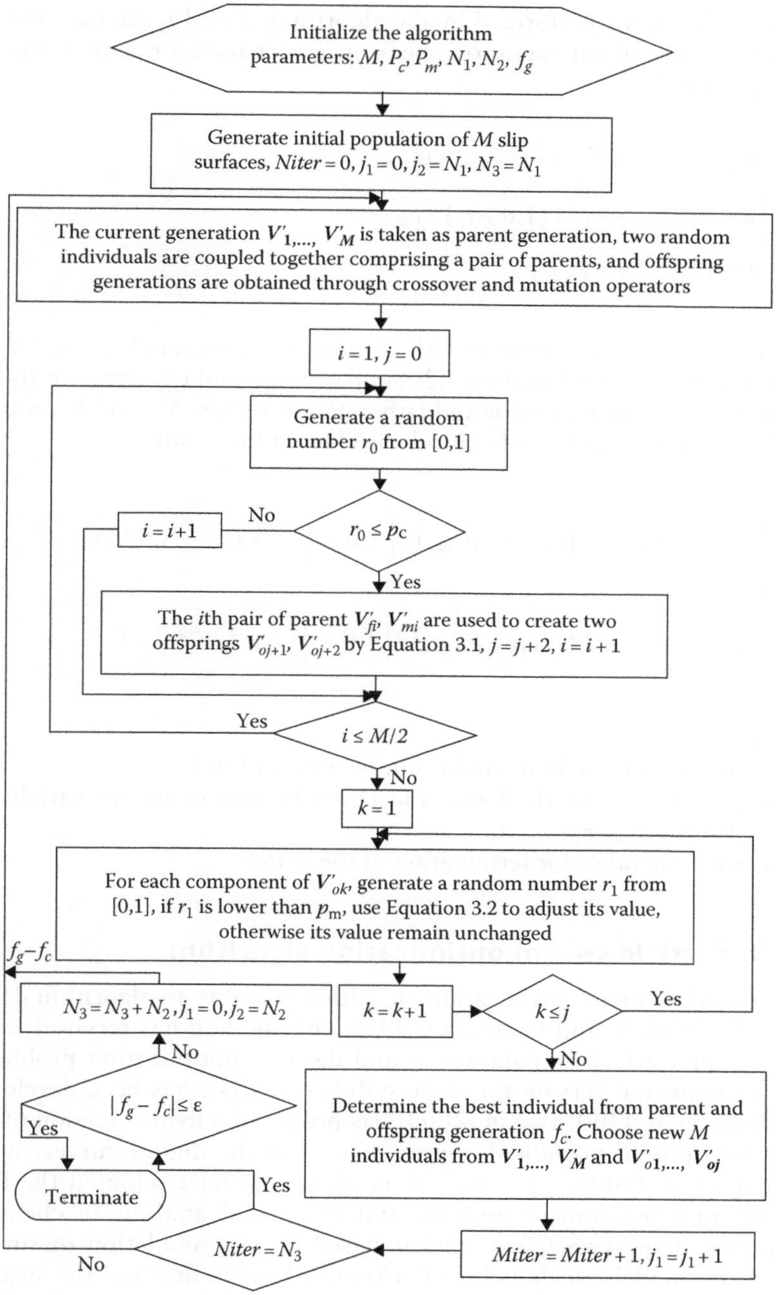

Figure 3.9 Flowchart for the GA.

the total iterations performed by the algorithm. j_1 and j_2 are used to perform the non-uniform mutation operator. The crossover operator is given by Equation 3.1:

$$\begin{cases} v_{oj+1,l} = v_{fi,l} \times r_c + (1.0 - r_c) \times v_{mi,l} \\ v_{oj+2,l} = v_{mi,l} \times r_c + (1.0 - r_c) \times v_{fi,l} \\ l = 1,2,\ldots,n+1 \end{cases} \tag{3.1}$$

where $v_{oj+1,l}$ and $v_{oj+2,l}$ mean the lth element of the vectors $\mathbf{V}_{oj+1}$ and $\mathbf{V}_{oj+2}$, respectively, given by Equation 3.2. Similarly, $v_{mi,l}$ and $v_{fi,l}$ represent the lth element of the mother parent and father parent vectors $\mathbf{V}_{mi}$ and $\mathbf{V}_{fi}$, respectively. $n+1$ is the number of control variables in this study.

$$\begin{cases} v_{oj+1,l} = v_{oj+1,l} - (v_{oj+1,l} - v_{l\min}) \times \left(1.0 - \dfrac{j_1}{j_2}\right)^2 \times r_m & r_{nd} \le 0.5 \\ v_{oj+1,l} = v_{oj+1,l} + (v_{l\max} - v_{oj+1,l}) \times \left(1.0 - \dfrac{j_1}{j_2}\right)^2 \times r_m & r_{nd} > 0.5 \end{cases} \tag{3.2}$$

where

r_m and r_{nd} are random numbers in the range of 0–1

$v_{l\min}$ and $v_{l\max}$ are the lower and upper bounds to the lth variable in $\mathbf{V} = (x_1, x_{n+1}, \sigma_2, \ldots, \sigma_n)$

ε is the tolerance for termination of the search

3.3.3 Particle swarm optimization algorithm

The particle swarm optimization algorithm (PSO) is an algorithm developed by Kennedy and Eberhart (1995). This method has received widespread applications in continuous and discrete optimization problems, and an improved version for slope stability analysis has been developed by Cheng et al. (2007a). Yin (2004) has proposed a hybrid version of the PSO for optimal polygonal approximation of the digital curves, while Salman et al. (2002) and Ourique et al. (2002) have adopted the PSO for the task assignment problem and dynamical analysis in chemical processes, respectively. The PSO is based on the simulation of simplified social models, such as bird flocking, fish schooling and the swarming theory. It is related to evolutionary computation procedures and has strong ties with the GAs. This method has been developed on a very simple theoretical framework, and it can be implemented easily with only primitive mathematical operators. Besides, it is computationally inexpensive in terms of both computer memory requirements and speed of the computation.

In the PSO, a group of particles (generally double the number of the control variables, M) referred to as the candidates or the potential solutions (as $\mathbf{V}$ described earlier) flown in the problem search space to determine their optimum positions. This optimum position is usually characterized by the optimum of a fitness function (e.g. FOS for the present problem). Each 'particle' is represented by a vector in the multidimensional space to characterize its position $\left(\mathbf{V}_i^k\right)$ and another vector to characterize its velocity $\left(\mathbf{W}_i^k\right)$ at the current time step k. The algorithm assumes that particle i is able to carry out simple space and time computations in order to respond to the quality environment factors. That is, a group of birds can determine the average direction and speed of flight during search for food, based on the amount of food found in certain regions of space. The results obtained at the current time step k can be used to update the positions of the next time step. It is also assumed that the group of particles is able to respond to environmental changes. In other words, after finding a good source of food in a certain region of space, the group of particles will take this new piece of information into the consideration to formulate the 'flight plan'. Therefore, the best results obtained throughout the current time step are considered to generate the new set of positions for the whole group.

In order to optimize the fitness function, the velocity $\mathbf{W}_i^k$ and hence the position $\mathbf{V}_i^k$ of each particle are adjusted in each time step. The updated velocity $\mathbf{W}_i^{k+1}$ is a function of three major components:

1. The old velocity of the same particle $\left(\mathbf{W}_i^k\right)$
2. Difference between the ith particle's best position found so far (called P_i) and the current position of the ith particle $\mathbf{V}_i^k$
3. Difference between the best position of any particle within the context of the topological neighbourhood of ith particle found so far (called P_g, its objective function value called f_g) and the current position of the ith particle $\mathbf{V}_i^k$

Each of the components 2 and 3 mentioned are stochastically weighted and added to component 1 to update the velocity of each particle, with enough oscillations that should empower each particle to search for a better pattern within the problem space. In brief, each particle employs Equation 3.3 to update its position.

$$\mathbf{W}_i^{k+1} = \omega \mathbf{W}_i^k + c_1 r_1 \left(\mathbf{P}_i - \mathbf{V}_i^k\right) + c_2 r_2 \left(\mathbf{P}_g - \mathbf{V}_i^k\right)$$

$$\mathbf{V}_i^{k+1} = \mathbf{V}_i'^k + \mathbf{W}_i^{k+1} \quad (3.3)$$

$$i = 1,2,\ldots,2n$$

where

c_1 and c_2 are responsible for introducing the stochastic weighting to components 2 and 3, respectively; these parameters are commonly chosen as 2, which will also be used in this study

r_1 and r_2 are two random numbers in the range [0,1]

ω is the inertia weight coefficient

A larger value of ω will enable the algorithm to explore the search space, while a smaller value of ω will lead the algorithm to exploit the refinement of the results. Chatterjee and Siarry (2006) have introduced a non-linear inertia weight variation for the dynamic adaptation in the PSO. The flowchart for the PSO in searching for the critical slip surface is shown in Figure 3.10.

The termination criterion for the PSO is not stated explicitly by Kennedy and Eberhart (1995) (same for other modern global optimization methods). Usually a fixed number of trials are carried out, with the minimum value from all the trials taken as the global minimum, and this is the limitation of the original PSO or other global optimization algorithms. Based on the termination proposal by Cheng et al. (2007a), if $\mathbf{P}_g$ remains unchanged after N_2 iterations are performed, the algorithm will terminate as shown by the following equation:

$$\left|f_{\mathrm{sf}} - f_g\right| \le \varepsilon \quad (3.4)$$

where

$\mathbf{V}_{\mathrm{sf}}$, f_{sf} mean the best solution found so far and its related objective function value

ε is the tolerance of termination

All global optimization methods require some parameters that are difficult to be established for general problems. Based on extensive internal tests, it is found that the PSO is not sensitive to the optimization parameters in most problems, which is an important consideration for recommending this method to be used for slope stability analysis.

3.3.4 Simple harmony search algorithm

Geem et al. (2001) and Lee and Geem (2005) developed a harmony search meta-heuristic algorithm that was conceptualized using the musical process

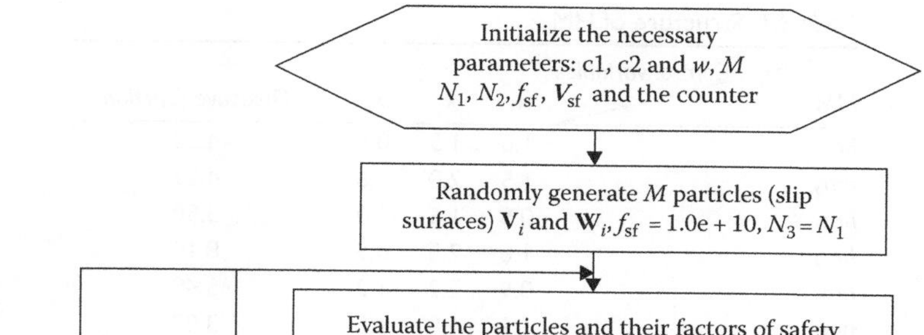

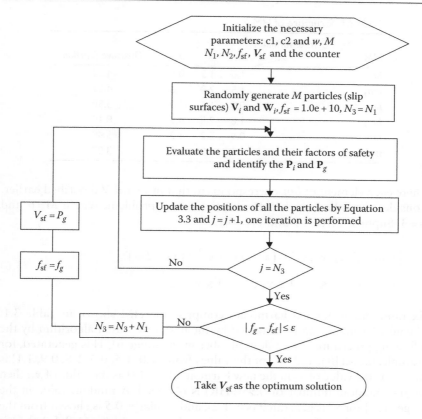

Figure 3.10 Flowchart for the particle swarm optimization method.

of searching for a perfect state of harmony. Musical performances seek to find pleasing harmony (a perfect state) as determined by an aesthetic standard, just as the optimization process seeks to find a global solution determined by an objective function. The harmony in music is analogous to the optimization solution vector, and the musician's improvisations are analogous to the local and global search schemes in the optimization process. The simple harmony search algorithm (SHM) uses a stochastic random search that is based on the harmony memory considering rate HR and the pitch adjusting rate PR, and it is a population-based search method. A harmony memory HM of size M is used to generate a new harmony, which is probably better than the optimum in the current harmony memory. The harmony memory consists of M harmonies (slip surfaces) and M harmonies are usually generated randomly. Consider $\mathbf{HM} = \{\boldsymbol{hm}_1, \boldsymbol{hm}_2, \ldots, \boldsymbol{hm}_M\}$:

$$hm_i = \left(v_{i1}, v_{i2}, \ldots, v_{im}\right) \tag{3.5}$$

Table 3.1 Structure of HM

HM \ *Control variables*	x_1	x_2	x_3	*Objective function*
hm_1	1.0	1.5	0.5	4.50
hm_2	1.5	2.0	1.8	4.29
hm_3	0.5	1.5	1.0	3.50
hm_4	1.8	2.5	0.9	8.10
hm_5	0.9	2.2	1.2	5.49
hm_6	1.1	1.9	1.5	3.87

where each element of hm_i corresponds to that in vector **V** described earlier. Consider the following function optimization problem, where $M=6$ and $m=3$. Suppose HR=0.9 and PR=0.1:

$$\begin{cases} \min \quad f(x_1,x_2,x_3)=(x_1-1)^2+x_2{}^2+(x_3-2.0)^2 \\ s.t. \quad 0 \le x_1 \le 2 \quad 1 \le x_2 \le 3 \quad 0 \le x_3 \le 2 \end{cases} \tag{3.6}$$

Six randomly generated harmonies comprise the HM shown in Table 3.1. The new harmony can be obtained by the harmony search algorithm by the following procedures. A random number in the range [0, 1] is generated, for example, 0.6(<HR), and one of the values from {1.0, 1.5, 0.5, 1.8, 0.9, 1.1} is chosen as the value of x_1 in the new harmony. Take 1.0 as the value of x_1; then another random number of 0.95(>HR) is obtained. A random value in the range [1, 3] for x_2 is generated (say 1.2), and similarly, 0.5 is chosen from the HM as the value of x_3; thus a coarse new harmony $hm_n' = (1.0,1.2,0.5)$ is generated. The improved new harmony is obtained by adjusting the coarse new harmony according to the parameter PR. Suppose three random values in the range [0, 1] (say 0.7, 0.05, 0.8) are generated. Since the former value 0.7 is greater than PR, the value of x_1 in hm_n' remains unchanged. The second value 0.05 is lower than PR, and so the value of 1.2 should be adjusted (say 1.10). The procedures mentioned proceed until the final new harmony $hm_n=(1.0,1.10,0.5)$ is obtained. The objective function of the new harmony is determined as 3.46. The objective function value of 3.46 is better than that of the worst harmony hm_4, and therefore, hm_4 is excluded from the current HM whereas hm_n is included in the HM. Up to this stage, one iteration step has finished. The algorithm will continue until the termination criterion is achieved. The iterative steps of the harmony search algorithm in the optimization of Equation 3.6 as given in Figure 3.11 are as follows:

Step 1: Initialize the algorithm parameters HR, PR and M and randomly generate M harmonies (slip surfaces) and evaluate the harmonies.

Step 2: Generate a new harmony (shown in Figure 3.11) and evaluate it.

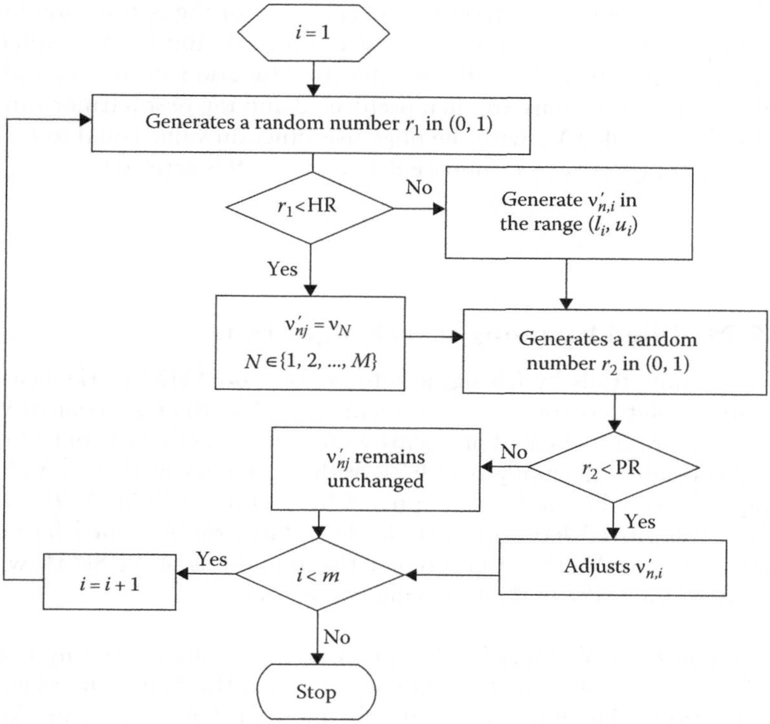

Figure 3.11 Flowchart for generating a new harmony.

Step 3: Update the HM. If the new harmony is better than the worst harmony in the HM, the worst harmony is replaced with the new harmony. Take the ith value of the coarse harmony h'_n, v'_{ni} for reference. Its lower bound and upper bounds are named as $v_{i\min}$ and $v_{i\max}$, respectively. A random number r_0 in the range [0, 1] is generated. If $r_0 > 0.5$, then v'_{ni} is adjusted to v_{ni} using Equation 3.7; otherwise, Equation 3.8 is used to calculate the new value of v_{ni}:

$$v_{ni} = v'_{ni} + \left(v_{i\max} - v'_{ni}\right) \times \text{rand} \quad r_0 > 0.5 \tag{3.7}$$

$$v_{ni} = v'_{ni} - \left(v'_{ni} - v_{i\min}\right) \times \text{rand} \quad r_0 \le 0.5 \tag{3.8}$$

where rand means a random number in the range [0, 1].

Step 4: Repeat Steps 2 and 3 until the termination criterion has been achieved.

In the original harmony search by Geem et al. (2001) and Lee and Geem (2005), an explicit termination criterion is not given. Cheng et al.

(2007b,d) have proposed a termination criterion for the optimization process. Suppose $M \times N_1$ iterations are first performed, and the best solution found so far is called $\mathbf{V}_{sf}$, with the objective function value equal to f_{sf}. Another $M \times N_2$ iterations are then preformed and the best harmony in the current HM is called $\mathbf{V}_g$, with the objective function value equal to f_g. The optimization process can terminate if Equation 3.9 is satisfied:

$$\left|f_{sf} - f_g\right| \leq \varepsilon \tag{3.9}$$

3.3.5 Modified harmony search algorithm

Based on many trials by Cheng, it is found that the SHM works fast and gives good solutions for simple problems with less than 25 control variables. For more complicated problems with a large number of control variables, the original harmony search algorithm becomes inefficient and can be trapped easily by the local minima. Cheng et al. (2007b, 2008c) have developed improved harmony search algorithms, the modified harmony search algorithm (MHM), to overcome the limitations of the SHM, which differs from the SHM in the following two aspects:

1. A uniform probability has been used in the original harmony search. Instead of that, a better approach may be: the better the objective function value of one harmony, the more probable will it be chosen for the generation of a new harmony. A parameter $\delta (0 < \delta \leq 1)$ is introduced and all the harmonies in HM are sorted in ascending order and a probability is assigned to each of them. For instance, $pr(i)$ means the probability to choose the ith harmony, which is given as

$$pr(i) = \delta \times (1 - \delta)^{i-1} \tag{3.10}$$

for $i=1, 2, \ldots, M$. From Equation 3.10, it can be seen that larger the value of δ, the more probable will the first harmony being chosen. An array $ST(i)$, $i=0, 1, 2, \ldots, M$ as given by Equation 3.11 should be used to implement the procedures mentioned earlier for choosing the harmony:

$$ST(i) = \sum_{j=1}^{i} pr(j) \tag{3.11}$$

where

$ST(i)$ represents the accumulating probability for the ith harmony

$ST(0)$ equals 0.0 for the sake of implementation

A random number r_c is generated from the range [0, $ST(M)$], and the kth harmony in HM is to be chosen if the following criterion is satisfied:

$$ST(k-1) < r_c \le ST(k), \quad k = 1,2,\ldots,M \tag{3.12}$$

2. Instead of one new harmony, a certain number of new harmonies (Nhm) are generated during each iteration step in the MHM. The utilization of the HM is intuitively more exhaustive by generating several new harmonies than by generating one new harmony during one iteration. In order to maintain the structure of the HM unchanged, the M harmonies with lower objective functions (for the minimization optimization problem) from $M+Nhm$ harmonies are included in the HM again, and the Nhm harmonies of the higher objective function values are rejected.

The HM shown in Table 3.1 is now reordered in increasing order and the new structure is illustrated in Table 3.2. Supposing $\delta=0.5$ and $Nhm=2$, the arrays pr and ST obtained are listed in columns 6 and 7, respectively, in Table 3.2. A random number in the range [0, 1] is generated, say 0.6($<HR$). One of the values from {1.0, 1.5, 0.5, 1.8, 0.9, 1.1} should be chosen as the value of x_1 in the new harmony. Given the value of r_c equals 0.4, for example, by using criterion (18), 0.5 is chosen to be the value of x_1. Another random number of 0.95(>HR) is obtained and a random value in the range [1, 3], 1.2, is generated. Similarly, a random number of 0.6 and $r_c=0.80$ are also obtained. 1.8 is chosen from the HM as the value of x_3, and thus, a coarse new harmony $hm'_n = (0.5,1.2,1.8)$ is generated. The fine new harmony is obtained by adjusting the coarse new harmony according to the parameter PR. Suppose two random values in the range [0, 1], say 0.7, 0.05, 0.8, are generated randomly. Since the former is greater than the PR, the value of x_1 in hm'_n remains unchanged. The latter value is lower than the PR, so the value of 1.2 should be adjusted. Suppose 1.10 is the

Table 3.2 Reordered structure of HM

HM \ Control variables	x_1	x_2	x_3	Objective function	$pr()$	$ST()$
hm_1	0.5	1.5	1.0	3.50	0.5	0.5
hm_2	1.1	1.9	1.5	3.87	0.25	0.75
hm_3	1.5	2.0	1.8	4.29	0.125	0.875
hm_4	1.0	1.5	0.5	4.50	0.0625	0.9375
hm_5	0.9	2.2	1.2	5.49	0.03125	0.9687
hm_6	1.8	2.5	0.9	8.10	0.01562	0.9843

Table 3.3 Structure of HM after first iteration in the MHM

HM \ Control variables	x_1	x_2	x_3	Objective function	pr( )	ST( )
hm_1	0.5	1.10	1.8	1.50	0.5	0.5
hm_2	0.9	1.5	1.3	2.75	0.25	0.75
hm_3	0.5	1.5	1.0	3.50	0.125	0.875
hm_4	1.1	1.9	1.5	3.87	0.0625	0.9375
hm_5	1.5	2.0	1.8	4.29	0.03125	0.9687
hm_6	1.0	1.5	0.5	4.50	0.01562	0.9843

new value of x_2; then the improved new harmony $hm_n = (0.5,1.10,1.8)$ is obtained. Similarly, the second new harmony $hm_n'' = (0.9,1.5,1.3)$ is also obtained. The objective functions of the two new harmonies are calculated as 1.5 and 2.75, respectively. So the six harmonies with lower objective functions $hm_1, hm_2, hm_3, hm_6, hm_n', hm_n''$ are introduced into the HM illustrated in Table 3.3 and one iteration is finished. The algorithm continues until the termination criterion is satisfied.

Based on extensive numerical tests by Cheng et al. (2007b, 2008c), it has been found that the MHMs shown in Figure 3.12 are more effective in overcoming the local minima compared with the original harmony search method for complicated problems. It is also more efficient than the original HM when the number of control variables is large, but is less efficient when there are only few control variables.

3.3.6 Tabu search algorithm

Tabu search (Glover, 1989, 1990) is not exactly an optimization algorithm, but a collection of guidelines to develop optimization algorithms. The basic idea of Tabu search is to explore the trial solutions for the problem, moving from a point to another point in its neighbourhood with solutions that have little differences from the point under consideration. Reverse moves and cycles are avoided by the use of a 'Tabu list' where the moves performed previously are memorized. In order to implement the Tabu search, the first step is the discretization of the problem space. Each dimension is divided into d elements and altogether, d^m hyper-cubes are obtained. If a solution is Tabu, it means the super-cube in which the solution locates is also Tabu. It is very difficult to directly generate a solution within the super-cubes that are not Tabu, and a trial procedure has been proposed by Cheng et al. (2007b). The procedure by which the new harmonies are obtained in the harmony search algorithm is used to obtain the trial solutions. If the super-cube of the new trial solution is Tabu, another trial solution will be tested until a trial solution that does not belong to the Tabu super-cubes is found. The flowchart for the Tabu search algorithm is shown in Figure 3.13.

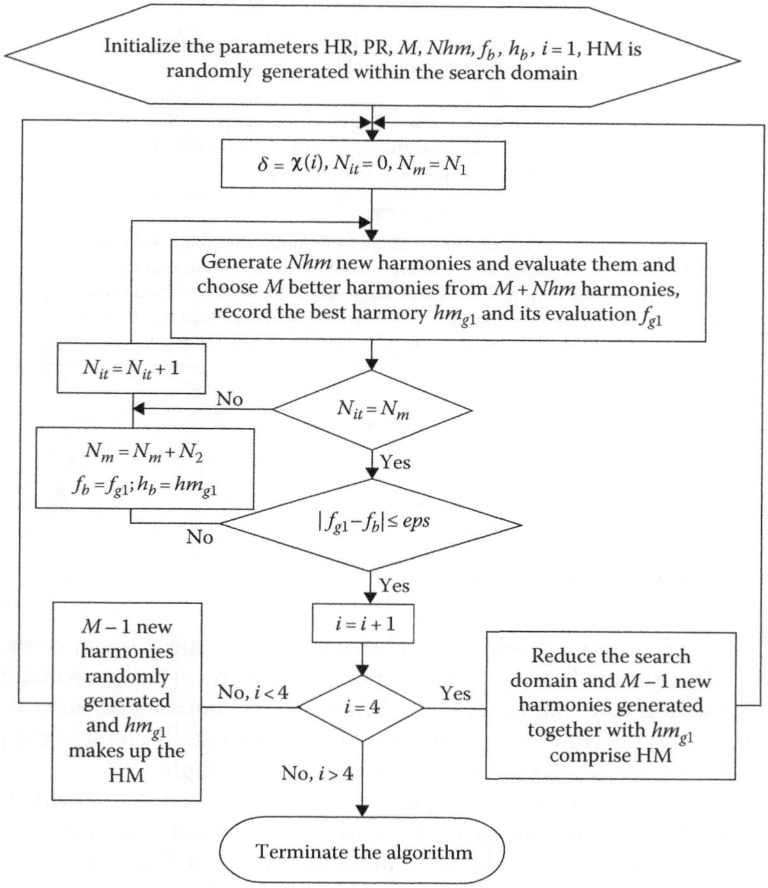

Figure 3.12 Flowchart for the MHM.

In Figure 3.13, f_g is the objective function value of the best harmony in the HM and the parameters of N_1, N_2, N_3 are used to terminate the algorithm. ε is the tolerance for the termination of the search.

3.3.7 Ant colony algorithm

The ant colony algorithm, which was developed by Dorigo (1992), is a meta-heuristic method that uses natural metaphors to solve complex combinatorial optimization problems. It is inspired by the natural optimization mechanism conducted by real ants. Basically, a problem under study is transformed into a weighted graph. The ant colony algorithm iteratively distributes a set of artificial ants onto the graph to construct tours corresponding

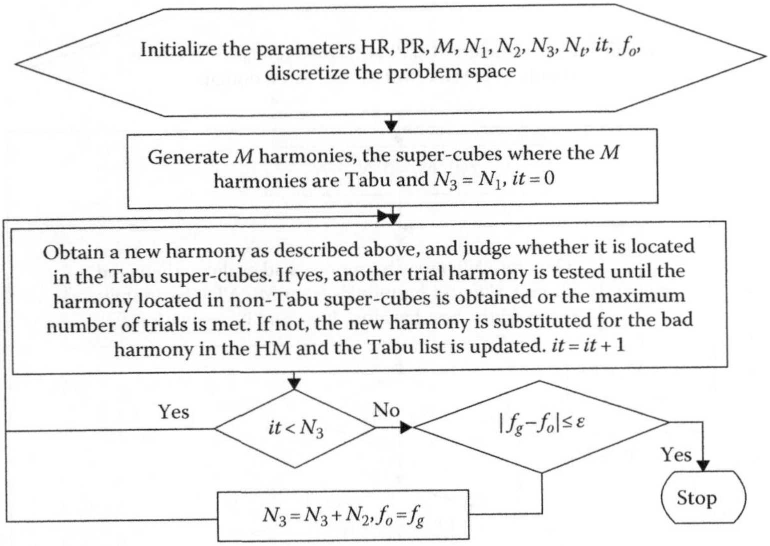

Figure 3.13 Flowchart for the Tabu search.

to the potential optimal solutions. The optimization mechanism of the ant colony algorithm is based on two important features: (1) the probabilistic state transition rule that is applied when an ant is choosing the next vertex to visit and (2) pheromone updating rule that dynamically changes the preference degree for edges that have been travelled through.

The continuous optimization problem should be first transformed into a weighted graph. In the case of locating the critical slip surface, each dimension is equally divided into d subdivisions and m dimension (5 and 3, respectively, in Figure 3.14) in the optimization problem. The solid circles located in adjacent columns are connected between each other.

An ant is first located at the initial point. Based on the probabilistic transition rule, one solid circle in the 'first variable' column is chosen

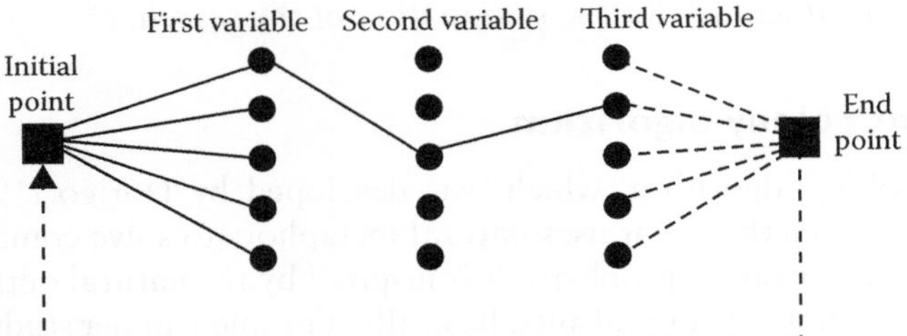

Figure 3.14 The weighted graph transformed for continuous optimization problem.

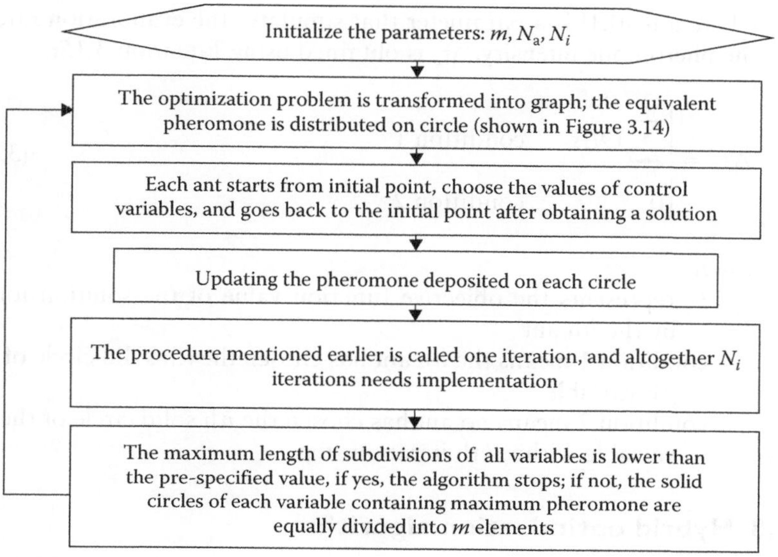

Figure 3.15 Flowchart for the ant colony algorithm.

and thus the value of the first variable is determined, and the procedures proceed to other variables. When an ant finishes determining the value of the end variable, it will go back to the initial point through the end point for the next iteration. Figure 3.15 shows the flowchart for the ant colony algorithm. In Figure 3.15, N_a means the total number of ants. The probabilistic transition rule and pheromone updating rule are described briefly as follows:

1. Probabilistic transition rule

$$\rho_{ij} = \frac{\tau_{ij}^{k}}{\sum_{i=1}^{d} \tau_{ij}^{k}}, \quad j = 1,2,\ldots,m \tag{3.13}$$

where

τ_{ij}^{k} represents the pheromone deposited on the ith solid circle of the jth variable within kth iteration step

ρ_{ij} means the probability of the ith solid circle of the jth variable to be chosen

2. Pheromone updating rule

$$\tau_{ij}^{k+1} = (1.0 - \mu) \times \tau_{ij}^{k} + \Delta\tau_{ij} \tag{3.14}$$

where $\mu \in [0,1]$ is a parameter that simulates the evaporation rate of the pheromone intensity. $\Delta\tau_{ij}$ is obtained using Equation 3.15:

$$\Delta\tau_{ij} = \begin{cases} \sum_{l=1}^{N_a} Q/fs_l & \text{condition 1} \\ 0 & \text{condition 2} \end{cases} \tag{3.15}$$

where

fs_l represents the objective function value of the solution found by the lth ant

condition 1 means the lth ant has chosen the ith solid circle of the jth variable

condition 2 means no ant has chosen the ith solid circle of the jth variable within kth iteration step

3.3.8 Hybrid optimization algorithm

While the heuristic algorithms as discussed can perform well over different conditions and appear to be stable and relatively insensitive to the optimization parameters, there are also cases where there are major fluctuations of the solutions near to the optimum solution. A simple example is Equation 3.16 over the solution domain from $x = 1$ to 6.

$$f(y) = \tan\left(\frac{y * \pi}{8}\right) \tag{3.16}$$

It is clear that the maximum and minimum of $f(y)$ will be given by 25.45 and 0.461, respectively, for Equation 3.16. Using the GA, harmony search, particle swarm search and ant colony method, as well as a starting point of $x = 1.0$ in the SA, point e (with value 0.727) is obtained as the minimum value of $f(y)$. If the global minimum point e is to be determined, a good initial trial is required for the SA, while specially tuned optimization parameters are used (obtained by trial and error). This situation occurs because the change from point e to f is relatively less extreme as compared with the section from f to g so that a trial within the region e to f is generated for the optimization analysis. For region in between points f, g and h, this region is too small and so, a trial may not be generated within this region. Even if a random trial is generated within this region by those optimization methods, the function $f(y)$ will still be high unless that random trial is very close to point g. As only limited trials will be generated within this special region, the absolute minimum is missed in the optimization search unless special treatment (may be problem-dependent) is adopted. For a simple one-dimensional problem, the absolute minimum can be obtained with

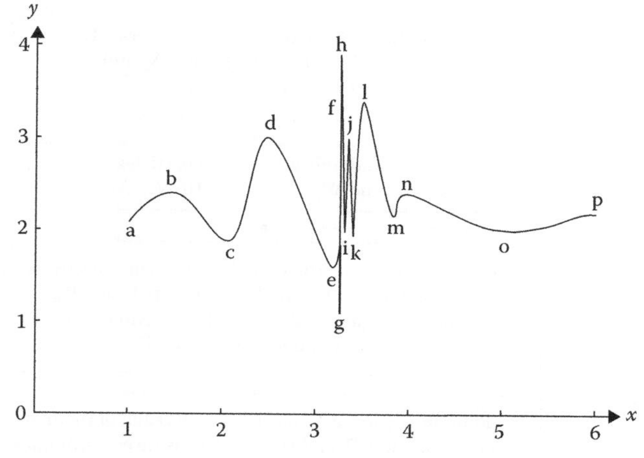

Figure 3.16 Equation 3.16 for $x = 1–6$.

ease by observation. For multi-dimensional problems, the situation will however be complicated. Cheng et al. (2012) have come across a practical case where the solution is similar to that shown in Figure 3.16. To overcome such difficult conditions, Cheng et al. (2012) have proposed a hybrid optimization algorithm based on the coupling of the harmony search (HS) and the PSO.

In the original PSO, the locations of the particles are updated by modifying the corresponding velocity vectors, and it is found that an incorrect value of ω may be trapped by the local minimum, which will be demonstrated in a later section. Generally speaking, a moderate value of 0.5 for ω is used for all the problems, or a larger value of ω can be applied at the initial analysis to search the solution space, which is then reduced linearly to a small value to find better results near the existing best position. There is another way to simulate the PSO procedure (Wang et al., 2008), as shown in Equation 3.3: the current positions of particles, the best position found so far, P_i, and the best position of any particle within the context of the topological neighbourhood of the ith particle found so far, P_g.

If we take the aforementioned positions (flights) from the PSO as the harmonies in the harmony memory in HS, a new position can also be obtained by the harmony search procedure. Similar to the modified PSO, N_a ($\leq M$) flights within each iteration step are allowed with different approaches. It is possible to choose N_a particles randomly from the total generation rather than based on the fitness of the particles in the modified PSO. In this way, the choice of flight is controlled by the procedure in HM rather than the original procedure as outlined in Equations 3.10 and 3.11. This is a minor and simple trick in combining the two methods.

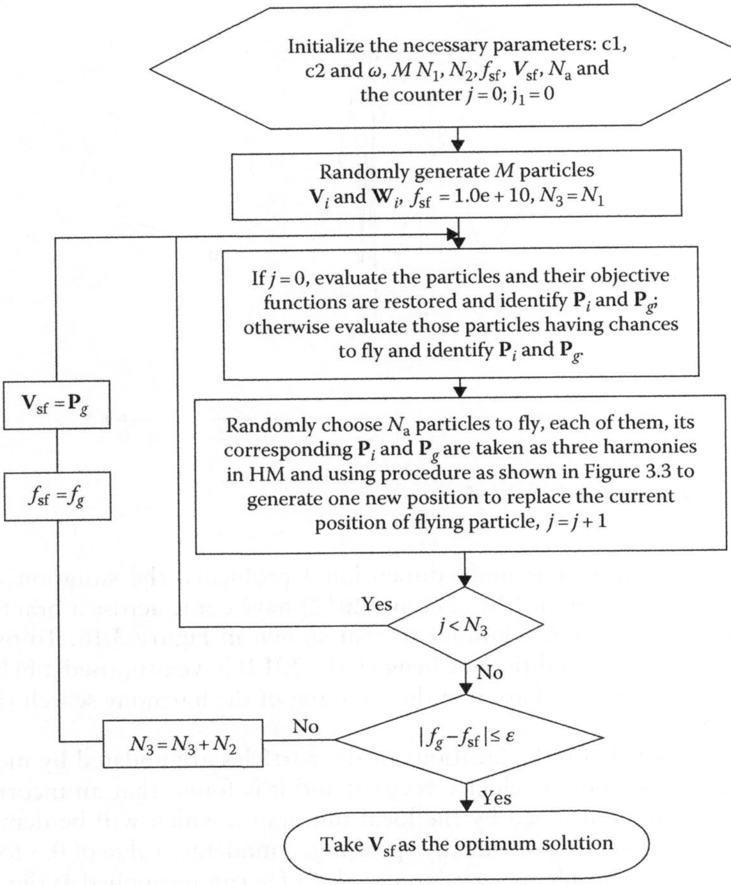

Figure 3.17 Flowchart of the coupled optimization method HS/PSO.

Cheng et al. (2007b) have tried the GA, the simulated annealing method, PSO, HM, Tabu search and ant colony search and have commented that no single method can outperform the other methods under all cases. Each optimization method has its own merits and limitations, and the combination of two optimization methods can possibly result in better performance under difficult cases.

The flowchart of the coupled PSO and HS, which is denoted as HS/PSO, is shown in Figure 3.17. Cheng et al. (2012) do not attempt to propose a highly complicated procedure in combining these two methods (for sake of simplicity). In Figure 3.10, the step on the updating of the positions of all the particles is replaced by the harmony search generation as shown in Figure 3.11. Such a minor change can retain the simplicity of both optimization methods so that the proposed algorithm is simple to use and does not require major computer memory. Cheng has come across several very complicated cases in some projects, and the algorithm proposed presently combines two optimization methods so that the coupled algorithm would be more stable and robust for very complicated problems. It is true that the present method will be less efficient, though effective, for simple methods and is not recommended for such purposes; the proposed algorithm is targeted towards complicated problems (discontinuous objective functions with multiple strong local minima and sudden major changes in material properties) for which the other algorithms may fail to perform satisfactorily.

Besides the coupling of PSO and HM, it is also possible to couple the Tabu search, the SA and the GA with the harmony search method. Cheng has also implemented these coupling methods successfully. As discussed by Cheng et al. (2007b), there is no single heuristic search method that can outperform other methods under all cases. Each method has its merits and limitations. By coupling two optimization methods, the resultant search algorithm is usually more effective in dealing with more complicated problems and is less likely to be trapped by the presence of local minima. It should be emphasized that the adoption of a coupling method usually requires more computation than the individual method for simple problems. In this respect, there is no simple way to achieve both effectiveness and efficiency under all cases. With advancement in computer technology and increasing complexities of problems, Cheng views that a more stable and effective algorithm is more important than a fast but less robust algorithm (keeping in mind that the increase in computation is not significant for the coupling method).

It should be noted that the flowchart in Figure 3.17 is a simple combination of that in Figures 3.10 and 3.11, and Cheng does not attempt to propose a highly complicated procedure in combining these two methods (for sake of simplicity). In Figure 3.10, the step on the updating of the positions of all the particles is replaced by the harmony search generation as given in Figure 3.11. Such a minor change can retain the simplicity of both optimization methods so that the proposed algorithm is simple to use and does not require major computer memory. Cheng has come across several very complicated cases in some projects, and the presently proposed algorithm combines two optimization methods so that the coupled algorithm will be more stable and robust for very complicated problem. It is true that the present method will be less efficient for simple method and it is not recommended for such purposes because it is less efficient (though effective) for such cases, and the proposed algorithm is targeted towards complicated problems (discontinuous objective function with multiple strong local minima and sudden major change in the material properties) for which the other algorithms may fail to perform satisfactorily.

3.4 VERIFICATION OF THE GLOBAL MINIMIZATION ALGORITHMS

The majority of the modern global optimization schemes have not been used in slope stability analysis in the past. The SA, SHM, MHM, PSO, Tabu and ant colony methods have been used first by Cheng (2003, 2007a) and Cheng et al. (2007b,d, 2008c) with various modifications to suit the slope stability problems. For the first demonstration of the applicability of these modern optimization methods, eight test problems are used to illustrate the effectiveness of Cheng's proposal on the modified SA (2003), and problems 4 and 8 are shown in Figures 3.18 and 3.19, respectively. Problems 1–3 are similar to problem 4 except for the external load. For problems 1–4, which are shown in Figure 3.18, there are totally two types of soils with a water table. In problem 1, there is no external load, whereas a horizontal load is applied in problem 2. A vertical load is applied in problem 3, whereas both vertical and horizontal loads are applied in problem 4. Problems 5–7 are also similar to problem 8, except for the external load. For problems 5–8, which are shown in Figure 3.19, there are totally three types of soils, a water table and a perched water table. In problem 5, there is no external load, while a horizontal load is applied in problem 6. A vertical load is applied in problem 7, while both vertical and horizontal loads are applied in problem 8. The cohesive strengths of the soils for problems 1–4 are 5 and 2 kPa, respectively, for soils 1 and 2, while the corresponding cohesive strengths for problems

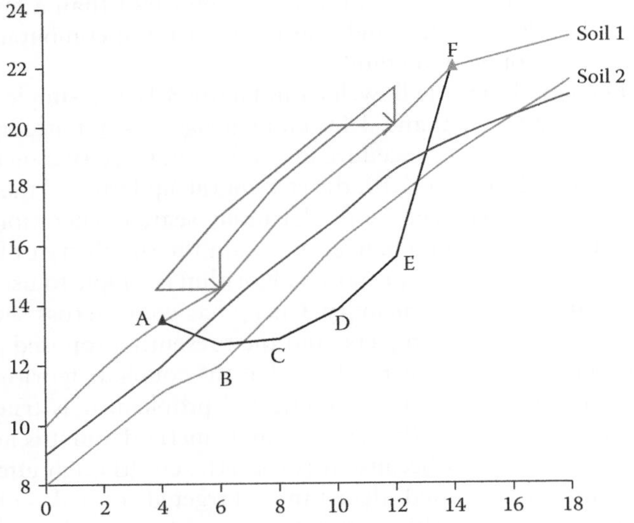

Figure 3.18 Problem 4 with horizontal and vertical loads (critical failure surface is shown by ABCDEF).

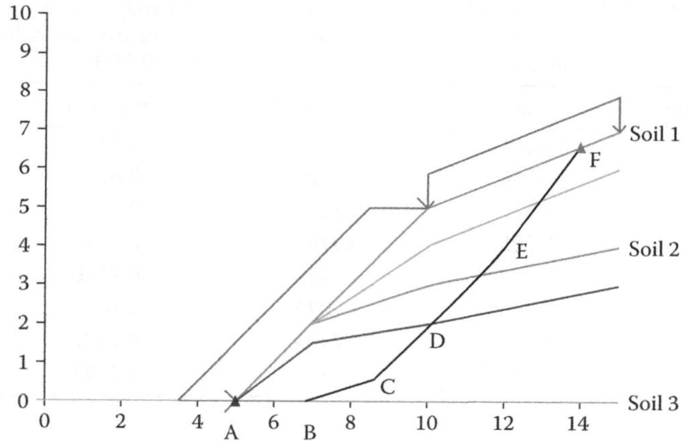

Figure 3.19 Problem 8 with horizontal and vertical loads (critical failure surface is shown by ABCDEF).

5–8 are 5, 2 and 5 kPa. The friction angles of the soils for problems 1–4 are 35° and 32°, respectively, for soils 1 and 2, while the corresponding friction angles for problems 5–8 are 32°, 30° and 35°. The unit weight of soil is kept constant at 19 kN/m^3 in all these cases. It is not easy to minimize the FOS for these problems by manual trial and error as the precise location of the failure surface will greatly influence the FOS. The minimum reference FOSs are determined by an inefficient but robust pattern search approach. To limit the amount of computer time used, the number of slices is limited to five in these studies and the slices are divided evenly.

The critical solution from the present study is shown in Figures 3.18 and 3.19, indicated by ABCDEF. The x-ordinates of the left exit end A (4.0 for problems 1–4 and 5.0 for problems 5–8) and the right exit end F (14.0 for all problems) of the failure surfaces are fixed so that only the y-ordinates of B, C, D, E are variables (x-ordinates of B, C, D, E are obtained by even division). There are hence four control variables in the present study. Based on the critical result BCDE obtained from the minimization analysis (rounded up to two decimal places), a grid is set up 0.5 m directly above and below B, C, D, E as obtained by simulated annealing analysis. The spacings between the upper and lower bounds are hence 1.0 m for all the four control variables. The grid spacing for each control variable is 0.01 m so that each control variable can take 101 possible locations. The present grid spacing is fine enough for pattern search minimization and all possible combinations of the failure surfaces are tried, which are 101 × 101 × 101 × 101 or 10,406,041 possible combinations.

The FOSs shown in Table 3.4 illustrate clearly that combined use of the failure surface generation and simulated annealing method is able to

Table 3.4 Comparison between minimization search and pattern search for eight test problems using the simulated annealing method (tolerance in minimization search = 0.0001)

Case	*Trials required in SA*	*FOS from SA*	*FOS from pattern search*
1	10,081	0.7279	0.7279
2	10,585	0.8872	0.8872
3	9,577	0.7684	0.7685
4	10,585	0.9243	0.9243
5	12,097	0.7727	0.7726
6	13,105	1.1072	1.1072
7	11,593	0.7494	0.7492
8	12,601	1.0327	1.0327

Table 3.5 Coordinates of the failure surface with minimum FOS from SA and pattern search for Figure 3.18

Point	*x-ordinate*	*y-ordinate from SA*	*y-ordinate from pattern search*
A	4	13.5*	13.5*
B	6	12.677	12.67
C	8	12.831	12.82
D	10	13.784	13.78
E	12	15.539	15.54
F	14	22.0*	22.0*

Note: Values with * are fixed and not control variables. SA, simulated annealing analysis.

minimize the FOSs with high precision, and the results are similar to those obtained by pattern search based on the 10,406,041 trials. The location of the critical failure surface obtained from simulated annealing analysis for problem 4 shown in Table 3.5 is very close to that obtained by pattern search and similar results are also obtained for all the other problems. The results in Tables 3.4 and 3.5 illustrate clearly the capability of the proposed modified SA in minimizing the FOSs, so that the burden on the engineers can be relieved by adoption of modern global optimization techniques. Besides the simulated annealing method, the other global optimizations as modified by Cheng's methods (2007b) can also be applied effectively for all the eight problems.

To illustrate the advantages of the present dynamic bound technique as compared with the classical static bounds to the control variables, the same problems are considered with static bounds analysis. The static bounds are defined as 0.5 m above and below the critical failure surface BCDE, and the results are shown in Table 3.6 (the same minimum values are obtained from the two analyses). It is clear that the present proposal can greatly reduce the

Table 3.6 Comparisons between number of trials required for dynamic bounds and static bounds in simulated annealing minimization

Case	*Trials for DBs*	*Trials for SBs*
1	10,081	19,823
2	10,585	21,023
3	9,577	17,234
4	10,585	22,131
5	12,097	23,968
6	13,105	25,369
7	11,593	23,652
8	12,601	25,104

SBs, static bounds; DBs, dynamic bounds.

time of computation as compared with the classical simulated annealing technique, which is highly beneficial for real problems. This advantage is particularly important when the number of control variables is great.

3.5 PRESENCE OF DIRAC FUNCTION

If there is a very thin soft band where the soil parameters are particularly low, the critical failure surface will be controlled by this soft band. This type of problem poses great difficulty as normal random number generation (uniform probability) is used within the solution domain, and this feature is difficult to be captured automatically. The thickness of the soft band can be so small that it can be considered a Dirac function within the solution domain. Such failures have been reported in Hong Kong and the slope failure at Fei Tsui Road is one of the famous examples in Hong Kong where failure is controlled by a thin band of soil.

For a thin soft band, the probability of the control variables falling within this region will be small with the use of the classical random numbers. From principle of engineering, the probability of the control variables falling within this soft band region should however be greater than that of falling within other regions. For this difficult case, Cheng et al. (2007) propose to increase the probability of the search within the soft band. Since the location of the soft band region is not uniform within the solution domain, it is difficult to construct a random function with increased probability within the soft band region at different locations. This problem can be solved by a simple transformation proposed by Cheng as shown in Figure 3.20 where a classical random function is used in simulated annealing analysis. In Figure 3.20, the actual domain for a control variable x_i ($N+1>i>2$ in present method) is represented by a segment AB, with a soft band CD in between

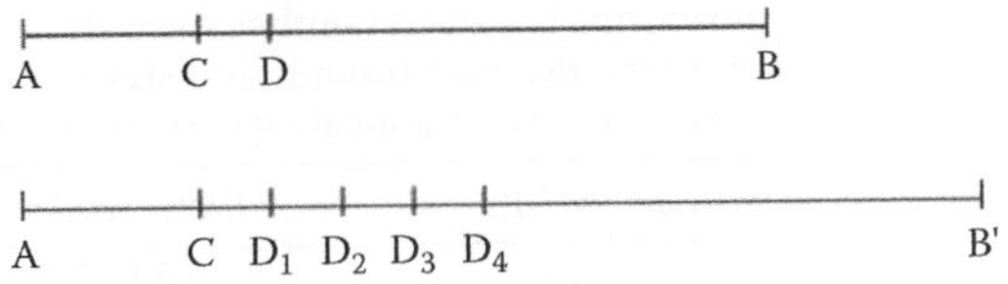

Figure 3.20 Transformation of domain to create special random number with weighting.

AB. For control variables x_j where $i \neq j$, the location of the soft band CD and the solution bound AB for a control variable x_i will be different from that for control variable x_j. For segment AB, several virtual domains with a width of CD for each domain are added adjacent to CD as shown in Figure 3.20. The transformed domain AB′ is used as the control domain of variable x_i. Every point generated within the virtual domain D1–D2, D2–D3, D3–D4 is mapped to the corresponding point in segment CD1. This technique is effectively equivalent to giving more chances to those control variables within the soft band. The weighting to the variables within the soft band zone can be controlled easily by the simple transformation as suggested in Figure 3.20. To the authors' knowledge, the search for critical failure surface with a 1 mm thick soft band has never been minimized successfully, but this has been solved effectively by the domain transformation proposed by Cheng (2007a). The transformation technique is coded into SLOPE 2000 and has been used to overcome a very difficult hydropower project in China where there were several layers of highly irregular soft bands. For that project, several commercial programmes had been used to locate the critical failure surface, without satisfaction.

3.6 NUMERICAL STUDIES OF THE EFFICIENCY AND EFFECTIVENESS OF VARIOUS OPTIMIZATION ALGORITHMS

The greater the number of the control variables, the more difficult will be the global optimization analysis. For the heuristic global optimization methods that have been discussed in the previous section, all of them are effective for simple cases with a small number of control variables. The practical differences between these methods are the effectiveness and efficiency under some special conditions with a large number of control variables. Consider example 1 as shown in Figure 3.21. It is a simple slope taken from the study by Zolfaghari et al. (2005). The soil parameters are as follows: unit weight 19.0 kN/m^3, cohesion 15.0 kPa and effective friction angle 20°. Zolfagahri (2005) used a simple GA and Spencer's method and obtained a minimum FOS of 1.75 for the non-circular failure surface.

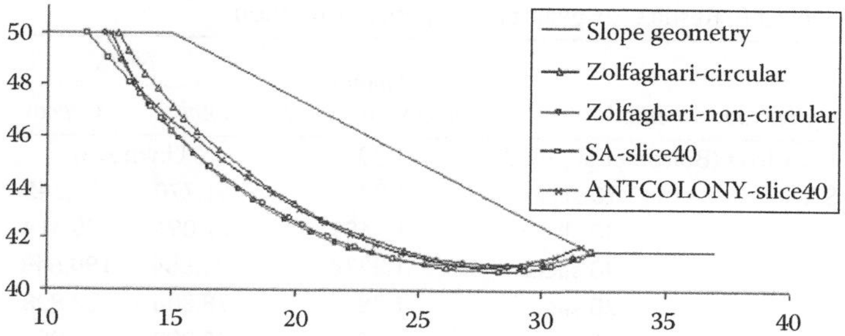

Figure 3.21 Example 1: Critical failure surface for a simple slope, example 1 (failure surfaces by SA, MHM, SHM, PSO and GA are virtually the same; failure surfaces by Tabu and Zolfaghari are virtually the same).

Table 3.7 Minimum FOS for example 1 (Spencer's method)

Optimization methods	*Minimum factors of safety*	*NOFs*	
		Total	*Critical*
Simple genetic algorithm by Zolfaghari et al. (2005)	1.75	Unknown	
SA	1.7267	103,532	102,590
GA	1.7297	49,476	49,476
PSO	1.7282	61,600	60,682
SHM	1.7264	107,181	98,607
MHM	1.7279	28,827	28,827
Tabu	1.7415	58,388	988
Ant colony	1.7647	83,500	16,488

NOFs, number of trials.

For this simple example, the results of the analyses are shown in Table 3.7 and Figure 3.21. All of the methods under consideration are effective in the optimization analysis. The SA and the SHM give the lowest FOSs, which are only slightly smaller than those by the other methods, but the number of failure surface trials (NOFs) are up to about 100,000, which is much greater than that for the other methods. MHM finds a minimum of 1.7279, which is slightly larger than that by the SA and the SHM, but only 28,827 trials are required, which is much more efficient in the analysis. The PSO and the GA give slightly larger factors of safety, which are 1.7297 and 1.7282, respectively, but the numbers of evaluations are modest when compared with those required by the SA and the SHM.

Example 2 is taken from the works by Bolton et al. (2003). There is a weak soil layer sandwiched between two strong layers. Unlike the previous example, the minimum FOS will be very sensitive to the precise location

Table 3.8 Results for example 2 (Spencer's method)

Optimization methods		*Minimum factors of safety*	*NOFs*	
			Total	*Critical*
Leap-frog (Bolton et al., 2003)		1.305	Unknown	
SA	20 slices	1.2411	51,770	51,745
	30 slices	1.2689	77,096	75,314
	40 slices	1.3238	190,664	190,648
GA	20 slices	1.2819	28,808	28,808
	30 slices	1.2749	39,088	39,088
	40 slices	1.2855	115,266	115,202
PSO	20 slices	1.2659	42,000	33,012
	30 slices	1.2662	64,800	55,810
	40 slices	1.2600	94,400	94,400
SHM	20 slices	1.3414	29,942	29,760
	30 slices	1.2784	118,505	97,055
	40 slices	1.2521	123,581	106,210
MHM	20 slices	1.2813	34,668	34,648
	30 slices	1.2720	26,891	26,891
	40 slices	1.2670	38,827	38,817
Tabu	20 slices	1.5381	30,548	1,148
	30 slices	1.5354	44,168	768
	40 slices	1.5341	58,188	788
Ant colony	20 slices	1.4897	43,500	4,721
	30 slices	1.5665	63,500	7,726
	40 slices	1.5815	83,500	1,501

of the critical failure surface. The soil parameters for soil layers 1–3 are, respectively, friction angle 20°, 10° and 20°; cohesive strength 28.73, 0.0, and 28.73 kPa; and unit weight 18.84 kN/m^3 for all three soil layers. The results of the analysis are shown in Table 3.8.

For this problem, all the methods are basically satisfactory except for the Tabu search and the ant colony method. The performance of the Tabu search and the ant colony method is poor for this example, which indicates that these two methods are trapped by the presence of the local minima in the analysis. Overall, the PSO is the most effective method for this problem, while the MHM ranks second with the least trials. The critical failure surfaces from the different methods of optimization are shown in Figure 3.22.

Example 3 is a case considered by Zolfaghari et al. (2005) where there is a natural slope with four soil layers as shown in Figure 3.23. Zolfaghari et al. (2005) has adopted the GA and Spencer's method for this example. The geotechnical parameters for this example are shown in Table 3.9.

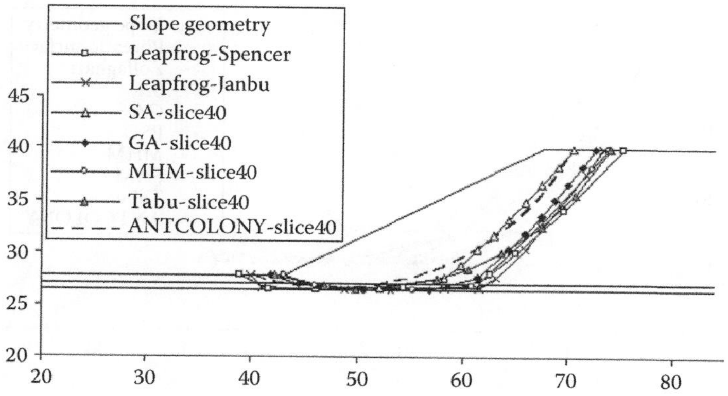

Figure 3.22 Critical slip surfaces for example 2 (failure surfaces by GA, PSO and SHM are virtually the same).

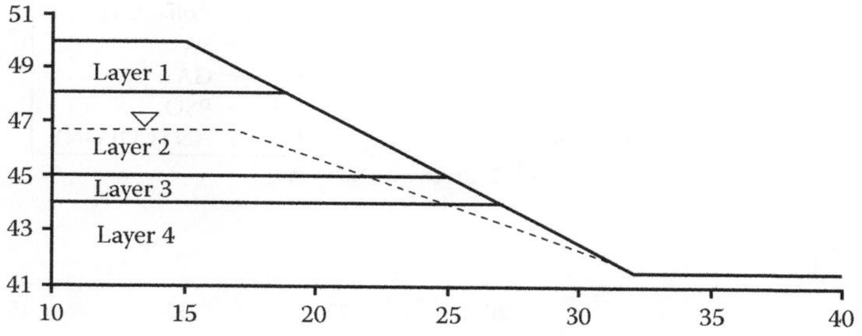

Figure 3.23 Geotechnical features of example 3.

Table 3.9 Geotechnical parameters of example 3

Layers	γ *(kN/m*^{3}*)*	*c (kPa)*	ϕ *(°)*
1	19.0	15.0	20.0
2	19.0	17.0	21.0
3	19.0	5.00	10.0
4	19.0	35.0	28.0

Four loading cases are considered by Zolfghari (2005): no water pressure and no earthquake loadings (case 1), water pressure and no earthquake loading (case 2), earthquake loading (coefficient = 0.1) and no water pressure (case 3) and water pressure and earthquake loading (case 4). The number of slices is 40, 41, 44 and 45 for case 1 to case 4. The critical failure surfaces are given in Figures 3.24 through 3.27 and Table 3.10. For this example, the Tabu search and the ant colony method are not good, while

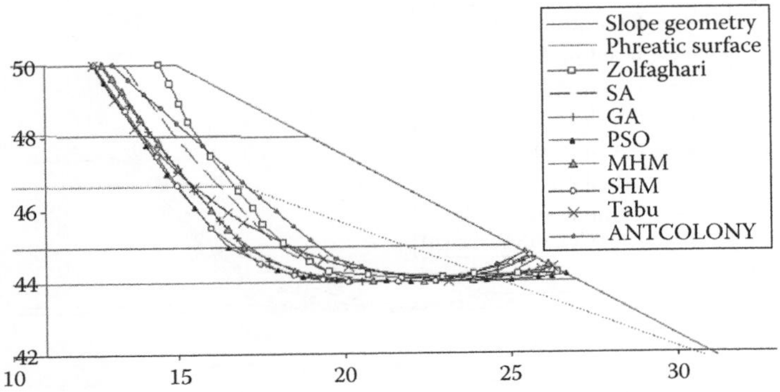

Figure 3.24 Critical slip surfaces for case 1 of example 3.

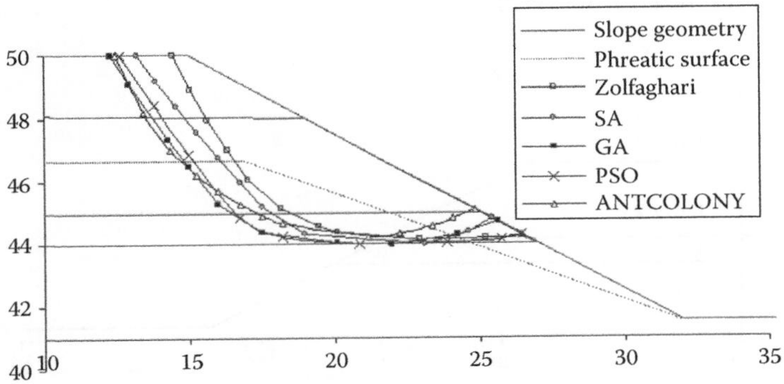

Figure 3.25 Critical slip surfaces for case 2 of example 3 (failure surface by GA, MHM, SHM and Tabu are virtually the same).

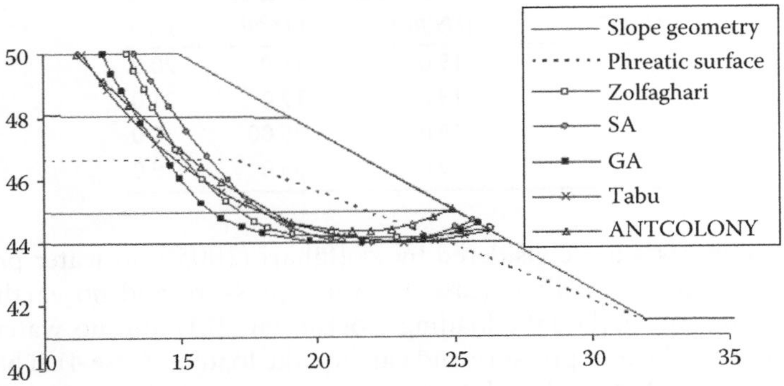

Figure 3.26 Critical slip surfaces for case 3 of example 3 (failure surfaces by GA, PSO, MHM and SHM are virtually the same).

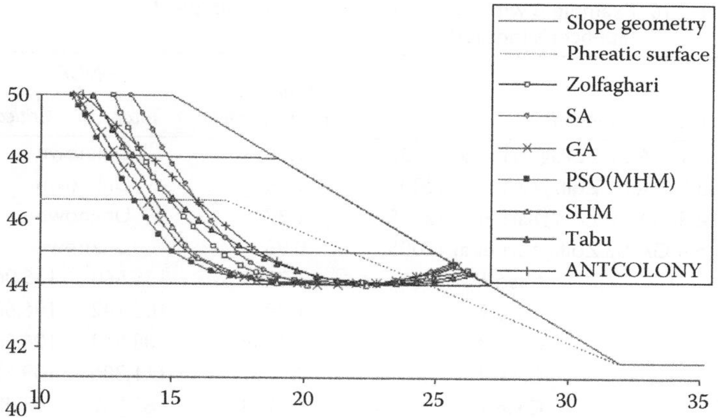

Figure 3.27 Critical slip surfaces for case 4 of example 3.

all the other methods are basically satisfactory. The PSO, GA and MHM are the most effective and efficient methods for this example.

Minimization of the FOS for a general slope stability problem is a difficult N–P hard-type problem because of the special features of the objective functions, which have been discussed before. Since there are many limitations in using the classical optimization methods in the slope stability problem, the current trend is the adoption of the modern global optimization methods for this type of problem. All these six types of heuristic algorithms can function well for normal problems, which are demonstrated in examples 1 and 2 and in some other internal studies by Cheng. For simple problems where the number of control variable is less than 25, it appears that the SHM and the MHM are the most efficient optimization methods. The SHM, Tabu search and ant colony method can perform well in many other applications, but they have been demonstrated to be less satisfactory for slope stability problems. Since the ant colony method aims at continuous optimization problems, it is not surprising that it is less satisfactory for slope stability problems where discontinuity of the objective function is generated by divergence of the FOS. Similarly, when there are great differences in soil properties between different soils, the SHM will be less satisfactory due to the use of uniform probability to individual harmony. The present study has illustrated the special feature of slope stability analysis during the optimization analysis, which is not found in other applications as production, system control or other similar disciplines.

On the other hand, for large-scale optimization problems or problems similar to example 3 with the presence of a thin layer of soft bands, which will create difficulties in the optimization analysis, the effectiveness and efficiency of the different heuristic optimization methods vary significantly between

Table 3.10 Example 6 with four loading cases for example 3 (Spencer's method)

Optimization methods		*Minimum factors of safety*	*NOFs* *Total*	*Critical*
Case 1 GA by Zolfaghari et al. (2005)		1.48	Unknown	
Case 2 GA by Zolfaghari et al. (2005)		1.36	Unknown	
Case 3 GA by Zolfaghari et al. (2005)		1.37	Unknown	
Case 4 GA by Zolfaghari et al. (2005)		0.98	Unknown	
SA	Case 1	1.3961	135,560	135,069
	Case 2	1.2837	106,742	106,662
	Case 3	1.1334	108,542	106,669
	Case 4	1.0081	111,386	109,667
GA	Case 1	1.3733	63,562	63,496
	Case 2	1.2324	77,178	77,114
	Case 3	1.0675	98,332	98,332
	Case 4	0.9631	84,272	84,272
PSO	Case 1	1.3372	62,800	33,116
	Case 2	1.2100	83,400	83,400
	Case 3	1.0474	69,600	69,600
	Case 4	0.9451	68,600	24,440
SHM	Case 1	1.3729	172,464	149,173
	Case 2	1.2326	126,445	100,529
	Case 3	1.0733	99,831	98,070
	Case 4	0.9570	212,160	186,632
MHM	Case 1	1.3501	32,510	32,500
	Case 2	1.2247	40,697	40,687
	Case 3	1.0578	40,476	40,440
	Case 4	0.9411	33,236	33,236
Tabu	Case 1	1.4802	58,588	1,188
	Case 2	1.3426	59,790	990
	Case 3	1.1858	63,796	796
	Case 4	1.0848	65,398	998
Ant colony	Case 1	1.5749	100,200	13,400
	Case 2	1.4488	102,600	1,801
	Case 3	1.3028	109,800	5,689
	Case 4	1.1372	112,200	18,436

different problems. The Tabu search and the ant colony method have been demonstrated to give poor results in some of the problems, while the PSO method appears to be the most stable and efficient, and is recommended for use in such cases. The presence of a thin soft band is difficult in the analysis, as a random number with equal opportunity in every solution domain is

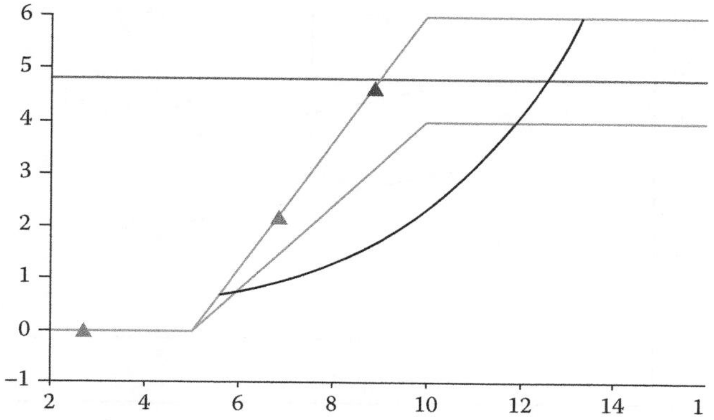

Figure 3.28 Slope with pond water.

used in the generation of the trial failure surface. In the present study, the domain transformation method has not been used. If the domain transformation technique as suggested by Cheng (2007a) is adopted, all six methods can work effectively and efficiently for problems with soft bands, with the MHM and the PSO being the best solution algorithms in terms of efficiency.

For Figure 3.28 where the water table is above the ground surface at the left-hand side of the slope, the SHM and the MHM cannot locate the global minimum using Spencer's method even when the optimization parameters are varied, unless the initial trial failure surface is close to the critical solution. Cheng et al. (2007b) have noted that many of the trial failure surfaces (20%) fail to converge in the optimization analysis, which is equivalent to the presence of discontinuity in the objective function. The SHM and the MHM are trapped by the local minima under such a case if the initial trial is not close to the critical solution, and there are major discontinuities in the objective function. The GA, Tabu search and ant colony method all suffer from this limitation. On the other hand, the PSO and the SA can locate the critical solution effectively.

Another interesting case is a steep slope in Beijing, where there is a thin layer of soft material, a tension crack, two soil nails, an external surcharge and water pressure at the tension crack (Figure 3.29). Only the SA method and the hybrid optimization method can work properly with Spencer's method for this case, while all the methods will fail to work properly unless a good initial is used. For this problem, Cheng et al. (2007b) noticed that all the initial 400–500 trials failed to converge with Spencer's method. Such a major discontinuity in the objective function creates great difficulty in determining the directions of search for all the global optimization methods, and no solution is obtained from all the optimization methods (complete breakdown) except for the SA and the hybrid method. This case

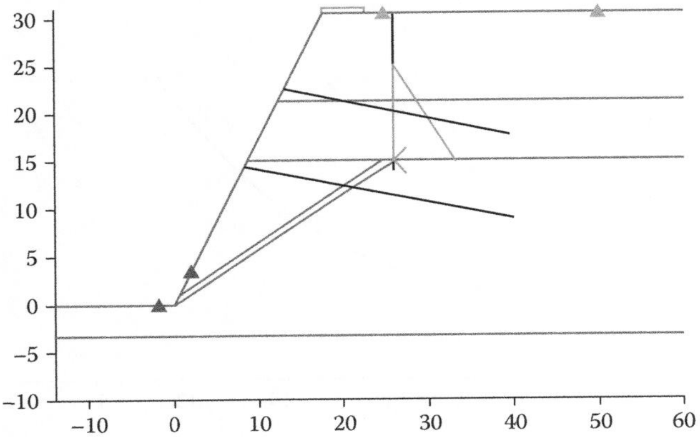

Figure 3.29 Steep slope with tension crack and soil nail.

is particularly interesting because all the optimization methods (except for the SA and the hybrid method) lose the direction of search and fail to find even one converged result before termination, unless the optimization parameters and the initial trial are specially tuned. This special example has also illustrated the difficulty in locating the critical failure surface for some special problems, where convergence is a critical issue. The SA and the hybrid method are less sensitive to the discontinuity of the objective function in general.

3.7 SENSITIVITY OF GLOBAL OPTIMIZATION PARAMETERS IN THE PERFORMANCE OF THE GLOBAL OPTIMIZATION METHODS

In all of the heuristic global optimization methods, there is no simple rule to determine the parameters used in the analysis. In general, these parameters are established by experience and numerical tests. It is surprising to find that the sensitivity of different global optimization methods with respect to different parameters has been seldom considered in the past, and the sensitivity of the parameters in slope stability analysis has not been reported. Cheng et al. (2007b) consider this issue to be important for geotechnical engineering problems as there are different topography, sub-soil conditions, ground water conditions, soil parameters, soil nails and external loads controlling the problem. It appears that many researchers have not appreciated the importance of the parameters used for the global optimization. The sensitivity of each parameter can be obtained through the nine numerical tests by the statistical orthogonal tests given in Tables 3.11 through 3.17

Table 3.11 Effects of parameters on SA analysis for examples 1 and 3

	λ	t_0	N	*Results*		*NOFs*	
	1 = 0.5 *2 = 0.8*	*1 = 10.0* *2 = 20.0*	*1 = 100* *2 = 300*	*Ex. 1*	*Ex. 3*	*Ex. 1*	*Ex. 3*
1	1	1	1	1.7256	1.3232	176,562	140,522
2	2	2	2	1.7241	1.2990	915,902	956,523
3	1	2	2	1.7235	1.2514	339,602	408,482
4	2	1	1	1.7264	1.2745	423,522	349,422
5	2	1	2	1.7258	1.2846	852,602	986,534
6	1	2	1	1.7239	1.3193	183,422	135,262
7	2	2	1	1.7262	1.3213	463,782	360,242
8	1	1	2	1.7268	1.2582	492,122	252,302

Note: $F_{\lambda} = 0.14, F_{t_0} = 0.47, F_N = 3.86, F_{0.05} = 7.7, F_{0.01} = 21.2.$

Table 3.12 Effects of parameters on GA analysis for examples 1 and 3

	ρ_c	ρ_m	*Results*		*NOFs*	
	1 = 0.85 *2 = 0.95*	*1 = 0.001* *2 = 0.1*	*Ex. 1*	*Ex. 3*	*Ex. 1*	*Ex. 3*
1	1	1	1.7273	1.2849	80,384	94,418
2	1	2	1.7266	1.2794	52,544	40,626
3	2	1	1.7272	1.2767	89,806	104,116
4	2	2	1.7266	1.2998	58,612	45,224

Note: $F_{\rho_c} = 0.18, F_{\rho_m} = 0.38, F_{0.05} = 161.4, F_{0.01} = 4052.$

Table 3.13 Effects of parameters on PSO analysis for examples 1 and 3

	c_1	c_2	Ω	*Results*		*NOFs*	
	1 = 1.0 *2 = 3.0*	*1 = 1.0* *2 = 3.0*	*1 = 0.3* *2 = 0.8*	*Ex. 1*	*Ex. 3*	*Ex. 1*	*Ex. 3*
1	1	1	1	1.7287	1.4430	59,200	45,200
2	2	2	2	1.7401	1.2671	59,200	45,200
3	1	2	2	1.7353	1.2692	59,200	94,800
4	2	1	1	1.7226	1.2368	108,400	231,800
5	2	1	2	1.7309	1.2545	59,800	46,400
6	1	2	1	1.7269	1.2405	75,600	57,600
7	2	2	1	1.7376	1.2747	59,200	45,200
8	1	1	2	1.7266	1.2479	59,200	58,200

Note: $F_{c_1} = 0.60, F_{c_2} = 0.37, F_{\varpi} = 0.52, F_{0.01} = 21.2.$

Table 3.14 Effects of parameters on SHM analysis for examples 1 and 3

	HR	PR	Results		NOFs	
	1 = 0.80 2 = 0.95	1 = 0.05 2 = 0.10	Ex. 1	Ex. 3	Ex. 1	Ex. 3
1	1	1	1.7330	1.2947	57,717	180,221
2	1	2	1.7438	1.3748	57,763	81,194
3	2	1	1.7231	1.2799	107,191	68,529
4	2	2	1.7259	1.2824	57,931	118,340

Note: $F_{HR} = 1.91, F_{PR} = 1.13, F_{0.05} = 161.4, F_{0.01} = 4052$.

Table 3.15 Effects of parameters on MHM analysis for examples 1 and 3

	HR	PR	Nhm	δ	Results		NOFs	
	1 = 0.80 2 = 0.95	1 = 0.05 2 = 0.10	1 = 0.1 2 = 0.3	1 = 0.3 2 = 0.8	Ex. 1	Ex. 3	Ex. 1	Ex. 3
1	1	1	1	1	1.7348	1.2838	7,654	13,547
2	1	1	1	2	1.7323	1.3523	13,654	9,545
3	1	2	2	1	1.7295	1.3102	31,446	34,436
4	1	2	2	2	1.7347	1.3025	31,446	45,235
5	2	1	2	1	1.7270	1.2976	17,159	27,053
6	2	1	2	2	1.7271	1.2874	16,986	26,591
7	2	2	1	1	1.7273	1.2989	9,640	15,219
8	2	2	1	2	1.7271	1.2878	19,621	13,128

Note: $F_{HR} = 0.97, F_{PR} = 0.07, F_{Nhm} = 0.10, F_{\delta} = 0.26, F_{0.05} = 10.1, F_{0.01} = 34.1$.

Table 3.16 Effects of parameters on Tabu analysis for examples 1 and 3

	d	N_t	HR	PR	Results		NOFs	
	1 = 2 2 = 5	1 = 30 2 = 50	1 = 0.80 2 = 0.95	1 = 0.05 2 = 0.10	Ex. 1	Ex. 3	Ex. 1	Ex. 3
1	1	1	1	1	1.7411	1.5714	58,388	44,168
2	1	1	1	2	1.7413	1.5391	58,188	44,168
3	1	2	2	1	1.7424	1.5661	58,188	44,168
4	1	2	2	2	1.7413	1.5661	58,188	44,168
5	2	1	2	1	1.7429	1.5661	58,588	44,168
6	2	1	2	2	1.7415	1.5427	58,188	44,168
7	2	2	1	1	1.7415	1.5470	58,188	44,368
8	2	2	1	2	1.7354	1.5561	59,188	44,168

Note: $F_d = 0.63, F_{N_t} = 0.17, F_{HR} = 0.49, F_{PR} = 1.43, F_{0.05} = 10.1, F_{0.01} = 34.1$.

Table 3.17 Effects of parameters on ant colony analysis for examples 1 and 3

	μ	Q	d	Results		NOFs	
	1 = 3 2 = 0.8	1 = 10.0 2 = 50.0	1 = 10 2 = 20	Ex. 1	Ex. 3	Ex. 1	Ex. 3
1	1	1	1	1.7447	1.5332	83,500	76,200
2	2	2	2	1.7404	1.9239	66,800	50,800
3	1	2	2	1.7636	1.8787	66,800	50,800
4	2	1	1	1.7717	1.7420	83,500	76,200
5	2	1	2	1.7377	1.9239	66,800	50,800
6	1	2	1	1.7538	1.5049	83,500	76,200
7	2	2	1	1.7569	1.7420	83,500	76,200
8	1	1	2	1.7591	1.8435	66,800	50,800

Note: $F\mu = 11.8, F_Q = 0.002, F_d = 39.7, F_{0.05} = 7.7, F_{0.01} = 21.2$.

for examples 1 and 2. If the F-value (Factorial Analysis of Variance after Fisher and Yates, 1963) of one parameter is larger than the critical value $F_{0.05}$ and smaller than $F_{0.01}$, it implies that the calculated result is sensitive to this parameter; otherwise, if F-value is smaller than $F_{0.05}$, the result is insensitive to this parameter. If F-value is larger than $F_{0.01}$, the result is hyper-sensitive to this parameter.

For the simple problem given by example 1, every method can work with satisfaction for different optimization parameters. For example 3, which is a difficult problem with a soft band (similar to a Dirac function), the Tabu search and the ant colony method are poor in performance (domain transformation technique is not used), while all the other optimization methods are basically acceptable, with F-value less than $F_{0.05}$. The efficiency of different methods for this case is, however, strongly related to the choice of parameters, unless the transformation technique by Cheng (2007a), which is equivalent to the use of a random number with more weighting in the soft band region, is adopted.

Every global optimization method can be tuned to work well if suitable optimization parameters or an initial trial are adopted. Since the suitable optimization parameters or the initial trial are difficult to be established for a general problem, the performance of a good optimization method should be relatively insensitive to these factors. Based on the numerical examples and the two special cases shown in Figures 3.28 and 3.29, and some other internal studies by the authors, the general comments on the different heuristic artificial intelligence–based global optimization methods are as follows:

1. For normal and simple problems, practically every method can work well. The harmony method and the GA are the most efficient methods when the number of control variables is less than 20. The Tabu search

and the ant colony method are sometimes extremely efficient in the optimization process, but the efficiency of these two methods fluctuates significantly between different problems and they are not recommended to be used.

2. For normal and simple problems where the number of control variables exceeds 20, the MHM and the PSO are the recommended solutions as they are more efficient in the solution, and the solution time will not vary significantly between different problems.
3. For more complicated problems or when the number of control variables is great, the effectiveness and efficiency of the PSO is nearly the best among all of the examples.
4. A thin soft band creates great difficulty in the global optimization analysis, and the PSO will be the best method in this case. However, using the domain transformation strategy of Cheng (2007a), all the global optimization methods can work well for this case.
5. For problems where an appreciable amount of trial failure surfaces will fail to converge, the simulated annealing method and the PSO are the recommended solutions.
6. Currently, there are many variants developed for each heuristic method in order to improve the robustness of the solution for more difficult problems. Improvements to the basic heuristic methods can usually improve the solution at the expense of more computations. The use of a hybrid optimization method represents another approach where a more stable solution can be obtained by combining the advantages of two optimization methods. The authors view that these two approaches are basically similar in many respects. More stable solutions will be obtained at the expense of more computation/time.
7. In view of the differences in the performance between different global optimization algorithms, different methods can be adopted for different conditions. For example, the PSO or the MHM can be adopted for normal problems, while the SA or the coupled method can be adopted when the 'failure to convergence' counter is high. Further improvement can be achieved by using the optimized results from a particular optimization method as a good initial trial, and a second optimization method adopts the optimized result from the first optimization algorithm for the second stage of optimization with a reduced solution domain for each control variable.
8. A programme should preferably allow users to change the optimization parameters and provide different optimization schemes for the solution. For experienced users, a faster scheme can be adopted for simple problems, while a more robust but time-consuming scheme can be chosen for more difficult problems. Users can also try different settings and optimization schemes to compare the results and increase

the confidence level of the solution. In the programme SLOPE 2000 designed by Cheng, the user is allowed to change the default settings and choose the optimization method for a specific problem, even though the default settings are virtually satisfactory for most of the practical engineering problems.

3.8 CONVEXITY OF CRITICAL FAILURE SURFACE

From Cheng's extensive trials, it can be concluded that most of the critical failure surfaces are convex in shape. The generation of convex failure surfaces as given in Section 3.2 is adequate enough for most cases. There are, however, some cases where the ground conditions may generate non-convex critical failure surfaces, which have been discussed by Janbu (1973), and this is further investigated in this section. Refer to the slope with a middle soft band shown in Figure 3.30. The soil properties for the three soils are: $c' = 4$ kPa, $\phi' = 33°$; $c' = 0$ kPa, $\phi' = 25°$ and $c' = 10$ kPa, $\phi' = 36°$, measured from top to bottom. This slope is analysed using the Janbu simplified method and the Morgenstern–Price (M–P) method (1965), and the critical failure surfaces from the analyses are shown in Figure 3.30. The two critical failure surfaces are controlled by the soft band and are similar in location except for section AB. Using the Janbu simplified method (1957), which satisfies only force equilibrium, the critical failure surface (FOS = 0.985 without correction factor) basically follows the profile of the soft band even though the kink AB should hinder the failure of the slope. On the other hand, if M–P method is used, the

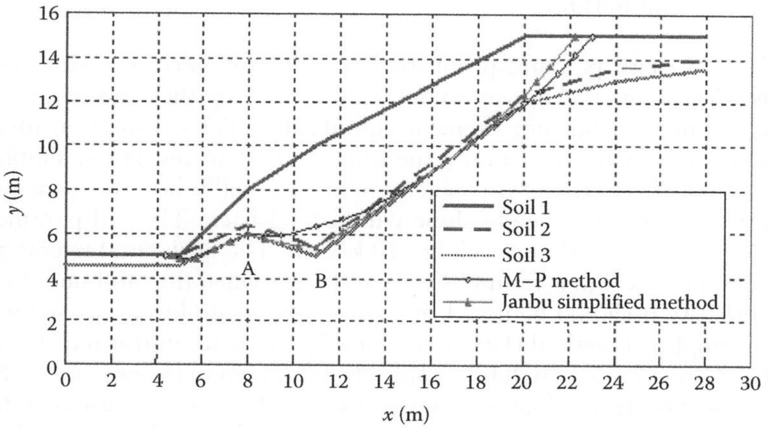

Figure 3.30 Critical failure surfaces for a slope with a soft band by Janbu simplified method and M–P method.

kink AB becomes important in the moment equilibrium and no kink is found for the critical failure surface unless the friction angle of soil 2 is lowered to 20° (a mild kink only). This result illustrates that different stability formulations may require different generations of the slip surface algorithm. These kinds of non-convex critical failure surfaces are not encountered commonly, but should be allowed in the generation of trial failure surfaces if necessary. The slip surface generation scheme outlined in Section 3.2 can achieve this requirement easily by simply eliminating the requirement on the lower bound of each control variable. This option is also available in SLOPE 2000, which can be chosen if required. For the acceptability of the solution, Cheng would like to point out that the simplified Janbu method (1957) can sometimes give FOSs more than 10% lower than those from the 'rigorous' methods if some surprising slip surfaces are accepted (with surprising internal forces usually). Towards this situation, the internal forces should be assessed before acceptance of the converged FOS for a proper analysis.

3.9 LATERAL EARTH PRESSURE DETERMINATION FROM SLOPE STABILITY ANALYSIS

Slope stability analysis methods based on the limit equilibrium approach are upper bound methods. In the Janbu simplified approach (1957), the basic assumptions are

1. Upper bound limit equilibrium approach
2. Mohr–Coulomb relation
3. Force equilibrium

In classical earth pressure problems, the Coulomb earth pressure theory is also based on the upper bound approach with the consideration of force equilibrium but not moment equilibrium. The assumption used in the Coulomb theory is actually the *same* as that in the Janbu simplified method (1957), so the two problems should actually be equivalent problems. Consider a 5 m height slope with a level back. The soil parameters are $c'=0$, $\phi'=30°$ and unit weight = 20 kN/m^3. The horizontal lateral pressure of 16.665 kN/m is found by trial and error when the minimum FOS is equal to 1.0. It should be noted that since force equilibrium is used in the Janbu simplified method (1957), the use of point load, uniform distributed load or triangular load will be completely equivalent in the present analysis. The correction factor f_0 should not be used in this case because only force equilibrium is considered in the Coulomb mechanism, whereas f_0 will correct for the inter-slice shear force and moment equilibrium for the Janbu simplified analysis (1957).

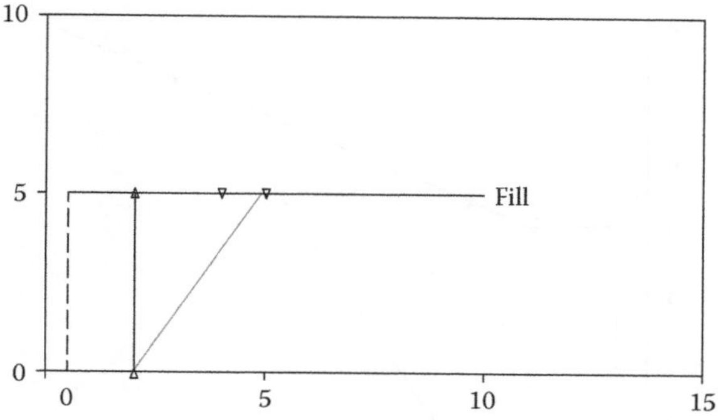

Figure 3.31 Critical failure surface by Janbu simplified without f_0 based on non-circular search, completely equal to Rankine solution.

The minimum FOS as found is 1.0001 and the critical failure surface as shown in Figure 3.31 is completely equivalent to the classical Rankine solution. The total load on this slope is 16.665 × 5 = 83.325 kN. The active pressure coefficient can be back-calculated as

$$83.325 = 0.5K_a \times 20 \times 5^2 \quad \Rightarrow \quad K_a = 0.3333 = \frac{(1-\sin\phi)}{(1+\sin\phi)}.$$

This result is exactly the same as the Rankine solution or the Coulomb solution. The critical failure surface from the non-circular search is also found to be a plane surface inclined at an angle of 60° with the horizontal direction, which is equivalent to $45° + \phi/2$ from the Rankine solution. The next example is the same as the previous one with a 20° slope behind a 5 m high slope. The total load on this slope for a minimum FOS of 1.0 is 110.25 kN. The active pressure coefficient can be calculated as

$$110.25 = 0.5K_a \times 20 \times 5^2 \quad \Rightarrow \quad K_a = 0.441.$$

This result is *exactly the same* as the Coulomb solution for a slope with 20° back, which is shown in Figure 3.32. The failure surface is also found to incline at an angle of 52°. The angle of inclination for this case can be found from Design Manual 7, which is given by

$$\tan\theta = \tan\phi + (1 + \tan^2\phi - \tan\beta/\sin\phi/\cos\phi)^{0.5}.$$

If we put $\beta = 20°$, θ is obtained as 52° and this is exactly the same as that obtained from the non-circular search as discussed before (see also the user guide of SLOPE 2000 for detailed results).

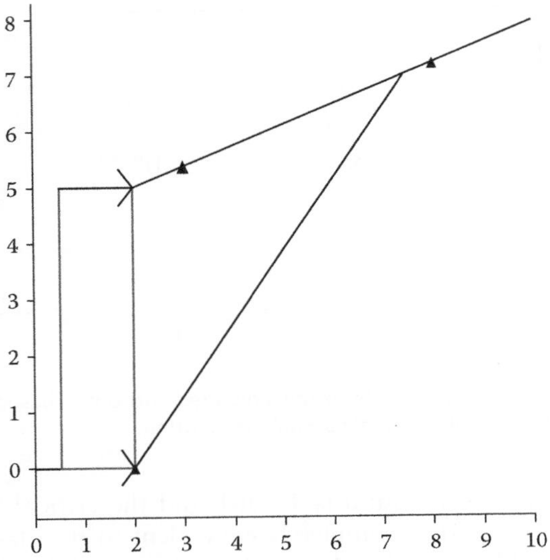

Figure 3.32 Critical failure surface by Janbu simplified without f_0 based on non-circular search, completely equal to Coulomb solution.

The point of application of the active pressure can be determined from the moment balance of the failure mass and is found to be at 1/3 height of the retaining wall. The present technique can be extended for a retaining wall with a non-homogeneous backfill, and the point of application of the active pressure can be determined by the moment equilibrium between the base normal and shear forces, the weight of the soil and the total active pressure from slope stability analysis.

3.10 CONVERGENCE PROBLEM DUE TO ITERATIVE SOLUTION OF FOS

The FOS function is a highly non-linear equation (Sarma, 1987). Currently, most of the slope analysis programmes are based on the iteration method, which requires an initial trial in analyses (commonly 1.0). Cheng et al. (2008a) have noticed that although the iteration method has been used for a long time, it has never been proven to be effective under all cases. Many engineers have experienced the problem of convergence with the M–P method or similar methods in determining the FOS, in particular, when there are soil nails or external loads in the problems. In all the slope stability programmes, if convergence is not achieved during evaluation of the FOS, an arbitrary large FOS is usually assigned to the trial failure surface.

If the phenomenon of 'failure to converge' is not a true phenomenon, the use of a large FOS (discontinuity of the safety factor function) can greatly affect the search for the critical failure surface, in particular, when the gradient-type method is used for the optimization analysis. This problem will be serious if convergence is important, and this will be demonstrated in a later section with some cases from Hong Kong.

In the search for the critical failure surface by manual trial-and-error approach, most of the engineers tackle the problem of 'failure to converge' by modifying the shape and location of the prescribed failure surface until a converged result is achieved. The minimum FOS will then correspond to the minimum value from the limited trial failure surfaces, which can converge by iteration analysis. It is also interesting to note that it has never been proven that a failure surface that fails to converge by iteration analysis is not a critical failure surface!

Failure to converge for 'rigorous' methods is experienced by many geotechnical engineers as iteration method is used by most of the commercial programmes. Cheng (2003) has formulated the slope stability problem in a matrix approach where the FOS and internal forces can be determined directly from a complex double QR matrix method without the need of an initial FOS. Cheng (2003) has proved that there are N FOSs associated with the non-linear FOS equation for a problem with N slices. In the double QR method, all N FOSs can be determined directly from the tedious matrix equation *without* using any iteration, and the FOSs can be classified into three groups:

1. Imaginary number
2. Negative number
3. Positive number

If all FOSs are either imaginary or negative, the problem under consideration has no physically acceptable answer, by nature. Otherwise, the positive number (usually 1–2 positive numbers left) will be examined for physically acceptability of the corresponding internal forces and a FOS will then be obtained. Under this new formulation, the fundamental nature of the problem is *fully* determined. If no physically acceptable answer exists from the double QR method (all results are imaginary or negative numbers), then the problem under consideration has no answer by nature, and the problem can be classified as 'failure to converge' under the assumption of the specific method of analysis. If a physically acceptable answer exists for a specific problem, then it will be determined by this double QR method. The authors have found that many problems, which fail to converge with the classical iteration method, actually possess meaningful answers by the double QR method. That means, the phenomenon of 'fail to converge' may come from the use of the iteration analysis and may be a false phenomenon

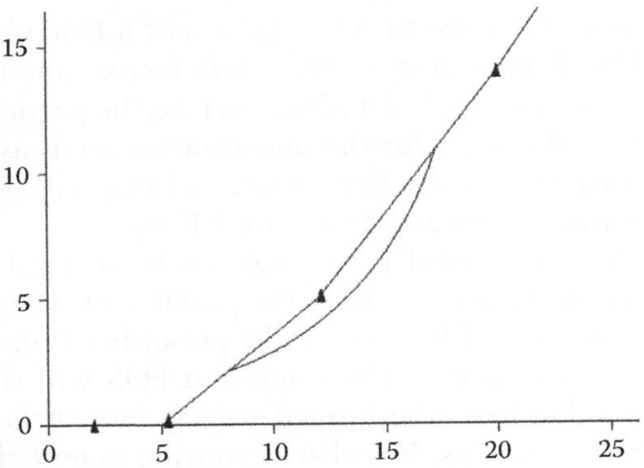

Figure 3.33 A simple slope fails to converge with iteration.

in some cases. Cheng et al. (2008a) have found that many failure surfaces, which fail to converge, are normal in shape and should not be neglected in ordinary analysis and design. This situation is usually not critical for slopes with simple geometry and no soil nail/external load, but convergence for 'rigorous' methods will be a more critical issue when soil nails/external loads are present in a problem. Since the use of soil nails is now very common in many countries, the problem of convergence, which is faced by many geotechnical engineers, should not be overlooked.

To evaluate the importance of the convergence in the analysis, specific problems and parametric studies using commercial programmes will be considered. Refer to Figure 3.33 for a simple slope with no water or soil nails; the soil parameters are $c' = 5$ kPa, $\phi' = 36°$ and unit weight = 20 kN/m^3. The prescribed circular failure surface fails to converge by iteration method (even when the correct FOS is used as the initial solution), but a FOS equal to 1.129 for M–P analysis is found by using the double QR method. The corresponding result for Sarma's analysis is 1.126, which is also similar to that by M–P analysis. This simple problem has illustrated that a failure surface that fails to converge by iteration method may actually possess physically acceptable answers. Since many Hong Kong engineers have encountered convergence problems with soil nailed slopes, the second problem shown in Figure 3.34 is considered where the soil nail loads are 30, 40 and 50 kN from left to right. The soil parameters for the top soil are $c' = 3$ kPa, $\phi' = 33°$ and unit weight = 18 kN/m^3, whereas the soil parameters for the second layer of soil are $c' = 5$ kPa, $\phi' = 35°$ and unit weight = 19 kN/m^3. The nail loads are applied at the nail head as well as on the slip surface for comparisons in this study. The results of analyses based on the iteration method by three commercial programmes and the double QR method by Cheng (2003) are shown in Table 3.17.

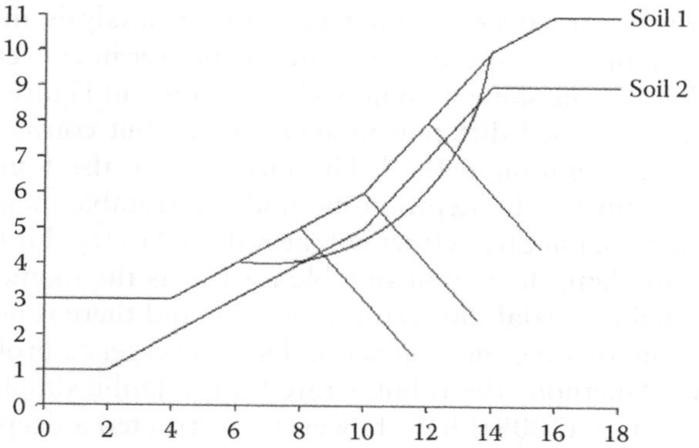

Figure 3.34 Slope with three soil nails.

Table 3.17 Performance of iteration analysis with three commercial programmes based on iteration analysis for the problem in Figure 3.32 (*c* means converged by iteration analysis, fail means failure to converge by iteration analysis)

	Soil nail 1–30 kN	Soil nail 2–40 kN	Soil nail 3–30 kN	No nail
Load applied at nail head				
Spencer	*c/c/c*	*c/c/c*	*c/c/c*	*c/c/c*
Load applied at slip surface				
Spencer	Fail/Fail/*c*	Fail/Fail/Fail	*c*/fail/*c*	*c/c/c*

	Soil nail 1 + 2	Soil nail 1 + 3	Soil nail 2 + 3	Soil nail 1 + 2 + 3
Load applied at nail head				
Spencer	*c/c/c*	Fail/Fail/Fail	*c/c/c*	*c/c/c*
Load applied at slip surface				
Spencer	Fail/Fail/Fail	*c/c/c*	Fail/Fail/Fail	Fail/Fail/Fail

For those in Table 3.17, the double QR method gives physically acceptable answers (or converged answers) for all the cases, while the iteration method fails to work for some of the M–P analyses. It is clear from Table 3.17 that the M–P method using the iteration method suffers from the 'failure to converge' problem, but answers actually exist for some of these problems. When only soil nail 3 is applied to the slope, it is noticed from Table 3.17 that one of the commercial programmes can converge while the other commercial programme fails to converge. It appears that convergence also depends on the specific procedures in the iteration analysis or the use of moment points in individual programme. Cheng et al. (2008a) have also found that the use of over-shooting in iteration analysis may generally improve the computation speed, but is slightly poorer in convergence (from internal study). For the slope with no soil nail shown in Figure 3.33, Cheng et al. (2008a) have tried different moment points but convergence is still not achieved with iteration method. The authors have also found that some problems may converge by iteration method if a suitable moment point is chosen for analysis, but great effort will be required to try this moment. For an arbitrary problem, the region suitable for use as the moment point has to be established by a trial-and-error approach, and there is no simple way to pre-determine this region in general. The convergence problem is critical for the M–P method (1965) but is rare for the Janbu simplified method (1957). Cheng et al. (2008a) have however constructed a deep-seated noncircular failure surface as shown in Figure 3.35 where $c' = 5$ kPa, $\phi' = 36°$ and unit weight = 20 kN/m^3. The Janbu analysis fails to converge with an initial FOS equal to 1.0, but convergence is possible with an initial trial of 2.0. The double QR method can work satisfactorily for this problem as no initial FOS is required and the FOS is obtained directly. For this problem, iteration method can work by changing the initial FOS, but this technique seldom works for those 'rigorous' methods. In general, convergence is important mainly for 'rigorous' methods, but is rare for those simplified methods. Up to present, Cheng et al. (2008a) could not construct a problem where Bishop's method fails to converge but possesses physically acceptable answers, and it appears that Bishop's method is virtually free from convergence problems. In fact, if the failure surface is circular in shape, Cheng et al. (2008a) are not able to construct any case for Bishop or Janbu

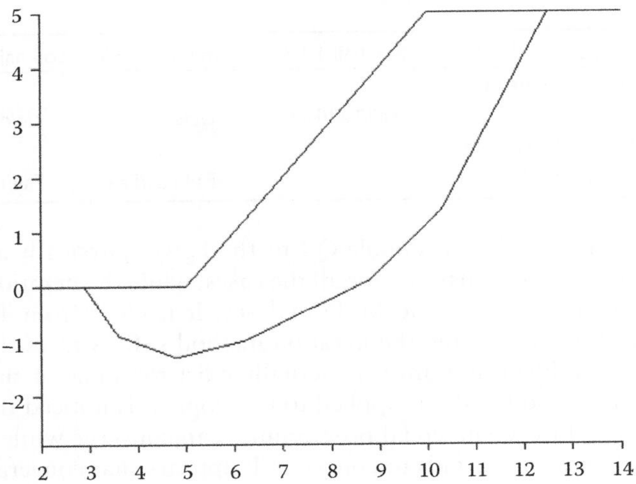

Figure 3.35 Failure to converge with Janbu simplified method when initial FOS = 1.0.

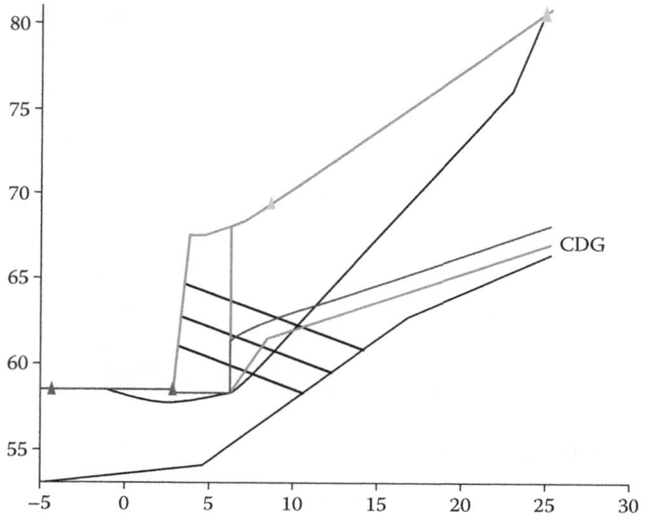

Figure 3.36 A problem in Hong Kong, which is very difficult to converge with iteration method.

analysis where iteration method fails to work, but physically acceptable answers can be determined with the double QR method.

Cheng et al. (2008a) have come across an interesting case in Hong Kong, which is shown in Figure 3.36. For this slope with a retaining wall, M–P method (1965) is adopted in the analysis. Engineers have experienced great difficulties in drawing suitable failure surfaces, which can converge in the analysis, and a minimum FOS of 1.73 is determined from several trials, which can converge. When iteration analysis is used by the authors using an automatic location of the critical failure surface, no solution can be found for the first 20,162 trials during the simulated annealing analysis, which has demonstrated that this problem is difficult to converge by iteration analysis. When Cheng et al. (2008a) re-consider this problem using the double QR method, 'failure to converge' is reduced greatly and a minimum FOS of 1.387, which is much lower than that by the engineers, is found. The critical failure surface shown in Figure 3.34 is re-considered by iteration analysis using some commercial programmes, but convergence cannot be achieved even if the correct answer is used as the initial FOS. This case has clearly illustrated the importance of convergence for some difficult problems.

3.10.1 Parametric study of convergence

To investigate the phenomenon of convergence, a systematic parametric study using M–P method is carried out for the simple slope shown in

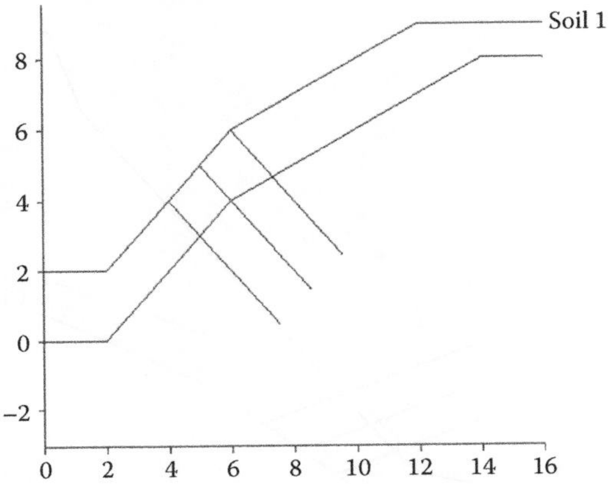

Figure 3.37 Slope for parametric study.

Table 3.18 Soil properties for parametric study

Case	*c′ (kPa)*	*ϕ′ (°)*	*Case*	*c′ (kPa)*	*ϕ′*	*Case*	*c′ (kPa)*	*ϕ′*	*Case*	*c′ (kPa)*	*ϕ′*
1	0	10	6	0	20	11	0	30	16	0	40
2	5	10	7	5	20	12	5	30	17	5	40
3	10	10	8	10	20	13	10	30	18	10	40
4	15	10	9	15	20	14	15	30	19	15	40
5	20	10	10	20	20	15	20	30	20	20	40

Figure 3.37 with only one soil and water. Twenty test cases with different c' and ϕ' are used in the parametric tests and the soil parameters are shown in Table 3.18. For the 20 test cases, three conditions are considered: no soil nail, three soil nails with each nail load equal to 30 kN and three soil nails with each nail load equal to 300 kN (maximum nail load in Hong Kong is 400 kN). A search for the critical circular failure surface is considered in generating trial failure surfaces by using a commercial programme. The x-ordinates of the left exit end of the failure surface are controlled within $x=0$ to $x=3.0$ m, while the x-ordinate of the right exit is controlled within $x=3.1$ m to $x=16$ m. Several thousands of failure surfaces are generated during the optimization search for each test case, and the percentage of 'failure to converge' is determined. Those cases that fail to converge with the iteration method are analysed individually by the double QR method, and many of these cases actually possess physically acceptable answers. The results of the analyses are shown in Figures 3.38 through 3.43.

There are two types of 'failure to converge' in the commercial programme adopted for the comparison, which are worth discussion. For type 1,

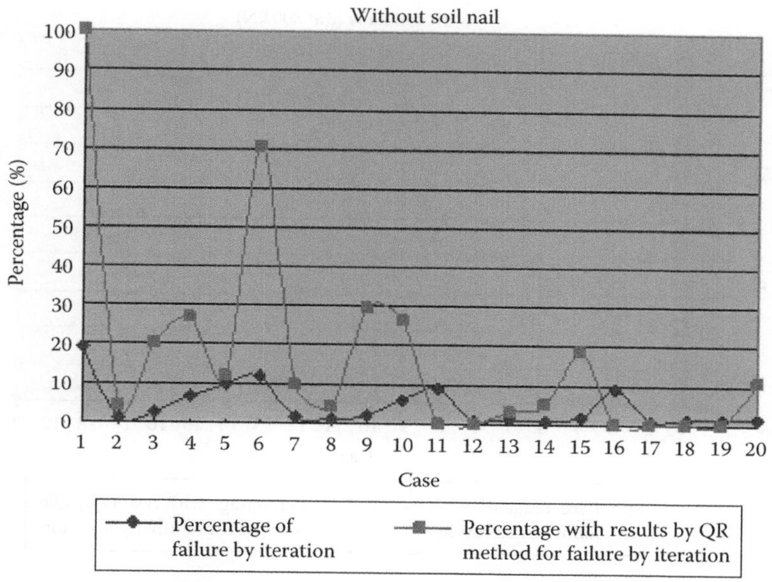

Figure 3.38 Percentage of failure type I for no soil nails.

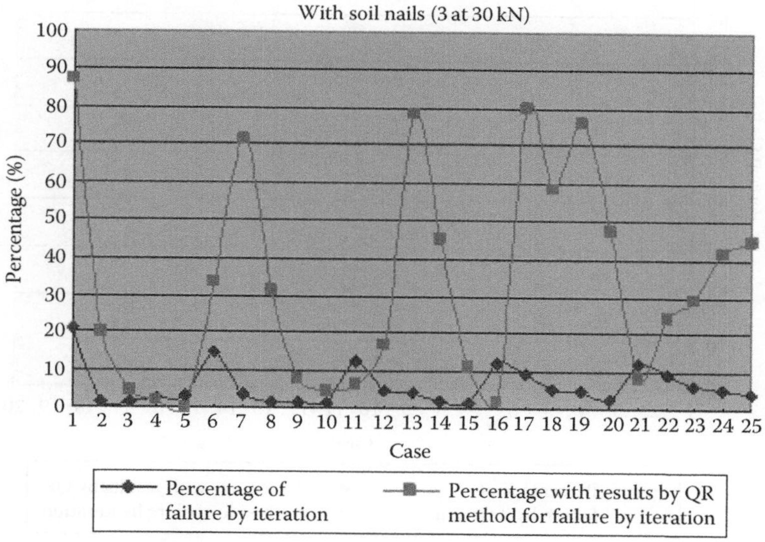

Figure 3.39 Percentage of failure type I for 30 kN soil nail loads.

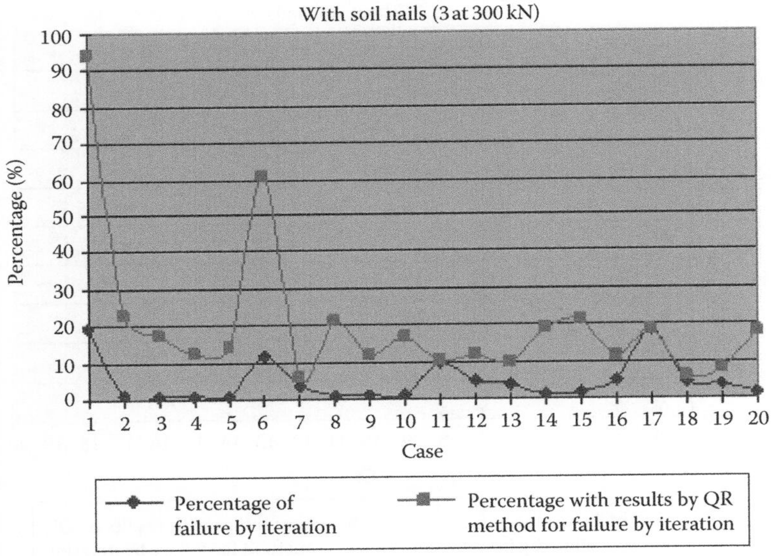

Figure 3.40 Percentage of failure type 1 for 300 kN soil nail loads.

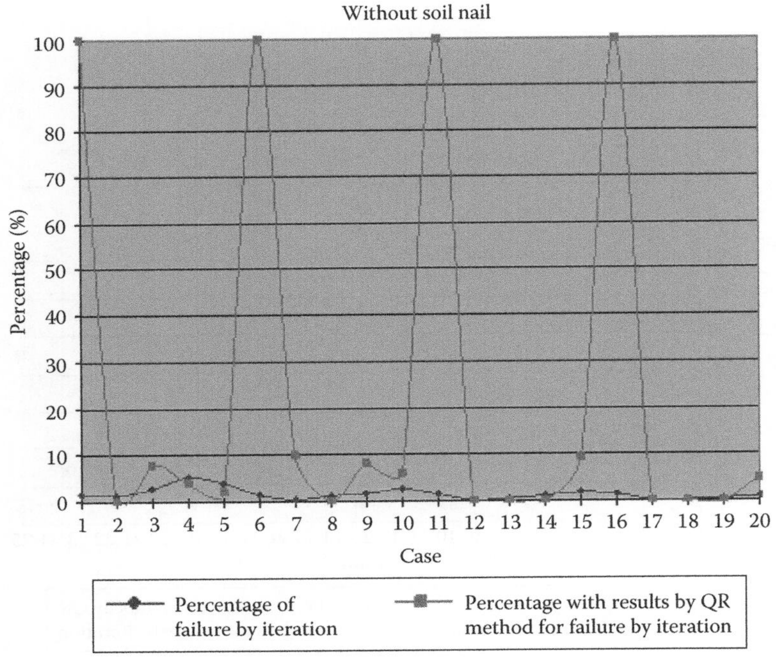

Figure 3.41 Percentage of failure type 2 for no soil nails.

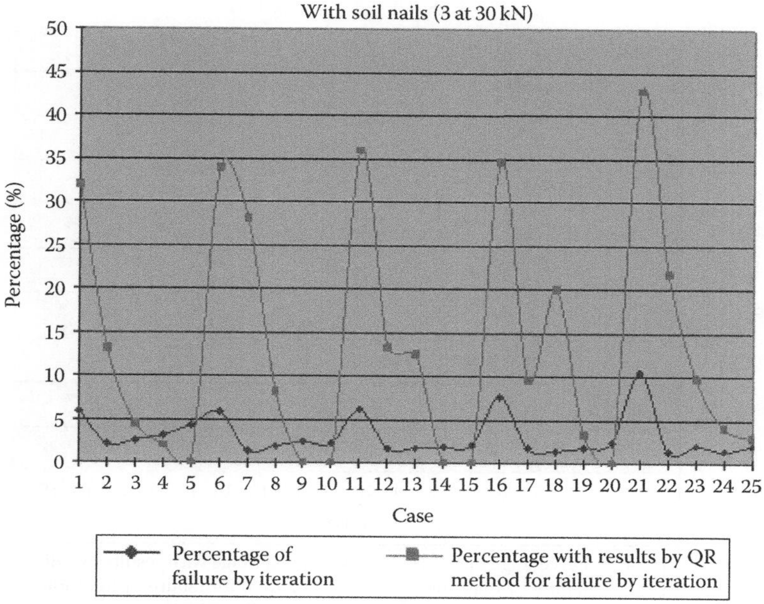

Figure 3.42 Percentage of failure type 2 for 30 kN soil nail loads.

converged result is not obtained with respect to the tolerance of the iteration analysis. For type 2, the 'converged' results are very small with unreasonable internal forces. The authors have also independently obtained very small 'converged' FOSs based on iteration analysis and the internal forces are all extremely large. If the FOS during the iteration analysis becomes very small, the difference between two successive trials can be less than the tolerance and a 'false' convergence is achieved. If the tolerance is further reduced towards 0, the FOS will also reduce further and tends to 0 while the internal forces will tend to infinity. Some commercial programmes will automatically screen out these kinds of results during analysis. When the double QR method is used, the small FOS is actually not the root of the FOS matrix, so use of the double QR method will not experience such 'false convergence' as in the iteration analysis.

As shown in Figures 3.38 through 3.40, the use of the iteration method by a commercial programme experiences 'failure to converge', with an interesting wave pattern for both type-1 and type-2 'failure to converge'. It is noticed that when ϕ' is 0 or very small, the use of the iteration method will experience more failure to converge while the double QR method is effective in determining meaningful answers for most of the cases. When the friction angle is high, iteration method appears to perform well. Besides that, the use of great soil nail forces will create great difficulties in

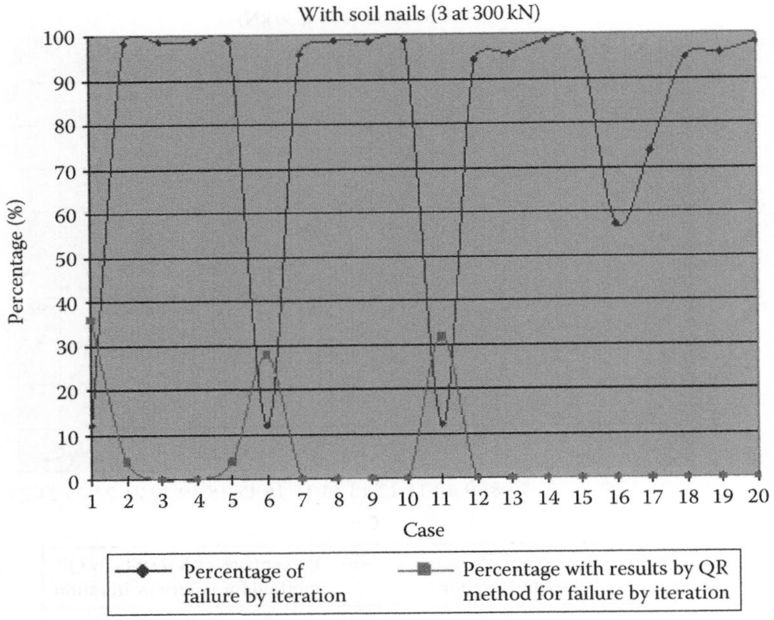

Figure 3.43 Percentage of failure type 2 for 300 kN soil nail loads.

convergence, which is also shown in Figures 3.40 and 3.43. The double QR method can however provide meaningful answers to many of the problems with great soil nail forces. The results shown in Figures 3.40 and 3.43 are particularly important to Hong Kong, as large-diameter soil nails with a maximum load of 400 kN for each bar are sometimes used in Hong Kong. Many engineers in Hong Kong have also experienced the problem of convergence with the presence of soil nails, and the problem in Figure 3.36 is a good illustration of the importance of convergence in slope stability analysis.

3.10.2 Combined impact of optimization and double QR analysis

The previous section has illustrated the importance of convergence in slope stability analysis. In this section, some cases from Hong Kong, which are analysed by experienced engineers, are re-considered by the authors. For the 13 cases shown in Table 3.19, all of them are analysed by engineers using the classical approach: manual location of the critical non-circular failure surface with 10–20 trials, while those failure surfaces that fail to converge will be neglected in the analysis. Cheng et al. (2008a) have used the double QR method in reducing convergence in the M–P analyses, and

Table 3.19 Impact of convergence and optimization analysis for 13 cases with M–P analysis

Case	*FOS by engineer*	*FOS by double QR*	*% Difference*	*Remark*
1	1.404	1.196	17.4	3 soil, no soil nail
2	1.458	1.152	26.6	4 soil, retaining wall, surcharge
3	1.5	1.18	27.1	2 soil, 12 soil nails
4	1.43	1.09	31.2	4 soil, 8 soil nails, surcharge
5	1.73	1.388	24.6	2 soil, retaining wall, 3 soil nails
6	1.406	1.253	12.2	3 soil, 7 soil nails
7	1.406	1.324	6.2	2 soil, 3 soil nails
8	1.4	1.293	8.3	3 soil, 4 soil nails
9	1.41	1.05	34.3	3 soil, 6 soil nails, steep slope
10	1.5	1.279	17.3	3 soil, 5 soil nails
11	1.408	1.328	6	2 soil, 3 soil nails
12	1.51	1.027	47	4 soil, 9 soil nails, steep slope
13	1.25	1.059	18	2 soil, 3 soil nails

the critical failure surfaces are located by the use of the simulated annealing method (Cheng, 2003). The results of analyses and comparisons are shown in Table 3.19.

In Table 3.19, the differences between the results by the engineers and those by Cheng et al. (2008a) are due to the optimization search and the convergence problems, but the individual contribution from these two factors cannot be separated clearly. Out of these 13 cases, the percentage differences are smallest for cases 7, 8 and 11 where the slope angles are not high and only small amounts of soil nails are required. Convergence problem is also less critical for these cases during optimization search. The critical failure surfaces for cases 2 and 5 are of deep-seated type, which lie below the retaining wall, and convergence is difficult for these deep-seated failure surfaces by iteration analysis with some commercial programmes. The slopes are steep for cases 3, 4, 9 and 12, and convergence with iteration analysis is difficult for these cases. It is to be noted that for steep slopes, deep-seated failure mechanisms or slopes stabilized with many soil nails, greater differences are found between the engineers' results and the refined results of the authors. For these cases, convergence is usually a problem, and only limited trials can be achieved by the manual trial-and-error approach.

3.10.3 Reasons for failure to converge

To investigate the reason behind 'failure to converge' even when the correct answer is used as the initial solution, the equilibrium equations shown in the following should be considered.

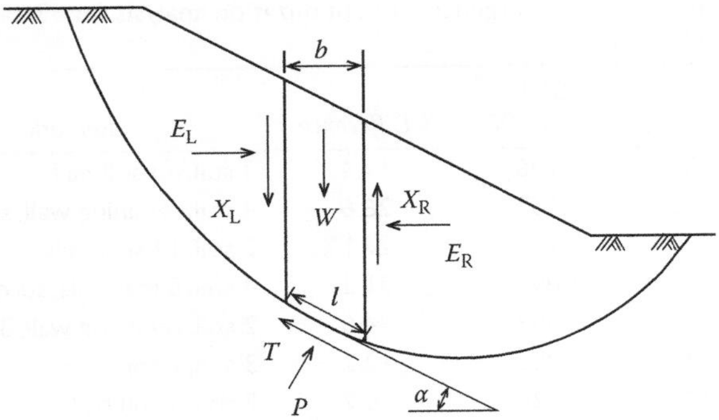

Figure 3.44 Forces acting on a slice.

For the slice shown in Figure 3.44, the base normal force P and the inter-slice normal forces E are given by

$$P = \frac{\left[W - (X_R - X_L) - 1/F\left(c'l\sin\alpha - ul\tan\phi'\sin\alpha\right)\right]}{m_\alpha} \tag{3.17}$$

$$E_R - E_L = P\sin\alpha - \left[c'l + (P - ul)\tan\phi'\right]\cos\alpha / F \tag{3.18}$$

where

$$m_\alpha = \cos\alpha\left(1 + \tan\alpha\frac{\tan\phi'}{F}\right)$$

In solving the equilibrium equations to determine the FOS F, the inter-slice shear forces X_R and X_L are usually assumed to be 0 in Equation 3.17 in the first step during the classical iteration method or the Newton–Rhapson method by Chen and Morgenstern (1983) or Zhu et al. (2001, 2005). When the correct FOS is used as the initial trial in Equation 3.17, N and hence F will not be correct even when F is correct on the right-hand side of Equation 3.17, as X_R and X_L are assumed to be 0, which is clearly not correct. The inter-slice normal force from Equation 3.18 will then be incorrect, which leads to the inter-slice shear force, which is computed based on $X=\lambda f(x)E$ to be incorrect. Equation 3.18, which is one step behind Equation 3.17, will not be correct as F is not correct from the left-hand side of Equation 3.17 in the first step. This iteration approach based on $X=0$ in the first step is used classically and is possibly the solution algorithm adopted in the commercial programmes. Referring to Figure 3.44, which is example 1 shown in Figure 3.33, the initial F_m based on the iteration analysis is close to the

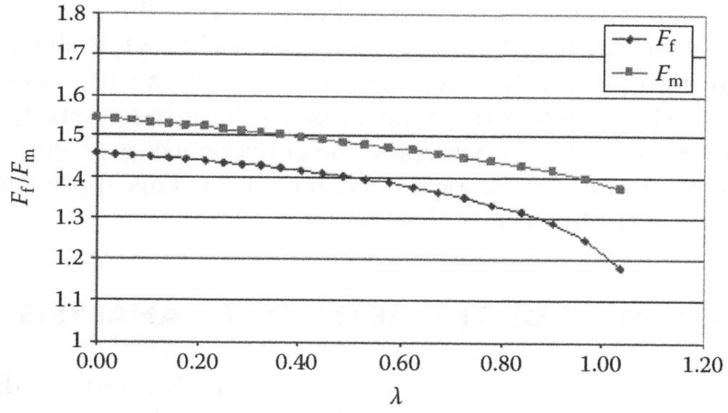

Figure 3.45 F_f and F_m from iteration analysis based on an initial FOS of 1.553 for example 1.

correct solution 1.553 when λ is 0 (FOS not sensitive to inter-slice shear force when λ is small). However, F_f is sensitive to the inter-slice shear force X when it is assumed to be 0 in the first step. When λ increases, X is no longer 0 but will deviate more and more from the correct value, and the effect on F_f becomes worse when an incorrect X is used in the iteration process. The results shown in Figure 3.45 will however be completely different when 10% of the correct inter-slice shear force is specified in the first step of the iteration analysis. As long as a constant ratio of 10% (or more) is specified for all the slices, F_f will be much closer to F_m initially by iteration analysis, and convergence can be achieved easily with $\lambda=0.71$ and $F=1.553$. While many problems are not sensitive to the inter-slice shear force so that the iteration method can work well, example 1 is very sensitive to the inter-slice shear force so that the iteration analysis leads to a wrong solution path during the non-linear equation solution, even when the correct F is used for the right-hand side of Equation 3.17 in the first step.

In Figure 3.45, the centroid of the soil mass is taken as the moment point for the analysis. If the moment point is varied, the results can be different and sometimes some problems, which fail to converge using default settings, may get converged using a different moment point. This practice is sometimes adopted by the engineers to overcome the convergence problem, and many commercial programmes allow the use of different moment points in the evaluation of FOSs. There is however no systematic and automatic way to change the moment point for general cases, and so the moment point will be kept constant during the optimization search in commercial programmes. From Cheng's experience, the use of a moment point above the slope surface is better than a moment point inside the slope in general.

Cheng et al. (2008a) have also tried to adopt the approach by Baker (1980) where the iteration method does not require $X=0$ in the first step.

It is found that convergence is improved by removing this requirement, but there are still cases where 'failure to converge' exists while the double QR method can find physically acceptable solutions. It can be concluded that the inter-slice shear force is the main cause for 'failure to converge' in classical iteration methods. To overcome the convergence problem, the extremum principle outlined in Section 2.9 can be used. This approach will give the FOS for practically every failure surface.

3.11 IMPORTANCE OF THE METHODS OF ANALYSIS

In general, different methods will give similar FOSs, and the differences between 'rigorous' and 'simplified' methods are small. Cheng has however come across many cases where there are noticeable differences in the FOSs, which is worth discussing. Consider the problem in Figure 3.46, which is a project in China. For the prescribed failure surface, the FOSs for the simplified methods are 2.358 (f_0 = 1.068), 2.159, 2.796 and 3.563 for Janbu simplified, Corps of Engineers, Lowe–Karafiath and load factor methods. The FOSs for the rigorous methods are 3.06, 3.38, 3.857, 3.361 and 3.827 for Sarma, M–P ($f(x)$ = 1.0 and sin(x)) and generalized limit equilibrium (GLE) methods ($f(x)$ = 1.0 and sin(x)). Since there is a wide scatter in the results, which means the effect of the inter-slice shear force is critical in the analysis, there is great difficulty in the interpretation of the results. The result by the Janbu simplified method (1957) is finally accepted by the engineers for sake of safety. As a good practice, the FOSs for complicated projects using different limit equilibrium methods (LEMs) should be determined and assessed before final interpretation.

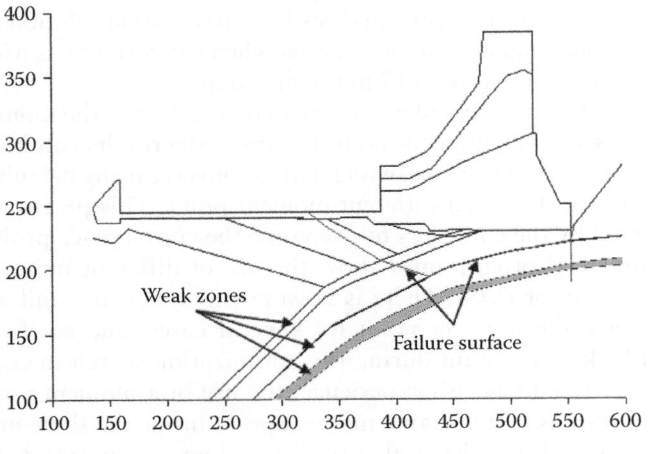

Figure 3.46 A complicated problem where there is a wide scatter in the FOS.

3.12 SOLUTION OF THE INTER-SLICE FORCE FUNCTION AND FUNDAMENTAL INVESTIGATION INTO THE PROBLEM OF CONVERGENCE

In Chapter 2 of this book, the basic formulation for the LEM is introduced. There are many limitations to the basic LEM: (1) requirement of assumption on inter-slice force function, thrust line or others; (2) inability to satisfy force and moment equilibrium locally and globally automatically; (3) different FOSs for different methods of analysis so that there is no unique answer to the problem and (4) failure to converge in the analysis. In this section, the fundamental limitations of the LEM will be investigated; Cheng et al. (2010) have proposed the extremum principle to overcome all these limitations.

One of the major outstanding questions of the LEM is the inter-slice force functions $f(x)$ or the thrust line in Janbu's rigorous method (1973) for which no rigorous formulation has ever been developed to determine this unknown. Various inter-slice force functions are available in most of the slope stability programmes. When $f(x)$ is equal to 1.0, which is the common practice used by most of the engineers (as the engineers do not know which option or parameters are used to define $f(x)$), the M–P method (1965) will reduce to the popular Spencer's method (1967). Although this option is adopted by many engineers in routine design and analysis works and appears to work well for most cases, there is no theoretical background to this option except for convenience. Sarma and Tan (2006) have implicitly assumed that the Mohr–Coulomb criterion is satisfied along the vertical interfaces between slices by using the limit analysis approach; and this approach is equivalent to enforcing a specific but unknown $f(x)$ in the analysis. In addition to the inter-slice force function, some researchers prefer the use of a low-order polynomial function for the base normal force (Zhu et al., 2003; Zheng et al., 2009), but this approach can also be viewed as being equivalent to an inter-slice force function.

Although $f(x)$ will exist for every problem, there is no theoretical background to determine it using the classical formulation. Fan, Fredlund and Wilson (1986) have proposed an error function for $f(x)$, which is based on elastic stress analysis for relatively simple problems. Cheng et al. (2010) believe that $f(x)$ based on elastic stress analysis may not be applicable to the ultimate limit state due to the stress redistribution when the system approaches the ultimate limit state. Morgenstern (1992), among others, has pointed out that for normal problems, the FOSs from different methods of analyses are similar, so that the assumptions on the internal force distributions are not important, except for some particular cases, but the validity of this observation has not been investigated fully for highly irregular $f(x)$. While the assumptions on $f(x)$ are not critical for normal problems, Krahn (2003) and Abramson et al. (2002) have pointed out that $f(x)$ may be critical for certain special cases.

Baker and Garber (1978), Baker (1980) and Revilla and Castillo (1977) have applied calculus of variation to determine the FOS of a slope that does not require any assumption on the internal force distribution. The variational formulation by Baker (1980) was criticized by De Jong (1980, 1981) as the stationary value may have an indefinite character rather than a minimum. The global minimum is not necessarily given by the gradient of the function as 0 if the global minimum lies at the boundary of the solution domain. This conclusion was also supported by Castilo and Luenco (1980, 1982), which was based on a series of counter-examples. Baker (2003) later incorporated some additional physical restrictions into the basic limiting equilibrium framework so as to guarantee that the slope stability problem has a well-defined minimum solution. Although the variational principle requires very few assumptions with no convergence problems during the solution, it is difficult to adopt when the geometry or the ground/loading conditions are complicated. As a variational principle suitable for general problems is not available, this approach has not been adopted in any commercial programme or used in any practical problem.

Cheng has worked on many complicated and large-scale slope stability problems in Hong Kong, China and Europe, and the LEM has been adopted in most of these projects. Some of the important questions about the fundamental problems of the LEM as raised by the engineers in these projects include the following:

1. The meaning of failure to converge during stability analysis – this is particularly serious for slopes with external loads and soil reinforcement, as the external loads and soil reinforcement may create local stress concentration, so that the problem may become difficult to be defined by a simple inter-slice force function.
2. Choice of $f(x)$ for some special problems where $f(x)$ is important in the analysis.
3. For cases where $f(x)$ is important, there will be a wide range of results based on different classical stability formulations, and acceptability of the results would be difficult, which is illustrated by the case shown in Figure 3.46.

These questions are important to both researchers and engineers for certain difficult problems, but there are very few previous studies devoted to these three questions.

Failure to converge with 'rigorous' methods is well known to many engineers, in particular, for complicated problems with heavy external loads or soil reinforcement. Cheng et al. (2008b) have carried out a detailed study on the convergence problem in stability analysis and have concluded that there are two reasons for failure to converge with rigorous methods. First, the iteration method that is commonly used to determine the FOS may fail to converge because the inter-slice shear force is assumed to be 0 in the first step of the iterative analysis. Cheng (2003) has developed the double QR method, which can evaluate the FOS and internal forces directly from a Hessenberg matrix. Based on this method, failure to converge in stability calculations due to the first reason can be eliminated. There are, however, many cases for which the double QR method determines that a physically acceptable answer does not exist for a given $f(x)$, which means no meaningful FOS will be available unless $f(x)$ can be varied. Actually, some engineers have questioned the meaning of 'no FOS available' for a given failure surface, as such a concept does not appear in structural engineering. So far, there has been little previous study on this type of failure to converge, and no rigorous method can guarantee convergence for the general case.

Since the critical failure surface may not necessarily converge according to the existing 'rigorous' methods of analysis (Cheng et al. 2008a), there is always a chance that the critical failure surface may be missed during optimization analysis. It is also interesting to note that it has never been proven that a slip surface that fails to converge in stability analysis is not a critical slip surface, but all commercial programmes will simply neglect those slip surfaces that fail to converge. Although this problem may not be critical in general, a failure surface with no FOS is still physically surprising. A system without FOS is not real and is just a human deficiency in making the wrong assumption, and this situation never appears in structural engineering or other similar disciplines. FOS always exists for a problem, but it is possible that we are not able to determine it simply because of the use of wrong assumptions, and this is supported by the study by Cheng et al. (2008a) that many smooth slip surfaces can also fail to converge using the popular Spencer's analysis.

Cheng et al. (2010) have investigated and addressed the three questions posed by the engineers using an approach that is equivalent to the variational principle. The upper and lower bounds to the FOS for any problem can be evaluated within a reasonable time for any complicated problem. The present approach can also be viewed as a lower bound approach, based on which $f(x)$ can be determined, and failure to converge will be virtually eliminated in the stability analysis. The present approach is also a practical and useful tool that has been used for several complicated projects in China.

3.12.1 Determination of bounds on FOS and $f(x)$

To determine the bounds of FOS and $f(x)$, the slope shown in Figure 3.47 can be considered. For a failure surface with n slices, there are $n - 1$ interfaces and hence $n - 1$ $f(x_i)$. $f(x)$ will lie within the range 0–1.0, while the mobilization factor λ and the objective function FOS based on M–P method (1965) will be determined for each set of $f(x_i)$. The maximum and minimum

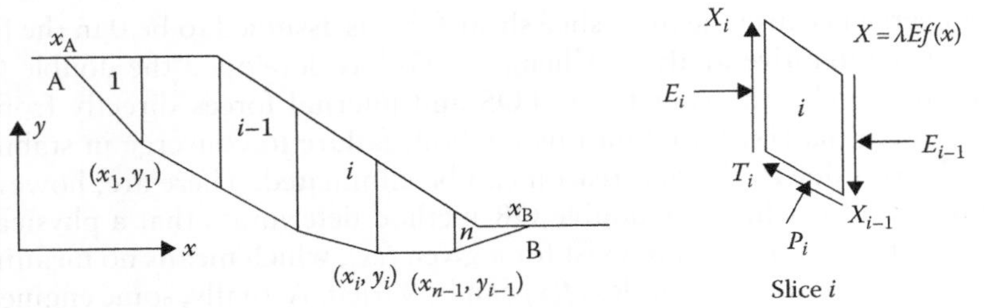

Figure 3.47 **Slip surface and slices.**

FOSs of a prescribed failure surface satisfying force and moment equilibrium will then be given by the various possible $f(x_i)$ satisfying Equation 3.19 and certain physical constraints that will be discussed later.

$$\text{Maximize (or minimize) FOS subject to } 0 \leq f(x_i) \leq 1.0 \text{ for all } i. \quad (3.19)$$

In carrying out the optimization analysis as given by Equation 3.19, the constraints from Mohr–Coulomb relation along the interfaces between slices as given by Equation 3.20 should be considered:

$$X \leq E \tan(\phi') + c'L \quad (3.20)$$

where L is the vertical length of the interface between slices. The constraint given by Equation 3.20 should also satisfy the requirement that the line of thrust of the internal forces lies within the soil mass, and Equation 3.20 can have a major impact on the FOS in some cases, which will be illustrated by numerical examples in the following section. Since other than $f(x)$ the M–P method (1965) is totally governed by force and moment equilibrium, the maximum and minimum FOSs found from varying $f(x)$ will provide the upper and lower bounds to the FOS of the slope that are useful for some difficult problems.

Since the objective function is highly discontinuous, the FOS is obtained by the double QR method by Cheng (2003). The simulated annealing method, which is more stable but less efficient, is used to determine the extrema (maximum or minimum FOS) with any given slip surface according to Equation 3.19. To evaluate the global minimum FOS of a slope, another global optimization analysis should be carried out for the FOS, which is an outer loop of the global optimization analysis. To ensure that 'false' failure-to-converge due to iteration analysis (Cheng et al. 2008b) is not encountered, so as to reduce the discontinuity of the objective function, the FOS is determined by the more time-consuming but robust double QR method. Since FOS is available for practically all of the failure surfaces, the more efficient harmony search method can be used for locating the critical failure surface. The complete process is computationally intensive, but the use of modern global optimization processes can make this process a reality on a personal computer within an acceptable computation time.

Pan (1980) has stated that the slope stability problem is actually a dual optimization problem to be not well known outside China. On the one hand, the soil mass should redistribute the internal forces to resist failure, which will result in a maximum FOS for any given slip surface, which is called the maximum extremum principle. On the other hand, the slip surface with the minimum FOS is the most possible failure surface, which is called the minimum extremum principle. The maximum and minimum extremum principles are actually equivalent to the lower and upper bound methods, which are well known and will not be discussed here. Mathematically, the solution from the use of variational principles is an extremum of a function, and this is also equal to the global maximum/minimum of the function, which can also be determined from an optimization process. The 'present proposal' can be viewed as a form of the discretized variational principle.

The maximum extremum principle is not new in engineering, and the ultimate limit state of a reinforced concrete beam is actually the maximum extremum state where the compressive zone of the concrete beam will propagate until a failure mechanism is formed. The ultimate limit state design of a reinforced concrete beam under application of a moment is equivalent to the maximum extremum principle. For any prescribed failure surface, the maximum 'strength' of the system will be mobilized when a continuous yield zone is formed, which is similar to a concrete beam. Pan's extremum principle (1980) can provide a practical guideline for slope stability analysis, and it is equivalent to the calculus of variation method used by Baker and Garber (1978), Baker (1980) and Revilla and Castillo (1977). This dual extremum principle is proved by Chen (1998) based on lower and upper bound analyses, and it is further elaborated upon with applications to rock slope problems by Chen et al. (2001a,b). The maximum extremum is actually the lower bound solution, and the present approach is actually a lower bound approach as well as a variational principle approach.

In the following sections, Cheng et al. (2010) will carry out the analysis in the following ways:

1. The FOS of a given failure surface with a given $f(x)$ will be determined by the use of the double QR method.
2. $f(x)$ will be taken as the control variable in the optimization process by the simulated annealing method, and the maximum FOS for a given failure surface and the associated $f(x)$ will be determined from an optimization analysis. The maximum extremum will be taken as the FOS for a prescribed failure surface; $f(x)$ can hence be obtained directly from the use of the optimization principle.

3. In addition to the maximum extremum FOS, Cheng et al. (2010) will also investigate $f(x)$ for the minimum extremum FOS, which will provide a lower bound of the FOS for comparison.
4. The problem of convergence will be investigated through the 'present proposal'.
5. A simple inter-slice force function $f(x)$ for simple problems will be proposed.
6. The relation between failure to converge and the shape of $f(x)$ will be investigated.

3.12.2 Numerical studies and comparisons with classical methods of analysis

Cheng et al. (2010) have applied the simulated annealing method (Cheng, 2003) complying with Equations 3.19 and 3.20 to evaluate the two extrema of the FOS, and the method is coded into a general-purpose programme SLOPE 2000 (available at Cheng's website: http://www.cse.polyu.edu.hk/~ceymcheng/). To determine the maximum and minimum extrema by the simulated annealing method, a tolerance of 0.0001 is used to control the optimization search and the FOS determination. This tolerance will terminate the search in a particular solution path during the optimization process (see also Cheng et al., 2007b for details of the heuristic optimization methods). Since Cheng et al. (2010) have adopted 15 slices in the computation, there are, in total, 14 $f(x_i)$ unknowns in the analysis, and the number of trials required to evaluate the two extrema ranges from 25,000 to 32,000, which is controlled by the tolerance during the optimization search. Based on this study, it was found that about 30%–80% of the trials can converge when $f(x)$ is varied, and those trials that fail to converge are controlled by either Equation 3.20, or no physically acceptable answer can be found from the double QR method. The number of trials, which fail to comply with Equation 3.20, is about three to five times the number when no physically acceptable answer can be found by the double QR method, and so compliance with Equation 3.20 (together with the requirement on the line of thrust), which has been neglected in M–P method (1965), is actually important if an arbitrary $f(x)$ is defined.

Consider a very simple 45°, 6 m high slope with a circular failure surface as shown in Figure 3.48. The unit weight of the soil is 19 kN/m^3, while c' and ϕ' vary as shown in Tables 3.20 and 3.21. The differences between the two extrema are less than 2% of the results given by Spencer's method (1967), which clearly indicates that the FOS is not sensitive to $f(x)$ and that Spencer's method (1967) gives a good result for this example (see Table 3.20). It is also interesting to find that whereas the FOS is not sensitive to Equation 3.20, λ is quite sensitive to Mohr–Coulomb relation along the interfaces as mobilization of inter-slice shear force to achieve maximum

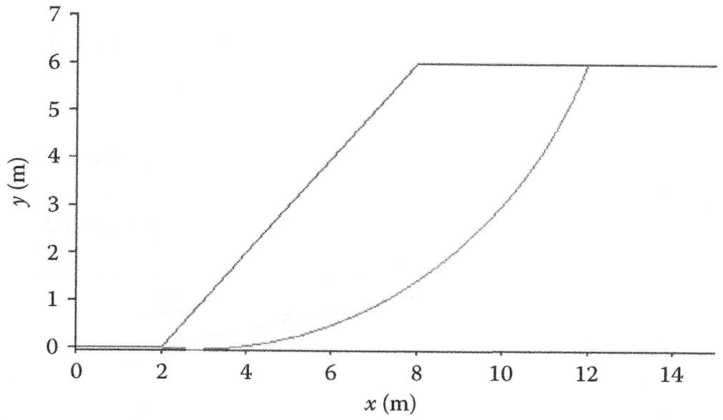

Figure 3.48 A simple slope with a circular failure surface.

Table 3.20 FOSs from lower bound and Spencer's analysis for Figure 3.48

FOS	*Max. FOS no Equation 3.20*	*Max. FOS with Equation 3.20*	*Min. FOS no Equation 3.20*	*Min. FOS with Equation 3.20*	*Spencer's*
$c'=0$ kPa, $\phi'=20°$	0.759	0.749	0.738	0.743	0.745[a]
$c'=0$ kPa, $\phi'=40°$	1.753	1.733	1.702	1.708	1.718
$c'=5$ kPa, $\phi'=20°$	1.017	1.012	1.002	1.003	1.007
$c'=5$ kPa, $\phi'=40°$	2.008	1.998	1.966	1.965	1.98
$c'=10$ kPa, $\phi'=20°$	1.280	1.277	1.268	1.267	1.272
$c'=10$ kPa, $\phi'=40°$	2.263	2.261	2.230	2.229	2.242

[a] Spencer's result violates Equation 3.20.

Table 3.21 λ from lower bound and Spencer's analysis for Figure 3.48

FOS	*Max. FOS no Equation 3.20*	*Max. FOS with Equation 3.20*	*Min. FOS no Equation 3.20*	*Min. FOS with Equation 3.20*	*Spencer's*
$c'=0$ kPa, $\phi'=20°$	1.867	1.0	1.888	1.0	0.522[a]
$c'=0$ kPa, $\phi'=40°$	1.82	1.151	1.886	0.901	0.522
$c'=5$ kPa, $\phi'=20°$	1.89	0.892	1.873	1.758	0.457
$c'=5$ kPa, $\phi'=40°$	1.857	1.208	1.896	1.903	0.491
$c'=10$ kPa, $\phi'=20°$	1.711	1.024	1.889	1.893	0.407
$c'=10$ kPa, $\phi'=40°$	1.855	1.432	1.846	1.907	0.464

[a] Spencer's result violates Equation 3.20.

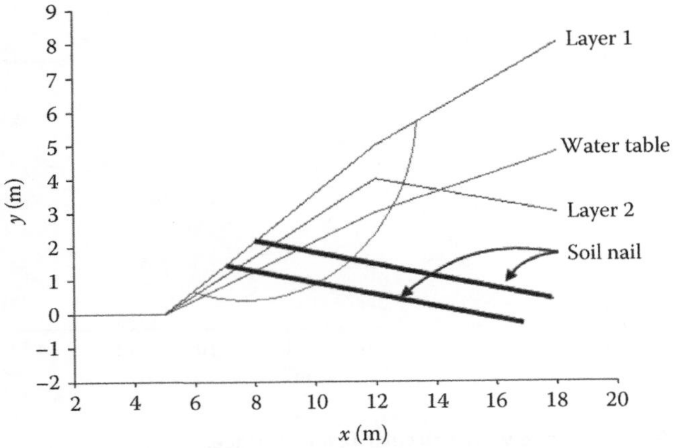

Figure 3.49 A problem with two soils, two soil nails and a water table.

and minimum resistance will involve a higher λ value. It is also observed that the values of λ from the two extrema are generally greater than that from Spencer's analysis (1967), and these observations also apply to many other examples.

For the slope shown in Figure 3.49 with the soil parameters given in Table 3.22, the various FOSs are given in Tables 3.23 and 3.24. With only the lower nail present, the differences between the two extrema as compared with Spencer's result (1967) are about 5.9% and 4.4% when Equation 3.20 is used or not used, respectively. The corresponding results/differences when the two soil nails are present are 10.5% and 4.1%. It can be observed

Table 3.22 Soil parameters for Figure 3.49

Soil	*Unit weight (kN/m³)*	*Saturated unit weight (kN/m³)*	*c′ (kPa)*	*ϕ′ (°)*
Top	18	20	5	36
Second layer	15	17	3	30

Table 3.23 FOSs from lower bound approach and Spencer's analysis for Figure 3.49

Case	*Max. FOS no Equation 3.20*	*Max. FOS with Equation 3.20*	*Min. FOS no Equation 3.20*	*Min. FOS with Equation 3.20*	*Spencer's*
Bottom nail	1.856	1.841	1.750	1.763	1.790
Two nails	2.661	2.600	2.398	2.498	2.515

Table 3.24 λ from lower bound approach and Spencer's analysis for Figure 3.49

Case	*Max. FOS no Equation 3.20*	*Max. FOS with Equation 3.20*	*Min. FOS no Equation 3.20*	*Min. FOS with Equation 3.20*	*Spencer's*
Bottom nail	1.149	0.944	0.924	1.902	0.488
Two nails	1.435	1.281	1.149	2.011	0.547

that when the soil nail or the external load is present, the choice of $f(x)$ has a noticeable impact on the results, and compliance with Equation 3.20 is also a critical issue, which should be considered in the determination of extrema. $f(x)$ for this problem is shown in Figure 3.50, and this inter-slice force function is clearly not similar to the functions commonly used. Based on the concept of extrema, the statically indeterminate nature of $f(x)$ is actually statically determinate for whatever complicated problem, and the procedure to determine $f(x)$ is completely independent of the difficulty of the problem.

3.12.3 Study of convergence by varying f(x)

Consider the slope with a steep failure surface as shown in Figure 3.51 with the soil parameters same as that for the problem in Figure 3.48. The FOSs are 1.542, 1.570, 1.526 and 1.550 based on Bishop's method, Janbu's simplified method (without the correction factor), the Swedish method and Sarma's method, respectively. The extrema are 1.602 and 1.547 if Equation 3.20 is not enforced and are 1.564 and 1.559 if Equation 3.20

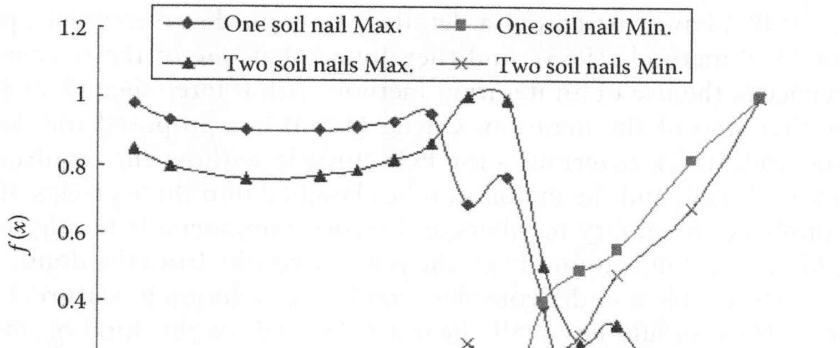

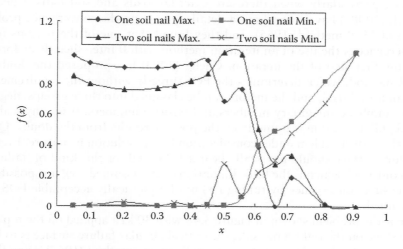

Figure 3.50 f(x) for the problem in Figure 3.49.

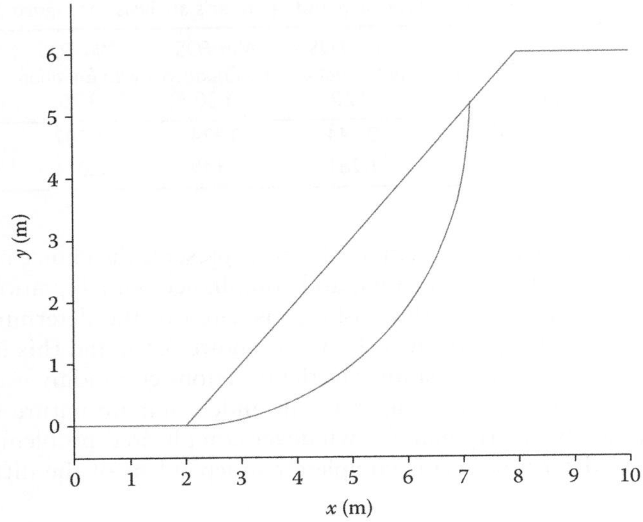

Figure 3.51 A steep failure surface for illustration.

is considered. No physically acceptable results can be found for Spencer's method (1967) using the double QR method, and 'failure to converge' is the fundamental problem in assuming $f(x) = 1.0$. For such a circular slip surface, the choice of $f(x)$ is the reason for the failure to converge, and an acceptable internal set of forces.

Failure to converge for 'rigorous' methods is well known to many engineers, particularly when there are external loads and soil nails. Cheng et al. (2008a) have carried out a detailed study on the convergence problem of M–P method (1965), and they found that one of the reasons for divergence is the use of an iteration method with 0 inter-slice shear force in the first step of the iteration. Cheng (2003) has proposed the double QR method, which determines the FOS directly without the requirement of an initial trial, and the factors can be classified into three groups: negative numbers, imaginary numbers and positive numbers. If no physically acceptable solution is found from the positive results from the double QR method, the problem under consideration has no solution by nature. Every failure surface should physically bear a FOS, and for this kind of 'failure to converge', which is the basic limitation of the assumed $f(x)$, it is possible to ensure convergence by tuning $f(x)$ until a physically acceptable FOS is obtained.

For the problem shown in Figure 3.48 with 30 kPa applied on the top of the slope on the right-hand side, the critical circular failure surface is to be determined. The minimum FOSs from Spencer's method (1967) using the double QR method and the iteration method are 0.995 and 0.989, respectively. Based on the harmony search, the percentages of surfaces that fail to converge are 6.2 and 24.6 from Spencer's method (1967) for this problem, based on the double QR method and the iteration method, respectively. It can be noted that there is a high percentage of failure by the classical iteration analysis, which has been investigated in detail by Cheng et al. (2008a). The double QR method has greatly overcome the limitations of the iteration method by direct evaluation of the FOS, but 6.2% of those failures still fail to converge due to enforcement of $f(x) = 1.0$ for the present simple problem. The minimum FOSs using the harmony search for the extrema are 1.013 and 0.85 if Equation 3.20 is not applied, and are 1.002 and 0.901 if Equation 3.20 is used; there is virtually no failure to converge based on the present approach. It can be observed that the lower bound of the FOS is relatively low as compared with the result by Spencer's method (1967), which means the choice of $f(x)$ is actually important for this case and is contrary to the comment by Morgenstern (1992) that $f(x)$ is not important except for isolated cases.

Based on various internal testing by Cheng, except for highly irregular zigzag failure surfaces or soils with very high cohesive strength and steep ends, which may fail to produce a result occasionally, the present method seldom fails to give a FOS, which is not possible with any other 'rigorous' method. Although the use of the simulated annealing method to determine the FOS for a single failure surface is time-consuming, the objective function FOS is practically a continuous function so that the more efficient harmony search method can be used to locate the critical failure surface. The present method is, hence, a practical solution for an engineer, and the time required to obtain the critical circular failure surface is about 10 min for the present problem, which can be considered acceptable. A further advantage of the present method is that the upper and lower bounds of the FOS for the critical failure surface can be evaluated for reference.

As shown earlier, $f(x)$ is actually determinate for all problems under the limit equilibrium formulation based on the time-consuming extrema determination. For simple problems where the use of $f(x) = 1$ may not be satisfactory, Cheng et al. (2010) have curve-fitted $f(x)$ by a simple formula, which requires only three variables. The simple formula given by Equation 3.21 is simple for optimization analysis, and Cheng et al. (2010) have found that the FOS from Equation 3.21 is very close to that from the present optimization analysis. Based on the internal tests by Cheng et al. (2010), Equation 3.21 should be adequate and sufficiently good for practical purpose:

$$f(x) = \cot^{-1}\frac{(ax+b)}{c} \tag{3.21}$$

Based on the concept of extrema determination, the upper and lower bounds of the FOS, which satisfy both the force and moment equilibrium, are determined from the tuning of $f(x)$ using the simulated annealing method and the double QR method. Based on the two extrema, $f(x)$ can be evaluated. Based on various case studies, Cheng et al. (2010) have found that the FOS from Spencer's method is a good approximation, which is well known among the engineers, as the differences between the upper and lower bounds are small. However, for more complicated problems, the differences between the upper and lower bounds are not negligible if Equation 3.20 is not considered, whereas the differences are still noticeable even when Equation 3.20 is applied. The present proposal provides the bounds to the FOS of a slope, which is clearly an advantage not possible with all the classical methods of analysis.

Cheng et al. (2008a) have demonstrated the importance of convergence in slope stability analysis based on 13 cases from Hong Kong. Since the extrema method is virtually free from convergence problems, the present method will be suitable for complicated problems, which are difficult to converge with the classical 'rigorous' methods. The previous studies on convergence by Baker (1980) or Cheng et al. (2008a) are mainly concerned with numerical results instead of investigating the fundamental importance of $f(x)$. For a problem with a set of consistent and acceptable internal forces, the FOS must exist as it can be determined explicitly if the internal forces are known. Failure to converge will not occur for this case if the double QR method is used, though the use of iteration method may fail to converge due to the limitation of the mathematical method. If no FOS can be determined from the use of double QR method in the FOS evaluation, this is equivalent to a consistent set of internal forces under the specified $f(x)$ cannot exist. The present approach in tuning $f(x)$ should not be viewed as an arbitrary process to generate a converged answer. The use of $f(x) = 1.0$ or any other $f(x)$ may not be associated with a set consistent and acceptable internal force (failure to converged cases), and the $f(x)$ that is commonly in use may not be correct from a physical point of view (though the error may be small). The present approach can establish a range of $f(x)$ with consistent and acceptable internal forces. If the maximum extremum is accepted as the FOS of the failure surface, then $f(x)$ is actually uniquely defined. For the two extrema from the present analysis, Cheng et al. (2010) view that the maximum extremum should be taken as the FOS of the prescribed failure surface. As discussed, the internal forces within the soil mass should redistribute until the maximum resisting capacity of the soil mass is fully mobilized. Beyond that, the soil mass will fail. The present definition also possesses an advantage that it is independent of the definition of $f(x)$. It is well known that there are also cases where $f(x)$ may have a noticeable influence on the FOS. There is not a clear guideline on the acceptance of the FOS due to the use of different $f(x)$. The use of the maximum extremum can also avoid this dilemma, which has been neglected in the past. The present approach is a typical lower bound approach as statically admissible forces associated with a prescribed $f(x)$ are considered. The selection of the maximum FOS is hence justified from the lower bound theorem.

Currently, there are more than 10 slope stability analysis methods available or equivalently, more than 10 sets of FOS under different assumptions for a problem. Although the differences between these FOSs are usually small, it is difficult to assess and compare these FOSs as the 'true' answer is not available. The authors view that the maximum extremum FOS can be taken as the FOS of the slope under consideration. Under this case, $f(x)$ can be determined from the present approach. The question of $f(x)$ and convergence can then be considered as settled under the present formulation. The difficulty in assessing the FOS for problems similar to that in Figure 3.46 can also be avoided, as there is only one FOS under the extremum formulation.

3.12.4 Validation of maximum extremum principle

The validity of the present approach can be illustrated further by a simple bearing capacity problem for clay soil, where the bearing capacity factor N_c is well known to be $\pi+2$ and is given by Prandtl (1920), Sokolovskii (1965) and Chen (1975). So far, there has not been any reported success in using the classical 'rigorous' slope stability methods to evaluate this factor, except for the approximate solution by Rocscience (2006), which is actually not good enough. Consider a smooth shallow foundation on clay as shown in Figure 3.52 where the failure mechanism can be obtained analytically. The stresses within the failure mass are determined by the iterative finite-difference solution of the slip line equations by Cheng and Au (2005), and a bearing capacity factor N_c of 5.14 ($\pi+2$) is determined by the authors from

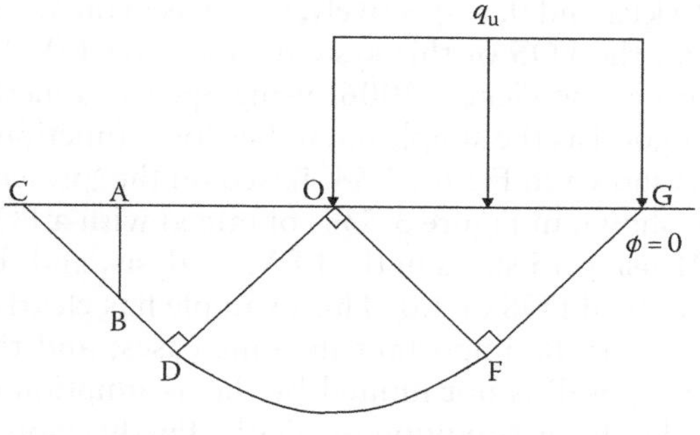

Figure 3.52 A simple smooth footing on clay.

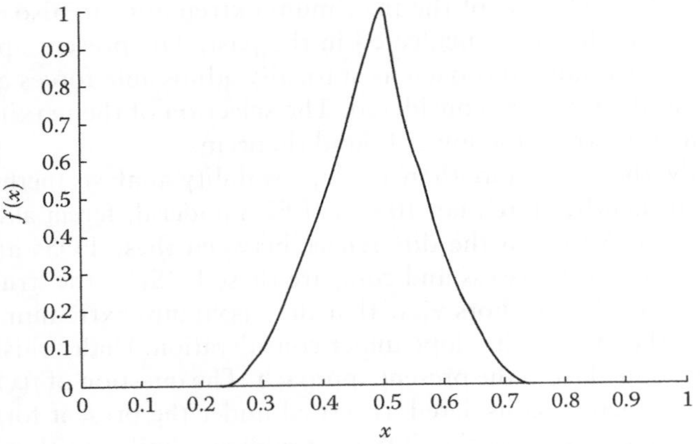

Figure 3.53 f(x) for the simple footing on clay as shown in Figure 3.52 based on slip line solution.

numerical analysis. Consider the vertical plane AB as shown in Figure 3.52; the normal and shear stresses on this plane have been determined, and the integration of the normal and shear stresses along AB will give the total normal and shear forces E and X on this surface. The ratios X/E [or $f(x)$] against different dimensionless distance x are hence obtained. Denote the largest ratio of X/E as λ; $f(x)$ is determined by the ratio between X/E and λ, and the corresponding $f(x)$ is shown in Figure 3.53, while λ is obtained as 0.388. For the left half of triangle CDO and the right half of triangle GFO in Figure 3.52, the principal stresses are along the horizontal and vertical directions so that $f(x)$ is equal to 0. Within zone ODF, there is a rapid change of the principal stress directions and $f(x)$ will no longer be 0. From D to O in Figure 3.52, $f(x)$ increases rapidly to 1.0 at $x=0.5$. For this problem, the inter-slice force function is determinate, but this kind of function appears to have never been used in the past. If the cohesive strength and unit weight of the clay are 20 kPa and 0, respectively, then based on an applied pressure q_u of 102.83 kPa, the FOS of this system should be 1.0. A FOS of 0.941 is however given by Rocscience (2006) using Spencer's method ($f(x)=1.0$). This result is not good as the simple inter-slice force function 1.0 is far from the actual $f(x)$ as shown in Figure 3.53. Based on the 'present proposal', an $f(x)$ close to that shown in Figure 3.53 is obtained with a FOS of 0.995 and λ equals to 0.391 using 15 slices in the LEM analysis, and the result is very close to the theoretical FOS of 1.0. This example has clearly demonstrated that a correct $f(x)$ can be important in some cases, and the applicability of the 'present proposal' is not limited by the assumption of $f(x)$ as there is no need to make this assumption explicitly. Furthermore, the result has also demonstrated that the maximum of the FOS as adopted in the 'present proposal' is correct as other $f(x)$ will give a smaller FOS different from the rigorous solution.

The inter-slice force function as shown in Figure 3.53 can be derived analytically, but such a shape cannot be described by Equation 3.21 and appears to have never been adopted in the past. This example has also illustrated that $f(x)$ can be important for some cases, and adoption of the 'present numerical proposal', which does not require $f(x)$ explicitly, can overcome this limitation.

3.13 VARIABLE FOS FORMULATION IN LEM

For the LEM, one of the basic assumptions common to all of the traditional soil and rock slope stability methods is a single FOS for the entire solution domain. Without this assumption, the slope stability problem will be statically indeterminate unless additional assumptions are used. The actual failure of a slope is however usually a progressive yielding phenomenon. If the shear strengths between adjacent blocks are fully mobilized, the unbalanced forces will distribute to the adjacent blocks until a failure mechanism is formed. This process is called the progressive failure of a slope. This phenomenon is well known but is difficult to be considered by the classical LEM. For a system with a FOS close to 1.0, the choices of the shear strength parameters become critical. The adoption of maximum shear strength or residual strength and the extent of adoption of different design parameters for the analysis are difficult to be decided, but the results of analysis will be greatly affected by the choice of the parameters. Under such case, the present proposal will be able to give an estimate of the extent to which residual strength is to be applied, which is not possible with other stability analysis methods.

Chugh (1986) presented a procedure for determining varying FOS along a failure surface within the framework of the LEM. Chugh (1986) predefined a characteristic shape for the variation of the local FOS along a failure surface, and this idea actually follows the idea of the inter-slice shear force function in M–P method (1992). The suitability of this varying FOS distribution function is however questionable, and there is no simple way to define this function for a general problem, as the local FOS should be controlled mainly by the local soil properties, topography and the shape of the failure surface. Sarma and Tan (2006) have recently proposed a new formulation with varying FOS based on the critical acceleration concept. No FOS distribution function is required in this formulation, and the varying FOS can be viewed approximately as an indication of the progressive yielding mechanism of the slope. The formulations by Sarma and Tan (2006) or Chugh (1986) are easy to implement with an estimation of the progressive failure, but cannot accept the post-peak strength in the analysis.

Lam et al. (1987) proposed a LEM for study of progressive failure in slopes under long-term conditions. Their main idea involved the recognition of the local failure and the operation of the post-peak strength. This concept is one of the progressive failure phenomena, which apply when deformation is very large with a major reduction in the strength of the soil, but this approach is not easy to implement and cannot be applied to the general progressive failure phenomenon. There is also no practical application of this method up to present.

3.13.1 Basic formulation for variable FOS formulation

Cheng et al. (2010) have proposed the use of the lower bound method to determine $f(x)$, but this formulation is also limited to a constant FOS. There is great difficulty in incorporating the concept of residual strength in the classical lower/upper bound method because a constant FOS is adopted in the whole domain. To overcome this limitation, the authors propose the concept of a varying local FOS together with a global FOS. A set of internal forces satisfying all equilibrium conditions but not violating the yield criterion is varied until the maximum strength of the system is achieved. The local equilibrium of each slice is controlled by a local FOS, while a global FOS is defined along the entire failure surface. This approach is effectively an approximate lower bound method, and the applicability of this method will be illustrated by an interesting engineering example.

Consider the slope shown in Figure 3.54; the soil mass between the potential slip surface and the ground surface is divided into n slices numbering from 1 to n. In Figure 3.54, F_0 represents the boundary thrust force and its value is usually equal to 0. Points A and B are the entrance and exit points of the sliding surface, respectively, with their x-coordinates denoted as x_A and x_B. The forces acting on a typical slice i are illustrated in Figure 3.55.

In Figure 3.55, the boundary between the i–1 slice intersects with the slip surface and the ground surface at point D_{i-1} and K_{i-1}, respectively.

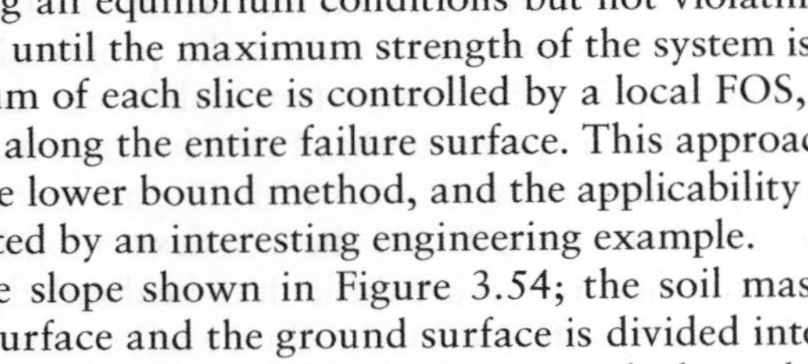

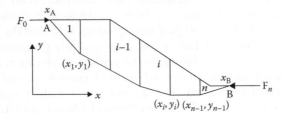

Figure 3.54 Slip surface and slices for variable factor of safety formulation.

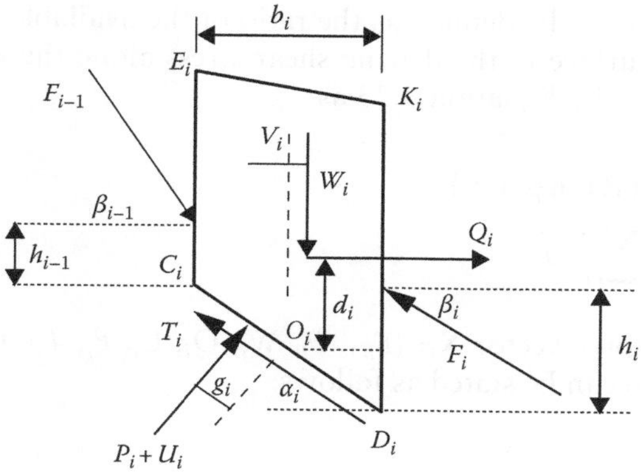

Figure 3.55 Forces acting on a typical slice.

F_{i-1} is the thrust force between the i–1 and i slices and h_{i-1} is the vertical distance from the thrust point of F_{i-1} to point D_{i-1}, while F_i represents the thrust force between the i slice and the $i+1$ slice, and h_i is the vertical distance from the thrust point of F_i to point D_i. β_{i-1} is the angle between F_{i-1} and the horizontal direction. W_i is the weight of slice i. v_i is the horizontal distance from the thrust line of W_i to the centre point of the slice base (used for moment arm) and can be taken as 0 for thin slices. P_i and U_i are the effective normal force and pore water pressure acting on the slice base, respectively. T_i is the mobilized shear force required to maintain the static equilibrium condition. Q_i is the external forces induced by earthquake. d_i is the distance from the thrust line of Q_i to O_i (midpoint of the base and is also used as the moment point). b_i is slice width. α_i is the inclination of the slice base, namely, the angle from horizontal to the slice base in clockwise direction. g_i is the distance from the base normal force P_i to point O_i along the slice base. Usually, P_i is assumed to be at point O_i, that is, $g_i = 0$, under the classical formulation, but can change during the optimization process under the present formulation, which is a more flexible arrangement. The local FOS for slice i is defined as the ratio of the available shear strength along a slice base to the driving shear stress along the slice as:

$$F_s{}^i = \frac{P_i \tan\phi_i + c_i}{T_i} \tag{3.22}$$

where

F_s^i is the local FOS for slice i
ϕ_i is the effective friction angle of the slice base
c_i equals $c_i' l_i$ and l_i is the base length of slice i

The global FOS can be defined as the ratio of the available shear strength along the slip surface to the driving shear stress along the whole slip surface, and is given by Equation 3.23 as

$$F_s = \frac{\sum_{i=1}^{n} (P_i \tan\phi_i + c_i)}{\sum_{i=1}^{n} T_i} \tag{3.23}$$

If we define a force vector $\mathbf{X}_i = (F_{i-1}, F_i, W_i, Q_i, U_i, P_i, T_i)$, then the maximum extremum can be stated as follows:

$$\begin{cases} \max F_s \left(\mathbf{X}_1, \mathbf{X}_2, \ldots, \mathbf{X}_n; h_1, \ldots, h_n; g_1, \ldots, g_n; \beta_1, \ldots, \beta_n\right) \\ \text{s.t.} \quad \sum_{i=1}^{n} \mathbf{X}_i\big|_x = 0 \qquad \sum_{i=1}^{n} \mathbf{X}_i\big|_y = 0 \\ \sum_{i=1}^{n} \mathbf{X}_i \times M_o = 0 \end{cases} \tag{3.24}$$

where

$\mathbf{X}_i$ represents the total forces imposed on slice i
$|_x$ means the projection of force vectors in x-direction
$|_y$ means the projection of force vectors in y-direction
$\mathbf{X}_i \times M_o$ is the moment of vector $\mathbf{X}_i$ about point O

This means the force vectors $\mathbf{X}_i$ ($i = 1, 2, \ldots, n$) and the variables h_i, g_i, β_i ($i = 1, 2, \ldots, n$) must satisfy the static equilibrium condition. There exist infinite groups of force vectors $\mathbf{X}_i$ and variables h_i, g_i, β_i that can satisfy the static equilibrium condition, and they will lead to different FOSs according to Equation 3.23. The FOS for a given slip surface will be the maximum value based on the lower bound principle, which can be determined from an optimization process.

The next step is to determine the control variables for the maximum extremum principle. If h_{i-1}, F_{i-1}, β_{i-1} are known, there will be six remaining variables h_i, β_{i-1}, P_i, T_i, F_i, g_i to be determined based on the static equilibrium. Since there are only three static equilibrium equations available for slice i, in order to make the problem determinate, three variables should be taken as the control variables in the optimization process. In this study, we assume that $g_i = 0$ in the initial trial and take h_i, β_i as the control variables. The boundary conditions give $h_0 = h_n = 0$ and $\beta_0 = \beta_n = 0$, so there are totally $2n-2$ variables ($h_1, \ldots, h_{n-1}; \beta_1, \ldots, \beta_{n-1}$) to be optimized.

Based on the boundary conditions, the recursive procedures will determine the local FOS and the related normal forces, the shear forces on the slice base and the thrust forces for all the slices. The global FOS is then determined from Equation 3.23. P_i is limited to positive value in the optimization analysis, which is also a constraint in the analysis.

Furthermore, during the implementation of the maximum extremum principle, there is the possibility that $F_s^i < 1.0$. Two approaches are adopted in this study. The first approach allows the occurrence of $F_s^i < 1.0$ and the other approach will assign $F_s^i = 1.0$ by transferring the unbalanced thrust forces to its adjacent slice in the sliding direction. The former approach is called the approach of instantaneous loading condition (Ailc) and the latter is called the approach of gradual loading condition (Aglc). The details of the Aglc are as follows (take slice i for example):

Step 1: The force equilibrium equations in the x- and y-directions give

$$\begin{cases} Q_i + F_{i-1}\cos\beta_{i-1} - F_i\cos\beta_i + P_i\sin\alpha_i - T_i\cos\alpha_i = 0 & x\text{-direction} \\ -W_i - F_{i-1}\sin\beta_{i-1} + F_i\sin\beta_i + P_i\cos\alpha_i + T_i\sin\alpha_i = 0 & y\text{-direction} \end{cases} \tag{3.25}$$

Assuming $g_i = 0$ initially, the moment equilibrium about point O leads to the following equation:

$$F_i = \frac{W_i v_i + Q_i d_i - F_{i-1}\sin\beta_{i-1}\dfrac{b_i}{2} + F_{i-1}\cos\beta_{i-1}\left(h_{i-1} + \dfrac{b_i}{2}\times\tan\alpha_i\right)}{\sin\beta_i\dfrac{b_i}{2} + \cos\beta_i\left(h_i - \dfrac{b_i}{2}\times\tan\alpha_i\right)} \tag{3.26}$$

In the following, we denote the whole denominator in Equation 3.26 as M^i for the sake of clearer interpretation. P_i, T_i can be obtained as Equation 3.27 by using the force equilibrium Equation 3.25 in x- and y-directions:

$$\begin{cases} P_i = -Ca_1\sin\alpha_i - Ca_2\cos\alpha_i \\ T_i = -Ca_2\sin\alpha_i + Ca_1\cos\alpha_i \end{cases} \tag{3.27}$$

where

Ca_1 equals $Q_i + F_{i-1}\cos\beta_{i-1} - F_i\cos\beta_i$.
Ca_2 equals $-W_i - F_{i-1}\sin\beta_{i-1} + F_i\sin\beta_i$.

Step 2: Calculate the global FOS F_s^i from Equation 3.26. If $F_s^i < 1.0$, local failure will occur, and T_i is adjusted to $P_i\tan\phi_i + c_i$ (or the residual strength if it is defined) with $F_s^i = 1.0$. If the local FOS of a slice is less than 1.0, it means this slice should have yielded, and either the maximum strength or the residual strength (if defined) should be used in the iteration process, while the local FOS should then be set to 1.0, as the strength is mobilized fully for this slice. The unbalanced thrust force will be distributed to F_i, P_i. P_i and F_i will be adjusted according to the following equation:

$$\begin{cases} P_i = \dfrac{-Cb_1 \sin\beta_i - Cb_2 \cos\beta_i}{Cd_1 \sin\beta_i + Cd_2 \cos\beta_i} \\ F_i = \dfrac{-Cb_2 Cd_1 + Cb_1 Cd_2}{Cd_1 \sin\beta_i + Cd_2 \cos\beta_i} \end{cases} \tag{3.28}$$

where

Cb_1 equals $Q_i + F_{i-1}\cos\beta_{i-1} - c_i \cos\alpha_i$
Cb_2 equals $-W_i - F_{i-1}\sin\beta_{i-1} + c_i \sin\alpha_i$
Cd_1 equals $\sin\alpha_i - \tan\phi_i \cos\alpha_i$
Cd_2 equals $\cos\alpha_i + \tan\phi_i \sin\alpha_i$

Residual strength can hence be considered easily by Step 2 in the present formulation. The moment equilibrium about an arbitrary moment point O will then be checked again. The variable h_i is calculated by the following equation:

$$h_{i-1} = \frac{M^i F_i + Q_i d_i + F_{i-1}\sin\beta_{i-1}\dfrac{b_i}{2}}{F_{i-1}\cos\beta_{i-1}} - \frac{b_i}{2}\tan\alpha_i \tag{3.29}$$

Pan (1980) has pointed out that the acceptable h_i should be in the range between $0.25|D_iK_i|$ and $0.5|D_iK_i|$, which is basically similar to the suggestion by Janbu in his 'rigorous' method (1973). $|D_iK_i|$ represents the vertical distance from point D_i to point K_i, which can be adjusted if necessary. If the constraint $0.25|D_iK_i| \le h_i \le 0.5|D_iK_i|$ (or any other similar range defined by the engineer) is not satisfied, the boundary value will be set to h_i. That is to say, if $h_i < 0.25|D_iK_i|$, then $h_i = 0.25|D_iK_i|$. The unbalanced moment induced by the violation of this constraint drives the force P_i to move along the slice base within a certain range. The maximum length for which P_i can move is set to $\dfrac{b_i}{\cos\alpha_i}\psi$, where $0 < \psi \le 0.5$. If g_i is lower than $\dfrac{b_i}{\cos\alpha_i}\psi$, then the computation for slice i will finish. ψ is set to 0.1 in the present study, otherwise, the centroids of the base normal forces will be close to the edges of the slice and are not acceptable. The effect of this value on the FOS can be considered by using the following equation:

$$g_i = \frac{M^i - F_i\left(\sin\beta_i \frac{b_i}{2} + \cos\beta_i\left(h_i - \frac{b_i}{2}\tan\alpha_i\right)\right)}{P_i} \tag{3.30}$$

The steps mentioned earlier are applicable to all the slices except for the last one. For the last slice, h_n, F_n are equal to 0 from the boundary condition, so F_n can be pre-determined as 0.0 instead of using Equation 3.26. By Equation 3.25, P_i and T_i are determined and the local FOS F_s^n is obtained. The moment equilibrium condition is maintained by varying P_i within the acceptable range between 0 (at the middle of the slice) and $\frac{b_n}{\cos\alpha_n}\psi$ (close to the edge of the slice).

For Ailc, only Step 1 is required as $F_s^i < 1.0$ is allowed. The optimization problem related to the maximum extremum principle (or lower bound method, Cheng et al., 2010) for a given slip surface Z′ is stated as follows:

$$\begin{cases} \max \quad g(h_1,\ldots,h_{n-1};\beta_1,\ldots,\beta_{n-1},\beta_-,\beta_+) \\ 0.25|D_iK_i| \le h_i \le 0.50|D_iK_i| \\ P_i > 0;\beta_- \le \beta_i \le \beta_+ \end{cases} \tag{3.31}$$

where β_-, β_+ are the minimum allowed angle and the maximum allowed angle, respectively. In the present study, the lower and upper limits of β_- and β_+ are −45° to 0° and 0° to 70°, respectively (it should be noted that β less than 0 is allowed by many commercial software). There are totally $2n$ variables to be optimized for the maximum extremum, and the global FOS will be obtained from this optimization procedure based on the mixed optimization algorithm, which will be discussed in the following.

In the Aglc formulation, progressive yielding can be considered approximately in two ways. If the global FOS exceeds 1.0, the system can redistribute the stresses for local yielding by Equations 3.28 and 3.29 while T_i is adjusted to $P_i \tan\phi_i + c_i$. That means, part of the failure surface can yield locally (with a local FOS of 1.0) while the whole soil mass is still maintained in a stable state by the remaining portion of the failure surface where the local FOSs exceed 1.0. Alternatively, if a residual strength is specified, the Aglc formulation can allow the use of the residual strength according to Step 2 mentioned earlier by replacing T_i with the residual stress during the stress redistribution, which will further extend the local yield zone in the analysis.

The present formulation is similar to that by Cheng et al. (2010), while a varying local FOS is defined with the explicit consideration of the local moment equilibrium of every slice, which is not possible with other classical formulation. In the formulation by Cheng et al. (2010), the violation of local moment equilibrium (actually thrust line) is enforced indirectly by rejecting the trial *f*(*x*), which gives a thrust line outside the soil mass (as M–P method does not consider local moment). The local moment equilibrium is however enforced automatically for every slice in the present formulations which is not possible with the other existing methods. On the other hand, this method requires the concept of local and global FOSs, which is different from the previous lower bound method of Cheng et al. (2010). The incorporation of residual strength is simple and direct in the present formulation, and the progressive yielding mechanism can be approximately estimated from the stress redistribution in the present formulation. With minor modification, the present formulation can also reduce to a special form of the classical Janbu's rigorous method (1973).

3.13.2 Analysis of variable FOS approach

Cheng et al. (2011b) have found that the present maximum extremum method of analysis gives results similar to the classical analysis methods for many prescribed failure surfaces (Cheng et al., 2011b). The FOS as well as the critical slip surface from this variable FOS approach are comparable to those obtained with the classical methods of analyses (Figure 3.56). There are however several aspects that are worth consideration. Sarma and Tan (2006) have implicitly assumed that the FOS along the interfaces between slices/blocks is unity at all interfaces. The limit analysis by Chen (1975, 2004) also implicitly assumes this FOS to be unity. Chen (2004) have found that this FOS is not unity by using the rigid element method. The authors view that there is not a strong theoretical background behind this assumption, and this assumption will be checked against the present formulation as well as Spencer's method.

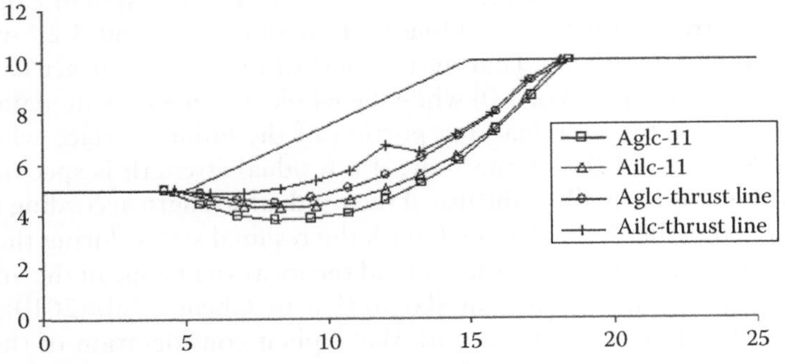

Figure 3.56 Critical slip surfaces and corresponding thrust lines by different methods for a simple slope.

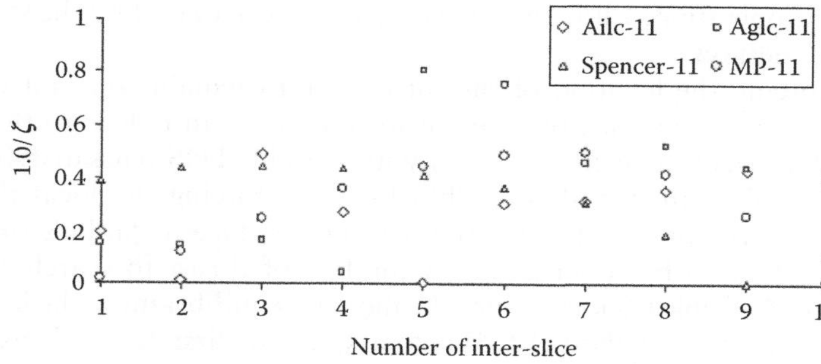

Figure 3.57 Distribution of the local FOS for example 1 (11 slices).

The local FOS along the interface between two adjoining slices is defined as $\zeta_i = F_i \cos\beta_i \tan\phi_{vi} + C_{vi}/F_i \sin\beta_i$, where ϕ_{vi} is the average friction angle along the ith inter-slice and C_{vi} is the average cohesion along the ith inter-slice. The distribution of $1/\zeta$ along the failure surface for critical failure surfaces by Ailc, Aglc and Spencer's method (classical and extremum principle) for examples 1 is shown in Figure 3.57. It is found that the FOSs are much greater than unity, which is greatly different from the assumption of Sarma and Tan (2006) or in the limit analysis of Chen (1975). Based on these results and the $f(x)$ as shown in Figure 3.53 ($f(x)$ are very small for the front and the end of the slope), it is clear that yielding along the vertical interfaces may not be a reasonable assumption for slope stability analysis.

3.13.3 Discussion on variable FOS approach

The variable FOS approach is based on a dual optimization formulation where the global FOS is obtained according to the maximum extremum principle (or equivalently the lower bound method) stated by Pan (1980). The numerical formulation of this approach is equivalent to the variational principle by Baker (1980) or Wu and Tsai (2005). The present formulation is however suitable to be used for general conditions, and the principle is still based on the classical limit equilibrium formulation, which is familiar to the engineers (but coupled with an optimization algorithm for the actual solution). It is demonstrated that the FOSs from the present formulation are similar to those by the classical methods. The authors have coded the present formulation into the programme SLOPE 2000, and a typical problem can be solved within 5 min, which is acceptable in terms of routine analysis and design. On the other hand, an equivalent problem considering residual strength with the strength reduction method takes at least 1 computer hour for computation (tested with Intel i920 with eight processes), and the effort to define the computer model is much longer than that. For routine analysis and design, the use of the present formulation has the advantage of fast and simple operation, while the answer will be close to that by the strength reduction method.

By enforcing the location of the thrust line to remain constant during the optimization analysis, the present formulation can reduce to the classical Janbu's rigorous method (1973) with variable FOS but satisfying the moment equilibrium for all slices/blocks. By enforcing the local FOS to be the same, the present formulation can also reduce to Janbu's rigorous method (1973) with the variation of the line of thrust to search for the extremum. A simple trick to ensure the moment equilibrium in the last slice of Janbu's rigorous method (1973) is to assume the first $N - 2$ thrust lines to be variables while the last thrust line ($N - 1$) is determined from a simple trial and error until the moment equilibrium of the last slice is satisfied. This is a modified Janbu's rigorous method (1973), which will satisfy the moment equilibrium for all the slices. While the equations as given earlier are developed for slices with vertical interfaces, they can be modified easily to suit problems with blocks (rock slope problem controlled by joints). The whole concept can be extended even further to general three-dimensional problems by the concept of lower bound principle-discretized variational principle as given here.

Based on some internal present studies, some conclusions are given as follows:

1. The Ailc approach usually gives a more conservative solution than the Aglc approach. The Ailc approach allows local failure without adjusting which may be slightly less satisfactory than the Aglc approach, which allows the redistribution of the internal forces.
2. The thrust lines given by the Aglc approach are generally smoother than those given by the Ailc approach.
3. The differences between the Aglc and Ailc approaches are small, except when FOS is close to 1.0. In this case, Aglc may not be applicable as this method allows local FOS to be smaller than 1.
4. The present formulation can consider progressive yielding in an approximate but useful way, which is particularly important for problems, which are controlled by a thin layer of weak zone with low residual strength.
5. The Aglc formulation can approximately consider progressive yielding in two ways. By nature, if the global FOS exceeds 1.0, the system can redistribute the stresses for local yielding. If a residual strength is specified, the Aglc formulation can allow the use of the residual strength during stress redistribution, which will extend further the local yield zone in the analysis.

Within the context of classical slope stability analysis where the FOS is defined in terms of the ultimate shear strength and mobilized shear strength, there should only be a single FOS for a problem. On the other hand, for a normal stable slope with overall FOS greater than 1.0, at least part of the system is not situated at the ultimate condition. The present study has demonstrated that for such cases, the FOSs based on a variable FOS and the classical approaches are similar. In this respect, the present formulation provides an alternative to the classical analysis methods. For normal and practical problems, the present formulation provides no special advantage over the classical analysis methods in terms of FOS, but a FOS together a consistent internal force system satisfying all kinematics is guaranteed from the present formulation. On the other hand, for those cases of stable slopes where the FOSs is slightly above 1.0 and part of the system may be controlled by the residual strength, the present formulation provides an estimation of the FOS, which is not possible with the classical analysis methods. For the two formulations shown earlier, Aglc is equal to Ailc if there is no local yielding, but it will be more appropriate than Ailc when local yielding occurs. In this respect, Aglc is a more general and appropriate formulation to be used in general.

3.14 USE OF INTERNAL/EXTERNAL VARIABLES IN SLOPE STABILITY ANALYSIS AND RELATION OF SLOPE STABILITY PROBLEM TO OTHER GEOTECHNICAL PROBLEMS

3.14.1 Basic methods in formulation

Broadly speaking, there are two major groups of 'rigorous' methods in the LEM analysis: (1) internal variables in the form of the direction or location of the inter-slice forces and (2) external variables (boundary stress) in the form of base normal forces (stress) acting on a potential slip surface.

For the first group of methods, M–P method (1965) and Janbu's rigorous method (1957, 1973) are the most common formulations. Since only the global moment equilibrium is used in the M–P formulation (1965), the back-calculated thrust line may lie outside of the soil mass, which is physically not possible. This situation is equivalent to the violation of the local moment equilibrium, and the classical M–P cannot enforce the local moment equilibrium automatically. In Janbu's rigorous method (1957, 1973), the distance between the thrust line and the base of the slip surface is assumed to be known, while the local moment equilibrium is used in the formulation. It should be noted that there are some differences between the international adaption of Janbu's method and the way Janbu himself intended (1957, 1973) it to be adapted. In the original computer implementation by Janbu and others in the Nordic countries, the moment equilibrium is taken on the interfaces of the slices but not on the actual slice itself.

Janbu (1973) has discussed 'the moment equilibrium for a slice of infinite small width', but this does not mean that the slices themselves would be very narrow. The method was developed during the time of hand calculations (Janbu et al., 1956), so relatively few slices had to be used. The trick was to take the moment equilibrium for an infinite extra-small slice at the intersection of normal slices. This averaging method resulted in quite good convergence for normal problems, but convergence problems began to appear as the slices got thinner. As the problem is actually over-specified by one unknown, the moment equilibrium of the last slice is not enforced in Janbu's rigorous method (1973). Thus, the true moment equilibrium is still not maintained in this method. Besides these two methods, there are many other variants of slope stability methods, which are usually based on these two important slope stability formulations. As long as a statically admissible stress field is defined over a domain, the solution will be a lower bound of the ultimate limit state (or an equivalent failure state for which no more external or internal loads can be added). In this respect, the LEM is an approximate, but not exact, lower bound solution (Chen, 1975), as force (lumping the stress over a finite length), instead of stress, is considered in the classical LEM.

In the second group of methods, the variational principle of Baker and Garber (1978) is the representative method. Baker and Garber (BG) method minimizes the safety functional with respect to both the potential slip surface $y(x)$ and the potential normal stress $\sigma(x)$ acting on the surface, using equilibrium requirements as the constraints. It should be noted that in the BG formulation (1978), the failure mass bounded by the potential slip surface and the ground surface is not divided into slices; complete equilibrium can be achieved using this group of methods, which is not possible with the first group of methods. It is important to realize that the variational technique of Baker and Garber (1978) is just one of the many different minimization procedures available. The variational technique is an analytical procedure that is convenient for the solution of simple slope stability problems, but it is difficult to adopt in cases when the layered geometry or the ground/loading conditions are complicated. Cheng et al. (2011b, 2013b) have demonstrated the equivalence between the variational principle and a global optimization analysis; the simpler global optimization analysis can be applied to general complicated cases without any problems.

Under the lower bound theorem in a limit analysis (Chen, 1975), the loads determined from the stress distribution alone, which satisfies (1) the equilibrium equations, (2) the stress boundary conditions and (3) nowhere violates the yield criterion, are not greater than the actual collapse load. Under the upper bound theorem, the loads determined by equating the external rate of work to the internal rate of dissipation, associated with a prescribed deformation mode (or velocity field) that satisfies (1) the velocity boundary conditions and (2) the strain and velocity compatibility conditions, are not less than the actual collapse load. The lower bound theorem, which does not involve energy dissipation, is also applicable to the limit equilibrium formulation. The major difference between the limit equilibrium and the limit analysis is the upper bound approach. In the limit analysis, the energy balance is considered in determining the critical solution; in the limit equilibrium formulation, the minimum resistance (force/moment) against failure is considered. Cheng et al. (2010) have demonstrated the equivalence of the ultimate limit and the maximum extremum of the system (whereby the maximum strength of a prescribed failure surface is utilized) by a simple footing on clay based on the slip line solution. The slip line solution corresponds to the instant of impending plastic flow, where both the equilibrium and the yield conditions are satisfied. Combining the Mohr–Coulomb criterion with the equations of equilibrium will provide a set of differential equations for plastic equilibrium, which can be solved with appropriate boundary conditions (Sokolovskii, 1965). Together with the stress boundary conditions, this set of differential equations can be used to investigate the stress under the ultimate condition. In the formulation by Cheng et al. (2010), which treats $f(x)$ as a variable to be determined, the overall moment equilibrium is used, while the local moment equilibrium of an individual slice is not directly enforced. The local moment equilibrium (or acceptability of the thrust line location by Cheng et al., 2010) is indirectly enforced by rejecting those $f(x)$ that are associated with the thrust line outside of the soil mass. Cheng et al. (2010) have pointed out that as long as an $f(x)$ is prescribed, the solution will be a lower bound to the ultimate limit state, which is the lower bound theorem. Under the ultimate limit state, where the strength of a system is mobilized fully, $f(x)$ is actually determined by this requirement, a boundary condition that has not been used in the past. Cheng et al. (2010) have applied a modern heuristic optimization algorithm to determine $f(x)$ for arbitrary problems, and have pointed out that every kinematically acceptable failure surface should have a FOS. Failure to converge in the classical stability analysis is caused by the use of an inappropriate $f(x)$ in the analysis.

Cheng's approach (2010) can be classified as a hybrid formulation of the first and second groups of methods. The adoption of the maximum extremum of the system is conceptually similar to the second group of methods, but there are two major differences between the formulations by Cheng et al. (2010) and Baker and Garber (1978). Baker and Garber (1978) minimize the FOS simultaneously with respect to the base normal forces as well as the locations of the failure surfaces, while Cheng et al. (2010) determine the maximum extremum of the system for any prescribed failure surface. The acceptability of the internal forces is not enforced in the BG approach (1978); however, a reasonable distribution of the internal stress is obtained by this approach (e.g., Baker, 1980, 2005). Cheng et al. (2010)

have enforced the acceptability of the internal forces during the extremum computation, which is different from the BG approach (1978).

In this section, Cheng et al. (2013a) will firstly use the well-known slip line solutions for a bearing capacity problem to determine $f(x)$ and the thrust line for a 'horizontal slope'. Based on the ultimate load, the failure surfaces and $f(x)$ or the thrust line from the slip line solutions, the FOSs will then be back-computed from M–P method (1965) and Janbu's rigorous method (1957, 1973), and the equivalence between the two stability methods under the ultimate condition will be illustrated. It is well known that difficulty is encountered in the convergence of thin slices with Janbu's rigorous method (1973), and computation can be very sensitive to the location of the thrust line; a method to improve convergence has been proposed by Cheng et al. (2013a).

Cheng et al. (2013a) will demonstrate that at the maximum extremum condition, there is no difference between the use of external or internal variables in specifying a problem. The authors believe that the use of internal variables is preferable over the use of boundary variables as the imposition of the acceptability of the internal forces can be easily enforced. Furthermore, the authors will demonstrate clearly that the classical LEMs with a prescribed internal/external force assumption will be a lower bound to the ultimate condition. The maximum extremum of the system from the LEM is also shown to be equivalent to the slip line solution in the present study.

3.14.2 Inter-slice force function *f*(*x*) and thrust line for horizontal slope problem

In this section, a 'horizontal slope' is considered under the action of an applied load. This case is used as the plasticity solutions (slip line solutions) are available for this ultimate 'horizontal slope'/bearing capacity problem. The slip line method is based on the theory of plasticity, and it considers the yield and equilibrium of a soil mass controlled by the Mohr–Coulomb criterion, which is a typical lower bound method. Combining the Mohr–Coulomb criterion with the equations of equilibrium gives a set of differential equations for the plastic equilibrium. Together with the stress boundary conditions, a set of differential equations given by Equations 3.32 and 3.33 can be used to investigate the stresses under the ultimate condition. Sokolovskii (1965), Booker and Zheng (2000), Cheng (2003), Cheng and Au (2005), Cheng et al. (2007c) and many others have provided solutions to slip line equations. In the present study, the slip line programme SLIP, developed by Cheng and Au (2005) (which has been compared and verified with the programme ABC by Martin 2004, as well as many published results), is used for slip line analysis of a bearing capacity problem:

α characteristics:

$$-\frac{\partial p}{\partial S_\alpha}\sin 2\mu + 2R\frac{\partial \theta}{\partial S_\alpha} + \gamma\left(\sin(\varepsilon + 2\mu)\frac{\partial y}{\partial S_\alpha} + \cos(\varepsilon + 2\mu)\frac{\partial x}{\partial S_\alpha}\right) = 0 \tag{3.32}$$

β characteristics:

$$\frac{\partial p}{\partial S_\beta}\sin 2\mu + 2R\frac{\partial \theta}{\partial S_\beta} + \gamma\left(\sin(\varepsilon + 2\mu)\frac{\partial y}{\partial S_\beta} + \cos(\varepsilon + 2\mu)\frac{\partial x}{\partial S_\beta}\right) = 0 \tag{3.33}$$

where

$p = \frac{\sigma_1 + \sigma_3}{2}$, $R = \frac{\sigma_1 - \sigma_3}{2} = p\sin\phi + c\cos\phi$

σ_1 and σ_3 are the major and minor principle stresses, respectively
S_α and S_β are the characteristic lines, as shown in Figure 3.58
c and ϕ are the cohesive strength and the friction angle of the soil, $\mu = (\pi/4 - \phi/2)$
γ is the unit weight of the soil
θ is the direction of the principal stress to the y-axis
ε is the angle between the body force and y-axis

For simplicity, effective soil parameters c' and ϕ' are represented by c and ϕ, respectively in the following section.

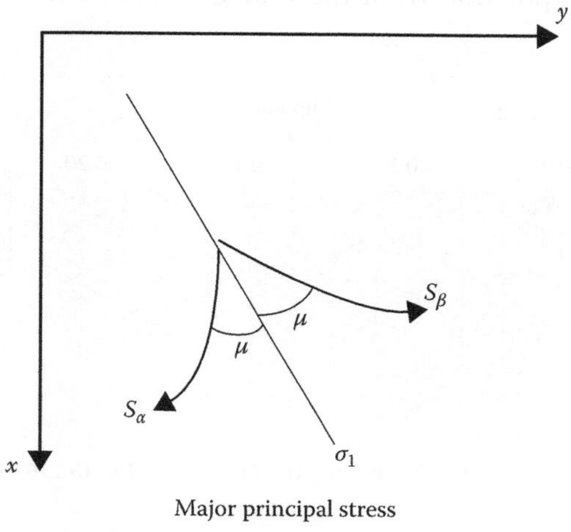

Figure 3.58 A typical slip line field.

Classically, bearing capacity factors N_c, N_q and N_γ are determined by a simple superposition principle. Michalowski (1997) and Cheng (2002) have demonstrated that this simple superposition is a good approximation and is only slightly conservative even for high-friction-angle conditions. A bearing capacity problem can be considered as a 'horizontal slope' where failure is induced by the bearing pressure from the foundation. This reasoning is physically obvious, but is practically not adopted for engineering use (due to poor results) because of its inability to specify a correct $f(x)$, as pointed out by Cheng et al. (2010). In this section, these three factors, which correspond to the ultimate condition, together with the corresponding $f(x)$, the thrust line and the base normal forces will be determined from slip line solutions. Once the stress field (p, R, θ) for the slip line field, shown in Figure 3.59, has been determined by Equations 3.32 and 3.33 using SLIP (Cheng and Au, 2005), the inter-slice force function and the thrust line location can then be determined as follows:

Step 1: Calculate the normal stress (σ_x, σ_y) and shear stress (τ_{xy}) at any grid point by SLIP using the following relations:

$$\sigma_1 = p + R \quad \sigma_3 = p - R \tag{3.34}$$

Hence, by Mohr–Coulomb relation

$$\sigma_x = p + R\cos 2\theta \quad \sigma_y = p - R\cos 2\theta \quad \tau_{xy} = R\sin 2\theta \tag{3.35}$$

Step 2: For any specified section (with a given x-ordinate), as shown in Figure 3.60, determine the normal stress and shear stress at equal vertical intervals by interpolation from the four grid points in the slip line field

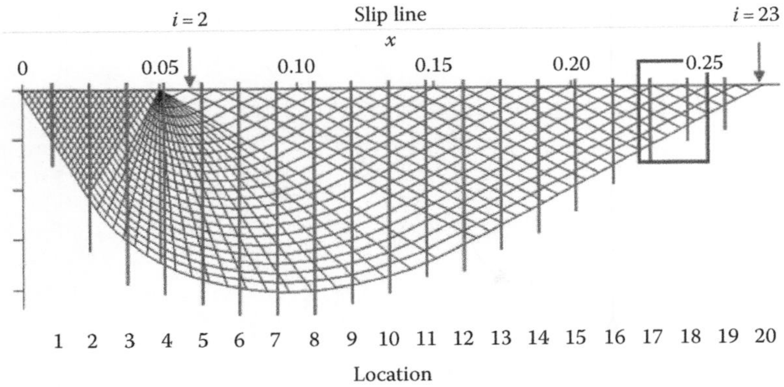

Figure 3.59 Determination of $f(x)$ and thrust line from slip line analysis.

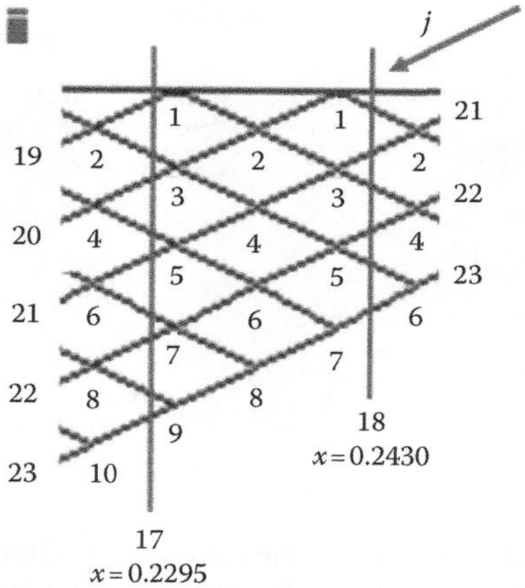

Figure 3.60 Interpolation to obtain stresses at any point from the grid points in the slip line field.

enclosing any given x and y coordinates using a bilinear equation similar to the four-node quadrilateral element used in the finite-element analysis.

Step 3: Calculate inter-slice normal force E and shear force X in each specified section by integrating the normal stress and shear stress at a vertical interval Δy in a vertical direction (as shown in Figure 3.59) by

$$E = \sum \sigma_x \Delta y \quad X = \sum \tau_{xy} \Delta y \tag{3.36}$$

Step 4: Determine the maximum ratio of X/E across all sections from Step 3, denoted as mobilization factor λ.

Step 5: Obtain $f(x)$ across the slip surface by

$$f(x) = \frac{X/E\left(\text{at each location } x\right)}{\lambda} \tag{3.37}$$

Step 6: Determine average normal stress σ_{ar} at each element along the vertical direction by stress σ_{x1} at the top and stress σ_{x2} at bottom of the element. Determine the lever arm h_t of the normal stresses above the base of the slip surface from

$$\sigma_{ar} = \frac{\sigma_{x1} + \sigma_{x2}}{2} \quad h_t = \frac{\sum\left[\sigma_{ar}\Delta y\left(h_i - y_i\right)\right]}{E} \tag{3.38}$$

where h_i and y_i are the average y-ordinate of the element and the y-ordinate of the slip surface at section i, respectively. In the present analysis, a very

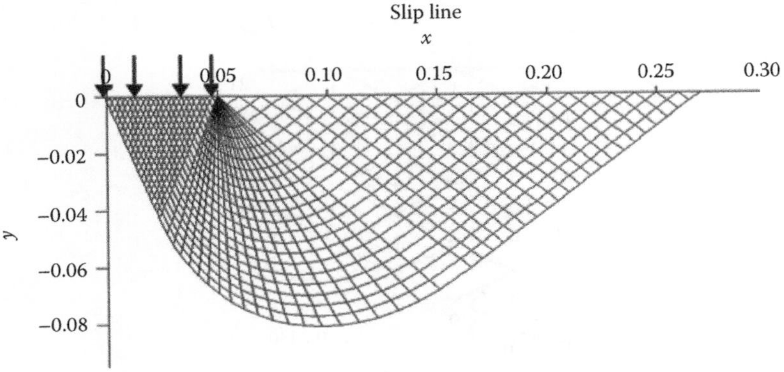

Figure 3.61 Slip line for the case of N_c when $\phi = 30°$, $N_c = 30.18$ using the programme SLIP and 30.14 from classical bearing capacity equation (using natural horizontal distance x instead of normalized distance x).

fine grid is used (1 mm for N_c and N_q and 0.1 mm for N_γ). With such a fine grid, simple interpolation within a sub-domain, as shown in Figure 3.60, and a simple trapezoidal rule, as used in Equation 3.38, is good enough for the analysis.

For assessment of the three bearing capacity factors, a direct superposition approach is assumed, which is also the basis for the determination of these three factors. For example, in determining N_γ, the surcharge and cohesive strength are assumed to be 0 in the slip line or the limit equilibrium analysis. From the results of SLIP, $f(x)$ is determined for different ϕ for cases associated with N_c, N_q and N_γ. A typical slip line field for the case of N_c is shown in Figure 3.61, where the pressure on the ground surface on the left-hand side is determined from a slip line analysis. The slip line field from SLIP is in accordance with the classical solution where active and passive wedges exist at the left- and right-hand sides of the problem, and the two wedges are connected by a log-spiral zone in between, as shown in Figure 3.61. On the other hand, for the slip line field for N_γ, as shown in Figure 3.62 with $\phi = 30°$, the active zone is actually curved, while the intermediate radial shear zone is not a true log-spiral zone with an inscribed angle less than 90°. When ϕ is further reduced to 10°, the radial shear zone becomes very small and the active zone will dominate the problem. As given in Figures 3.62 and 3.63, the bearing capacity factors from SLIP are very close to the slip line solutions of Sokolovskii (1965).

$f(x)$, for the cases associated with the determination of N_c, N_q and N_γ, is given in Figures 3.64 through 3.66. In Figures 3.64 and 3.65, $f(x)$ is symmetrical about $x = 0.5$ when $\phi = 0$, which is consistent with the classical plasticity solution. $f(x)$ is 0 at the left- and right-hand sides of the wedge, as shown in Figures 3.64 and 3.65. These results are consistent with the slip line results

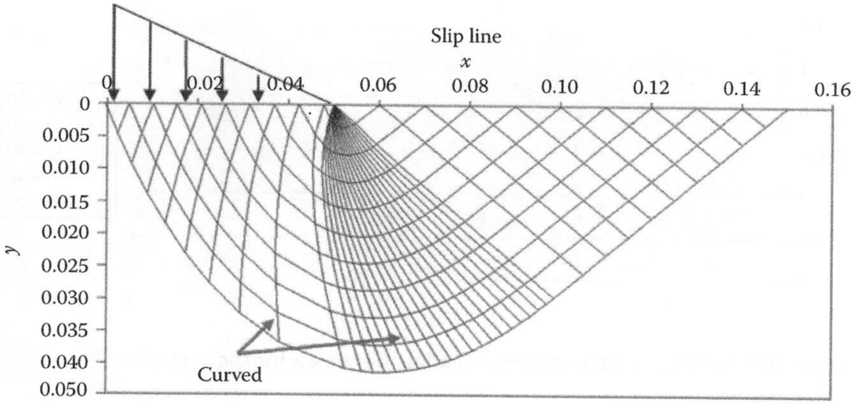

Figure 3.62 Slip line for the case of N_γ when $\phi = 30°$, $N_\gamma = 15.32$ using the programme SLIP and 15.3 by Sokolovskii (1965) (using natural horizontal distance x instead of normalized distance x).

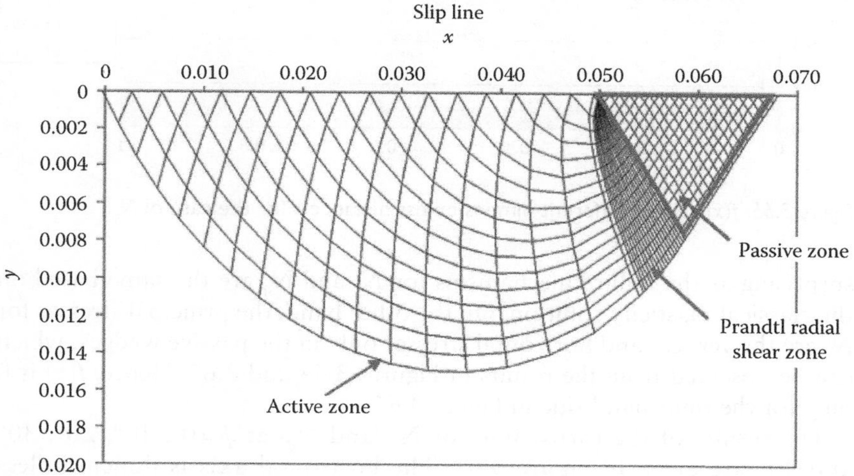

Figure 3.63 Slip line for the case of N_γ when $\phi = 10°$, $N_\gamma = 0.54$ from the programme SLIP and 0.56 by Sokolovskii (1965) (using natural horizontal distance x instead of normalized distance x).

in which the principal stresses are the vertical and horizontal stresses in the active and passive wedges. When $\phi > 0$, $f(x)$ moves towards the left-hand side of the figure with increasing ϕ. These results are also obvious because the failure zone will become longer with increasing ϕ. It is also interesting to note that $f(x)$ is the same for factors N_c and N_q. These results are not

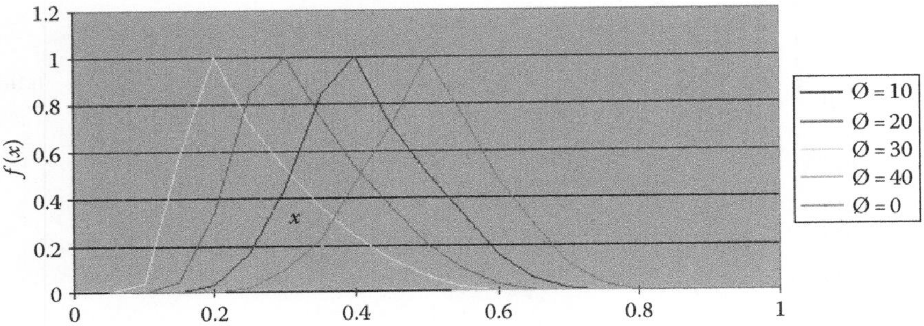

Figure 3.64 f(x) against different dimensionless distance x for the case of N_c.

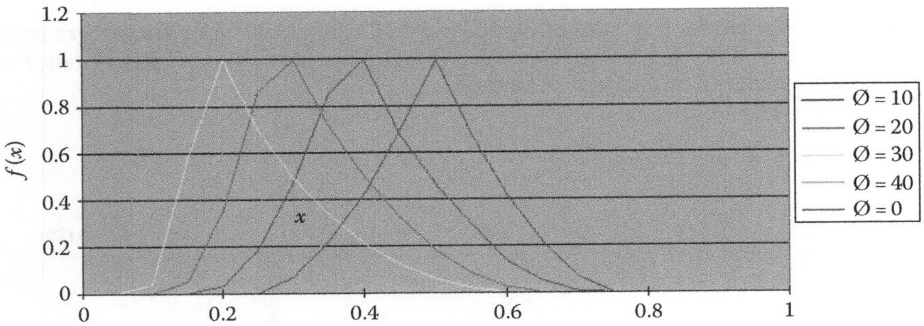

Figure 3.65 f(x) against different dimensionless distance x for the case of N_q.

surprising as the failure mechanisms for N_c and N_q are the same based on the classical plasticity solution. On the other hand, the principal stresses for N_γ are the vertical and horizontal stresses only in the passive wedges, which can be observed from the results in Figures 3.64 and 3.65. Hence, $f(x)$ is 0 only for the right-hand side in Figure 3.66.

The results of the thrust line for N_c (and N_q) at $\phi=0°$, 10°, 20°, 30° and 40° are given in Figure 3.67. The horizontal axis is dimensionless distance x in the range of 0–1. In the passive wedge region on the right-hand side, the thrust line ratio (LOT) is always 0.5. In the very beginning of the active wedge, the thrust line ratios are very close to 0.5, but deviate slightly from 0.5 due to the minor error arising from the iteration analysis in the slip line analysis. Outside the foundation, the LOT will go below 0.5 and then gradually rebound to 0.5. The fluctuation in the LOT outside the foundation is mainly caused by the radial shear zone. It should be noted that while the ultimate bearing capacity factors from SLIP are relatively insensitive to the grid sizes used in the analysis, the thrust line

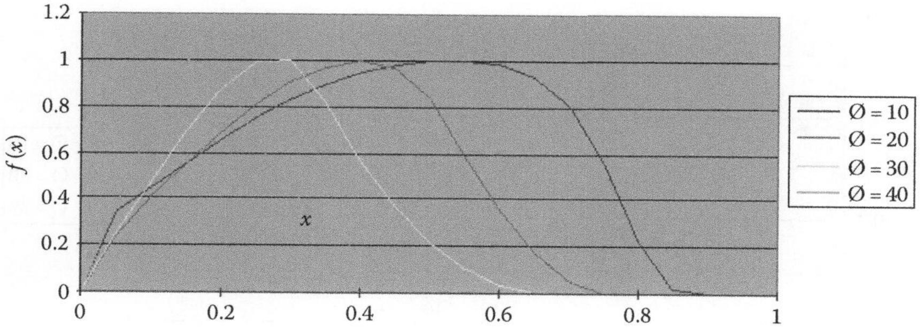

Figure 3.66 $f(x)$ against different dimensionless distance x for the case of N_γ.

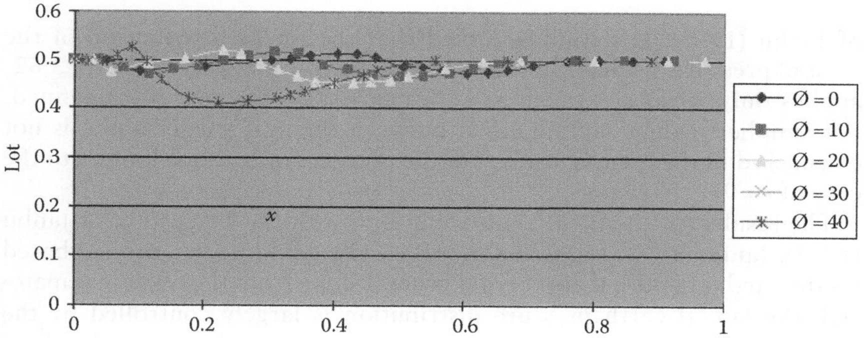

Figure 3.67 Thrust line for different ϕ angles for the case of N_c (x coordinate is x ratio) (same result for N_q).

is more sensitive to the size of the grid. This situation is particularly important for the case of N_γ, and thus, a very fine grid is adopted for the case of N_γ or else there would be a larger fluctuation in the location of the thrust line.

The results of the thrust line ratio for N_q are the same as those for N_c and are also given in Figure 3.67. These results are in line with those for which $f(x)$ is the same for the cases of N_q and N_c.

For the case of N_γ, the results are different from those of the previous two cases. Just beneath the foundation, the stresses are controlled mainly by the ground pressure so that the thrust line is slightly less than 0.5, which is shown in Figure 3.68. At the passive zone, where there is no imposed pressure, the vertical pressure is totally controlled by the weight of the soil. Therefore, the thrust line ratio is 1/3, which implies a triangular pressure distribution; this is consistent with the recommendation

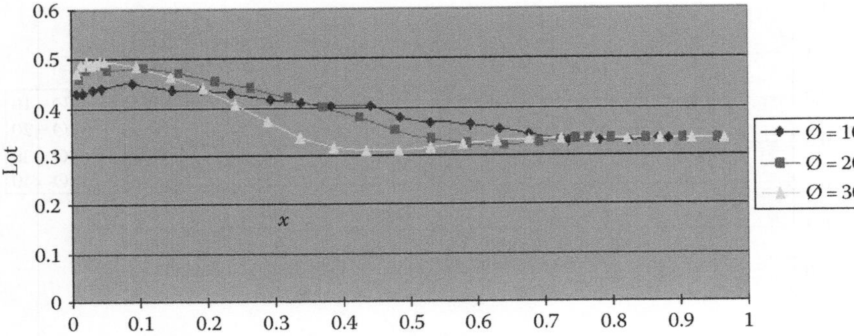

Figure 3.68 Thrust line for different ϕ angles for the case of N_γ (x coordinate is x ratio).

of Janbu (1973). It should be noted that the linear distribution of the ground pressure, as determined from the slip line analysis in Figure 3.62, applies only when c is taken as 0, which is also the way N_γ is defined. For simplicity, the coupling effect between the unit weight and c is not considered in the present study, but the present study is not limited to the case of $c = 0$.

The results for the thrust line are in line with the suggestion of Janbu (1973). Janbu (1973) suggested that the LOT could be determined based on the earth pressure theory. For a general slope from the frictional material, the lateral earth pressure distribution is largely controlled by the unit weight of the soil and will be close to a triangular shape; hence, a generally referred value of 1/3 for the LOT is suggested. In the present study, for both N_c and N_q, where the unit weight of the soil is 0, the horizontal and vertical pressure under half of the footing will be constant, and thus, the LOT should be exactly 0.5. The later part of the slip surface represents a passive earth pressure state in which the earth pressure distribution is similarly constant, and again, LOT = 0.5. For N_γ, there is a triangular-shaped earth pressure distribution in the passive zone, and hence, LOT = 1/3.

When $f(x)$ or the thrust line is defined, the problem can be back-analysed in the following ways. For the case of N_c, a uniform pressure corresponding to cN_c is applied on ground surface without any surcharge outside the foundation. The unit weight of the soil is set to 0 in the slope stability analysis. For the case of N_q, a surcharge of 1 unit is applied outside the foundation, and the uniform foundation pressure is given by unit N_q, while the unit weight of the soil and the cohesive strength are set to 0 in the analysis. For the case of N_γ, a triangular pressure (see Figure 3.62) with a maximum equal to $\gamma_{BN\gamma}$ is applied on the ground surface (average pressure is 0.5 $\gamma_{BN\gamma}$), while the cohesive strength and the surcharge outside the foundation are set to 0. Based on the $f(x)$ shown in Figures 3.64 through 3.66, and the failure surfaces given by the slip line solutions, Cheng et al. (2013a) have back-computed the FOSs to be nearly 1.0 (only 0.001–0.002 less than 1.0) for all the cases. These results are obvious as the solutions from the slip line equations are the ultimate solutions of the system. On the other hand, when the thrust line ratios shown in Figures 3.67 and 3.68 are used in Janbu's rigorous method (1973), there are major difficulties in convergence with the international adaption of Janbu's rigorous method (1973) (equations by Janbu are approximations only). The majority of the analysis using Janbu's rigorous method (1973) cannot converge using the exact thrust line location from the slip line solution. After a series of investigations, Cheng et al. (2013a) found that the solutions can be very sensitive to the thrust line location, and they have finally proposed another procedure that can truly satisfy the moment equilibrium (instead of the approximations by Janbu, 1973).

For the slice shown in Figure 3.69, P and T are the base normal and the shear forces, respectively; l is the base length of the slice; E_L and X_L are the inter-slice normal and shear forces at the left, respectively, while E_r and X_r are the inter-slice normal and shear forces at the right, respectively; h_t is the height of the thrust line above the base of the slice to the right; W is the weight of the soil mass; and α is the base angle of the slice base. Based on the Coulomb relation applied to force, which is the common approach for slope stability analyses, we obtain

$$T = \frac{1}{F}\left(cl + \left(P - ul\right)\tan\phi\right) \tag{3.39}$$

For the vertical force equilibrium,

$$P\cos\alpha - T\sin\alpha = W + \left(X_r - X_L\right) \tag{3.40}$$

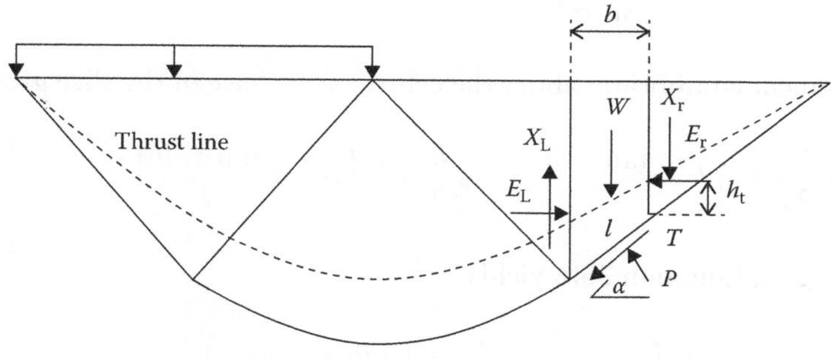

Figure 3.69 Slip surface and slices for Janbu's rigorous method.

Rearranging and substituting for T gives

$$P = \frac{1}{m_\alpha}\left[W + (X_r - X_L) + \frac{1}{F}(cl\sin\alpha - ul\tan\phi\sin\alpha)\right] \tag{3.41}$$

where

$$m_\alpha = \cos\alpha - \sin\alpha\frac{\tan\phi}{F}$$

For the horizontal force equilibrium,

$$-P\sin\alpha - T\cos\alpha = E_r - E_L \tag{3.42}$$

Rearranging and substituting for T gives

$$P = \frac{1}{i_\alpha}\left[-(E_r - E_L) - \frac{1}{F}(cl\cos\alpha - ul\tan\phi\cos\alpha)\right] \tag{3.43}$$

where

$$i_\alpha = \sin\alpha + \frac{1}{F}\tan\phi\cos\alpha$$

For the force equilibrium, resolving the forces parallel to the base of the slice along the base shear force direction yields

$$-T - (E_r - E_L)\cos\alpha = \left(W + (X_r - X_L)\right)\sin\alpha \tag{3.44}$$

Rearranging Equation 3.44 gives

$$X_r - X_L = -W - \frac{1}{\sin\alpha}\left(T + (E_r - E_L)\cos\alpha\right) \tag{3.45}$$

The moment equilibrium about the centre of the base of the slice gives

$$X_L\frac{b}{2} + E_L\left(h_j - \frac{\tan\alpha\cdot b}{2}\right) + X_R\frac{b}{2} = E_r\left(h_{j+1} + \frac{\tan\alpha\cdot b}{2}\right) \tag{3.46}$$

Rearranging Equation 3.46 yields

$$X_L + X_r = -E_L\left(2\frac{h_j}{b} - \tan\alpha\right) + E_r\left(2\frac{h_{j+1}}{b} + \tan\alpha\right) \tag{3.47}$$

For the overall force equilibrium in the horizontal and vertical directions, in the absence of surface loading, the internal forces will balance out and produce

$$\Sigma(E_r - E_L) = 0 \quad \Sigma(X_r - X_L) = 0. \tag{3.48}$$

From Equations 3.44 and 3.48

$$\Sigma(X_r - X_L) = \Sigma\left(-W - \frac{1}{\sin\alpha}\left(T + (E_r - E_L)\cos\alpha\right)\right) = 0 \tag{3.49}$$

Hence, the FOS is given by

$$F = \frac{\Sigma\left(cl + (P - ul)\right)\tan\phi}{\Sigma\left(-W\sin\alpha - (E_r - E_L)\cos\alpha\right)} \tag{3.50}$$

The iteration solution starts from a good estimate of the initial FOS and the first inter-slice normal force E_L. From Equations 3.43 and 3.45, P and X_L for the first slice are then computed. Once X_L is known, Equations 3.44 and 3.45 can then be used to compute X_r and E_r. The process is continued until all internal forces and the FOS have been computed. This solution procedure is advantageous in that no finite-difference scheme is required. Here are some notes about this new modified method:

1. This solution procedure is still sensitive to the location of the thrust line, but is better than Janbu's rigorous method (1973) for thin slices.
2. A good initial choice for the FOS has to be defined. In general, the FOS by Janbu's simplified method can be used as the initial solution.
3. A good initial guess for the first inter-slice normal force should be supplied in the beginning, and this value can be estimated from M–P solution.

Based on these procedures, all the problems can now converge nicely with FOSs close to 1.0 for all cases (only 0.001–0.002 less than 1.0). It should be noted that for normal slopes, the convergence of Janbu's rigorous method (1973) is actually not too bad, although it is not very good either. On the other hand, if a more rigorous consideration of the moment equilibrium as suggested is adopted, then the FOS is 1.0, while the $f(x)$ and internal forces obtained are virtually the same as those from the slip line analysis. That means a correct thrust line will correspond to a correct $f(x)$, and the choice of internal or external variables is not important under the ultimate condition.

It is interesting to note that all FOSs for the three bearing capacity factors are very close to 1.0 (0.001–0.002 different from 1.0) using either $f(x)$ or the thrust line from the ultimate condition. That means as long as the ultimate condition is given consideration, there is no difference between the uses of $f(x)$ or the thrust line in defining a problem. In this respect, M–P method (1965) and Janbu's method (1973) are judged to be equivalent methods when specifying a problem under the ultimate condition. Other than the ultimate condition, the choice of $f(x)$ or the thrust line will give different FOSs (well known in limit equilibrium analyses) as the solutions are only typical lower bound solutions. Since iteration analyses are sensitive to the thrust line location, the use of $f(x)$ for normal routine engineering analyses and designs is advantageous in that it is easier to achieve convergence for normal cases.

3.14.3 Boundary forces in LEM

Baker and Garber (1978) have proposed to use the base normal forces as the variables in the variational principle formulation of slope stability problems. For N_q where $\phi = 30°$, the base normal stresses under an external surcharge of 1 kPa outside the foundation are determined by the slip line method, which are shown in Figure 3.70.

Based on the stresses determined from the slip line analysis, the base normal stresses and hence, the forces for the slices can be determined correspondingly. Once P is known, based on $\Sigma(E_r - E_L) = 0$ and using Equation 3.42, the FOS can then be computed by force equilibrium as

$$F = \frac{\Sigma(cl + P\tan\phi)\cos\alpha}{\Sigma{-P}\sin\alpha} \tag{3.51}$$

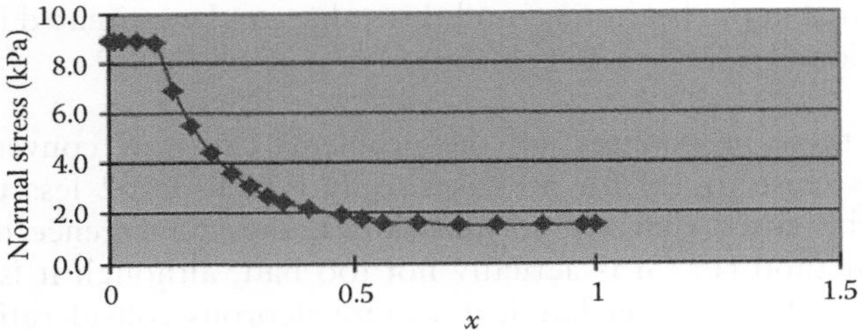

Figure 3.70 Base normal stress distribution along the slip surface for N_q when $\phi = 30°$ and $q = 1$ kPa.

Based on the base normal stress in Figure 3.70, which has been tested against different grid sizes used for the slip line analysis, the FOS from Equation 3.51 is exactly equal to 1.0, which is as expected. Using *P* or the thrust line as the control variables is, however, less satisfactory as compared with the use of the inter-slice force function in the optimization analysis. When the thrust line is defined, the moment equilibrium of the last slice is not used in the analysis, so that true moment equilibrium cannot be satisfied (the well-known problem for Janbu's rigorous method [1973]). It should be pointed out that for the original Janbu's moment equilibrium (1973), the moment equilibrium of the slice interface instead of the slice is considered so that there is no problem for the last slice, but the moment equilibrium for each slice is not strictly enforced. The use of *P* also suffers from this limitation. Once *P* is prescribed, *F* will be known from Equation 3.51. Based on force equilibrium in horizontal and vertical directions as well as the moment equilibrium, the inter-slice normal and shear forces as well as the thrust line will be defined for the first slice. These results can then be used to compute the internal forces between slices 2 and 3. The computation progresses up till the last slice for which both the force and moment equilibrium cannot be enforced automatically. To apply the base normal forces as the control variables in the extremum evaluation, the base normal forces for $N-1$ slices (N=total number of slices) are taken as the variables, while the base normal forces for the last slice will be determined by a trial-and-error process when the equilibrium of the last slice is satisfied. The same principle can also be applied to the thrust line, where thrust lines for only $N-2$ interfaces are prescribed and the thrust line for the last interface is obtained by a trial-and-error process until moment equilibrium is achieved. The use of $f(x)$ is simpler in that the majority of the back-computed thrust lines are acceptable, so that the solution for the prescribed $f(x)$ can be adopted directly without a trial-and-error process. If the thrust line is not acceptable, then the solution is simply rejected and another trial $f(x)$ can be considered.

3.14.4 Lower bound solution and the maximum extremum from limit equilibrium analysis

For the previous problems where $f(x)$ or the thrust line from the ultimate limit state is used, the FOSs of the system will be very close to 1.0. As discussed by Cheng et al. (2010), whenever an $f(x)$ is prescribed, the solution will always be the lower bound, which can be illustrated by the results in Table 3.25. The maximum extremum corresponds to the state that a system will exercise its maximum resistance before failure, and this condition is simply the ultimate condition of the system (Cheng et al., 2010). It is interesting to note that as ϕ increases, the rate of decrease of the FOS increases, and the FOSs corresponding to N_c and N_q can be considered to be the same.

Table 3.25 FOS corresponding to the three bearing capacity factors based on *f*(*x*) = 1

ϕ (°)	Nc	Nq	Nγ
0	0.941	0.941	—
10	0.941	0.941	1.0
20	0.937	0.936	0.989
30	0.925	0.924	0.968
40	0.894	0.892	0.939

The maximum difference for the FOSs corresponding to N_c and N_q with the ultimate limit state solution (1.0) is about 10%. On the other hand, the FOS corresponding to N_γ is close to 1.0 (bearing in mind that the failure surface is not the classical wedges/log-spiral mechanism).

Based on the stresses at yield at the two ends of a failure surface, Chen and Morgenstern (1983) have established the requirement that the inclination of the internal forces at the two ends of a failure surface must be parallel to the ground slope. For the previous problems, $f(x)$ is 0 at the two ends and it satisfies this requirement. Consider a bearing capacity problem (equivalently, a slope stability problem) for a soil with $c = 0$ and $\phi = 30°$, and that the ground is sloping at an angle of 15°. A solution to this problem is given by Sokovloskii (1965) as

$$N_c = \left[\frac{1+\sin\phi}{1-\sin\phi}e^{(2\alpha-\pi)\tan\phi} - 1\right]\cot\phi \tag{3.52}$$

where $\alpha = 165°$ in the present problem. If Spencer's method (1967) is used, a FOS of 0.951 with $\lambda = 0.281$ is obtained for this slip surface, which should bear a FOS of 1.0 by the classical plasticity solution. Based on the slip line solution by SLIP, $f(x)$ for this problem is given in Figure 3.71, and $\lambda f(x)$ at the end of the failure surface is 0.268, which is exactly tan 15°, and the result has clearly satisfied the requirement by Chen and Morgenstern (1983). If Spencer's method (1967) is used in the global minimum analysis, the minimum FOS is 0.825 with $\lambda = 0.219$. The critical failure surface based on Spencer's method (1967) as shown in Figure 3.72b is slightly deeper than the classical slip line solution as shown in Figure 3.72a. More importantly, the critical solution based on Spencer's method (1967) appears to be a wedge-type failure, which is different from the classical solution. The low FOS from the Spencer's method (1967) has illustrated the importance of $f(x)$ in the analysis.

Based on the previous problems and the present problem, it is established that the maximum extremum of the system is very close to the ultimate

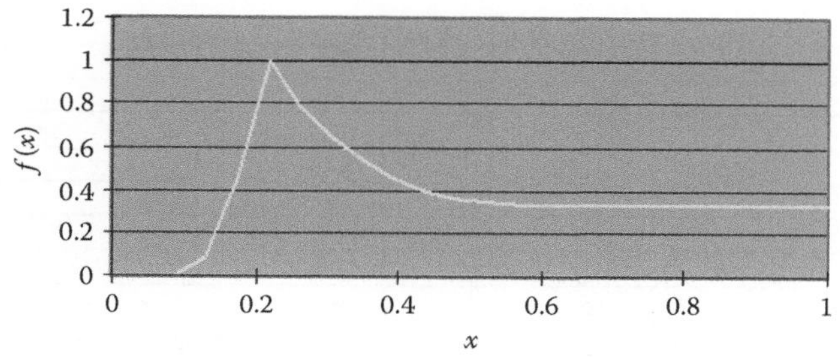

Figure 3.71 $f(x)$ for N_c with a sloping ground of 15°.

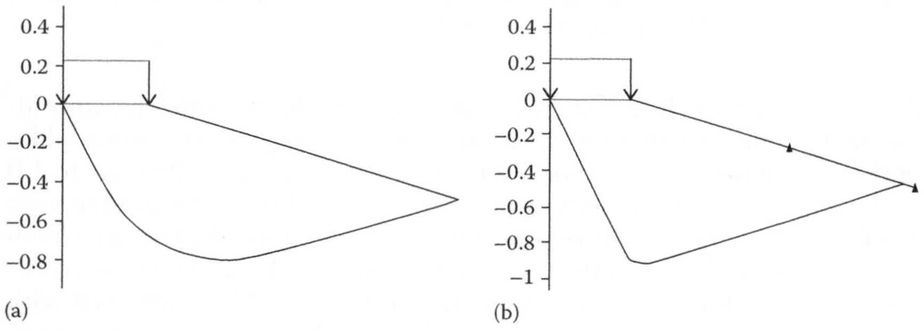

Figure 3.72 (a) Failure surface based on the classical plasticity solution using $f(x)$ from Figure 3.71 ($\lambda = 0.79$, FOS = 1.0). (b) Critical failure surface based on Spencer's method ($\lambda = 0.219$, FOS = 0.825).

limit state of a system. For a prescribed failure mechanism, the system will exercise its maximum strength before failure, and this is conceptually the lower bound theorem. Either $f(x)$ or the thrust line corresponding to the ultimate limit state will be sufficient to define the system, and $f(x)$ and the thrust line can be determined if the method of Cheng et al. (2010) is adopted.

The results in Table 3.25 and Figure 3.72 have clearly illustrated the concept of lower bound analysis, and every prescribed $f(x)$ with acceptable internal forces will gives a lower bound solution to the problem. On the other hand, if unacceptable internal forces are accepted for any prescribed set of $f(x)$, thrust line or base normal forces, it is actually possible to obtain a low FOS or a value higher than 1.0, which is in conflict with the assumption of lower bound analysis. In this respect, the acceptability of the internal forces, which is not imposed explicitly in the methods of Morgenstern and Price (1965), Janbu (1973) or Baker (1978), is actually important if arbitrary internal or external variables are imposed in a stability analysis. The lower bound concept can be visualized clearly from

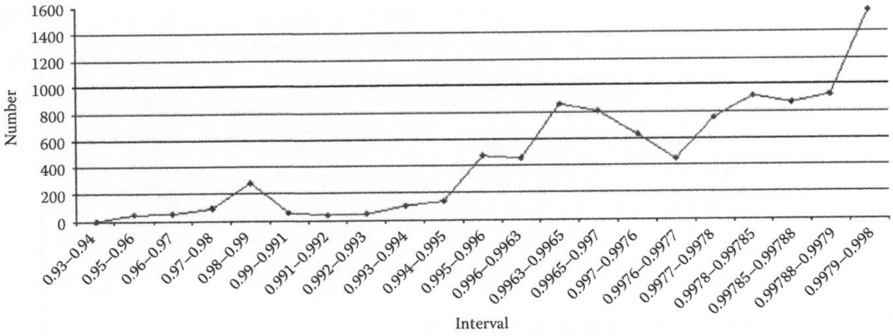

Figure 3.73 Distribution of acceptable FOS during simulated annealing analysis for N_q with $\phi = 30°$ using $f(x)$ as the variable.

Figure 3.73, which shows all the temporary FOSs during simulated annealing analysis in searching for the maximum FOS using $f(x)$ as the variables in the extremum determination for factor N_q when $\phi = 30°$. $f(x)$ is set to 1.0 as the initial trial in the optimization analysis, and the FOSs are far from 1.0 initially. As the global optimization analysis proceeds, the extremum of the system will tend to the theoretical value of 1.0, and no FOS exceeding 1.0 can be found. The results shown in Figure 3.73 comply well with the assumption of the lower bound analysis, and they further support the adoption of the maximum extremum as the lower bound solution. A further demonstration of the extremum principle is shown for a very thin slice for a 30° slope with $\phi = 30°$. According to the classical soil mechanics, the FOS for this thin slice should be 1.0. From the extremum principle, a factor of 1.0014 is obtained for the slip surface (minor difference from 1.0 as a circular arc is actually used). From the results in Figures 3.72 and 3.73, it is clear that the maximum resistance of the system has been mobilized in the extremum analysis, and the extremum principle will not over-predict the FOS of the system. Besides the use of $f(x)$, thrust line and the base normal forces from slip line analysis, Cheng et al. (2013a) have also adopted the extremum principle and have obtained FOSs close to 1.0 for the three bearing capacity factors. Using the extremum principle, all the FOSs corresponding to the three bearing capacity factors are very close to 1.0 by using either $f(x)$, thrust line or base normal forces. These results further demonstrate that the results from the extremum principle practically correspond to the ultimate condition of a system from a limit equilibrium viewpoint.

For the bearing capacity problem, the classical plasticity solutions based on the lower bound slip line method are well known and accepted. Since the stresses are determinate for this problem, all internal or external variables under the ultimate condition can be evaluated. Cheng et al. (2013a) have proved that a bearing capacity problem is equivalent to a horizontal slope stability problem if $f(x)$, thrust line or the external boundary forces P are known under the ultimate condition. The use of $f(x)$, thrust line or the external boundary forces are different ways to specify a limit equilibrium slope stability problem, and under the ultimate conditions, the uses of $f(x)$, thrust line or base normal forces as the control variables are actually equivalent. It has been demonstrated in the present study that $f(x)$ are the same for N_c and N_q factors from both the slip line methods and the extremum principle, and this result is not surprising as the failure mechanisms for N_c and N_q are actually the same. It is also demonstrated that $f(x)$ for N_c and N_q are 0 at the two ends and assume the maximum value at an x ratio less than 0.5. On the other hand, $f(x)$ for N_γ assumes the maximum value at a different x ratio. For N_c and N_q, the thrust line ratio clusters around 0.4 to 0.5, while for N_γ, the thrust line ratio starts at a value between 0.4 and 0.5 and decreases gradually to 1/3 outside the foundation. Based on $f(x)$, thrust line or base normal forces as obtained, the FOSs using M–P (1965), Janbu's rigorous method (1973) or BG method (1978) are very close to 1.0. That means as long as the ultimate condition is considered, there is no practical difference between the uses of internal or external variables in defining a problem. Different formulations should give the same results under the ultimate condition, which is clearly illustrated in this section.

The inter-slice force functions obtained from the slip line method in this study are not simple functions, and they have fully complied with the requirement by Chen and Morgenstern (1983), and the inclination of the internal forces at the two ends of a failure surface is always parallel to the ground slope in the present study. On the other hand, $\lambda f(x)$ is set to 1.0 in the popular Spencer's method (1967), which can be considered an approximate lower bound of the true failure mechanism. Poor results are obtained in the bearing capacity problem if Spencer's method (1967) is used, which implies that the precise values of the internal or external variables can be very important in some problems. Cheng et al. (2013a) have also clearly illustrated that as long as an assumption is prescribed, the solution will be a lower bound of the ultimate condition; hence, the classical LEMs are practically lower bounds to the ultimate condition.

It should be pointed out that the extremum principle of Cheng et al. (2010) can be viewed as a form of the variational principle (Sieniutycz and Farkas, 2005; Cheng et al., 2011a, 2013). The maximum FOS as determined may however deviate slightly from the true ultimate limit state, as the Coulomb relation is applied as a constraint along the vertical interface. For a true ultimate limit state, Mohr–Coulomb relation should be applicable throughout the whole medium instead of applying to a lumped global inter-slice normal and shear force. In this respect, the extremum principle as proposed by Cheng et al. (2010) is only a good approximate to the ultimate condition as the yield condition is checked globally at each interface instead of being enforced at each infinitesimal domain. Nevertheless, the method of

Cheng et al. (2010) provides a good solution with minimum effort in providing a practical solution to a problem without complete discretization of the solution domain. In the strength reduction analysis of a slope, the FOS is varied until the system cannot maintain stability. Stress will redistribute during the non-linear elasto-plastic analysis, and as long as the stress can redistribute without violating yield and equilibrium, the trial FOS can be increased further. It should be noted that the concept of the extremum is actually in line with the concept of the strength reduction method in this respect. It is hence not surprising that the FOSs from strength reduction analysis (ultimate condition) are always greater than those from the lower bound Spencer's analysis (Cheng et al., 2007a), provided that the global minima from Spencer's analysis are used for comparisons.

M–P method (1965) and Janbu's rigorous method (1973) can be considered as the same under the ultimate condition. These two methods are demonstrated to be equivalent under the ultimate condition, as they are controlled by both yield and equilibrium equations, except for ease of mathematical manipulation. Besides that, it is also demonstrated that use of base normal forces can be an alternative to M–P method (1965) and Janbu's rigorous method (1973). The choice of the assumption is hence physically not important under the ultimate condition. On the other hand, if the ultimate condition is not considered, there are practical differences between the use of $f(x)$, thrust line or base normal forces in the LEM, and every prescribed assumption giving acceptable internal forces will be a lower bound to the ultimate limit state of the problem, which is clearly illustrated in Figure 3.73. It is also interesting that the present study has demonstrated the equivalence between the LEM and the slip line solution for a medium that is fully under the plastic condition, provided the extremum from the LEM is used in the comparison. The hyperbolic partial equations governing the α and β characteristic lines as given by Equations 3.32 and 3.33 can be well approximated by tuning $f(x)$ in slope stability analysis until the critical solution is obtained with simple force and moment equilibrium. This is an interesting and useful application of the lower bound concept to a more general problem where the classical slip line methods fail to work. For a homogeneous problem with a continuous stress field, the conclusion from the present study will be valid. It should be kept in mind that for a non-homogeneous problem with a discontinuous stress field, the present conclusion will however be no longer valid, but the difference between the use of $f(x)$, thrust line and base normal forces should be small under the ultimate condition.

Chapter 4

Finite-element methods for slope stability analysis and comparisons with limit equilibrium analysis

The limit equilibrium method (LEM) and the strength reduction method (SRM) based on finite-element/finite-difference methods are currently the most popular methods among the engineers. For the limit analysis (including the rigid element) and distinct element methods, they are still not popular among the engineers, and comparisons with the two methods will not be included in this chapter. Some of the engineers responsible for the design of dams in China even have doubt on the activation of energy balance along the vertical interfaces in limit analysis, which is shown to be not valid in Figure 2.20.

4.1 COMPARISONS BETWEEN SRM AND LEM

The LEM, which is based on the force and moment equilibrium, is a popular method among the engineers. Besides the LEM, which has been introduced in Chapter 2, the use of finite-difference/finite-element methods has also attracted the engineers in recent times, which are introduced in Section 2.8. This approach is currently adopted in several well-known commercial geotechnical finite-element programmes. The SRM by finite-element analysis was used for slope stability analysis as early as 1975 by Zienkiewicz et al. Later, the SRM was applied by Naylor (1982), Donald and Giam (1988), Matsui and San (1992), Ugai and Leshchinsky (1995), Dawson et al. (1999), Griffiths and Lane (1999), Song (1997), Zheng et al. (2005) and others. In the SRM, the domain under consideration is discretized and the equivalent body forces are applied to the system. The yield criterion adopted is usually the Mohr–Coulomb criterion, but the use of other yield criteria is also possible. Different researchers and commercial programmes have adopted different definitions to assess the factor of safety (FOS). The most popular definitions for the FOS include (1) a sudden change in the displacement of the system; (2) failure to converge after a pre-determined number of iterations have been performed and (3) formation of a continuous yield zone.

Many researchers have compared the results between the SRM and the LEM and have found that the two methods will generally give similar FOSs. Most of the studies are, however, limited to homogenous soil slopes, and the geometry of the problems is relatively regular, with no special features (e.g., the presence of a thin layer of soft material or special geometry). Furthermore, there are only limited studies, which compare the critical failure surfaces from the LEM and the SRM as the FOSs appear to be the primary quantity of interest. In this chapter, the two methods are compared under different conditions and both the FOS and the locations of the critical failure surfaces are considered in the comparisons. In this chapter, both a non-associated flow rule (SRM1 and dilation angle = 0) and an associated flow rule (SRM2 and dilation angle = friction angle) are applied in the SRM analyses. To define the critical failure surface from the SRM, both the maximum shear strain and the maximum shear strain increment definition can be used. Cheng et al. (2007a) have found that these two definitions will give similar results under most cases and maximum shear strain increment is chosen for the present study.

In this chapter, the LEM is considered using the Morgenstern–Price method with $f(x) = 1.0$ (equivalent to Spencer's method). It is also found that the differences of the FOS and the critical failure surfaces from $f(x) = 1.0$ and $f(x) = \sin(x)$ are small for the present study. In performing the SRM analysis, many soil parameters and boundary conditions are required to be defined, which are absent in the corresponding LEM analysis. The importance of the various parameters and the applicability of the SRM under several special cases, are considered in the following sections.

4.2 STABILITY ANALYSIS FOR A SIMPLE AND HOMOGENEOUS SOIL SLOPE USING LEM AND SRM

To investigate the differences between the LEM and the SRM, a homogeneous soil slope with slope height equal to 6 m and slope angle equal to 45° (Figure 4.1) is considered in this section. For the three cases in which friction angle is 0, since the critical slip surface is a deep-seated surface with a large horizontal extent, the models are larger than the one shown in Figure 4.1 and have a width of 40 m and height of 16 m. In the parametric study, different shear strength properties are used and the LEM, SRM1 and SRM2 analyses are carried out. The cohesive strength c' of the soil varies between 2, 5 and 10–20 kPa, while friction angle ϕ' varies between

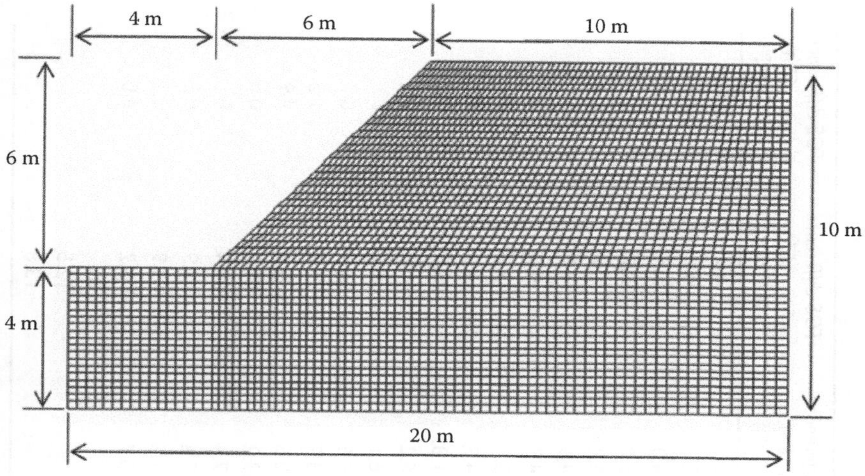

Figure 4.1 Discretization of a simple slope model.

5°, 15°, 25° and 35°–45°. The density, elastic modulus and Poisson's ratio of the soil are maintained at 20 kN/m^3, 14 MPa and 0.3, respectively, in all the analyses. As shown in Figure 4.1, the size of the domain for the SRM analyses is 20 m in width and 10 m in height and there are 3520 zones and 7302 grid points in the mesh for analysis. Based on limited mesh refinement studies, it was found that the discretization shown in Figure 4.1 is sufficiently good so that the results of analyses are practically insensitive to a further reduction in the element size. For the LEM, Spencer's method, which satisfies both the moment and the force equilibrium, is adopted and the critical failure surface is evaluated by the modified simulated annealing technique proposed by Cheng (2003). The tolerance for locating the critical failure surface by the simulated annealing method is 0.0001, which is sufficiently accurate for the present study.

From Table 4.1 and Figures 4.2 and 4.3, it is found that the FOS and the critical failure surfaces determined by the SRM and the LEM are very similar under different combinations of soil parameters for most cases, except when $\phi' = 0$. When the friction angle is greater than 0, most of the FOSs by the SRM differ by less than 7.4% with respect to the LEM results, except for case 16 ($c' = 20$ kPa, $\phi' = 5°$) where the difference is up to 13.2%. When the friction angle is very small or 0, there are relatively major differences between the SRM and the LEM for both the FOS and the critical slip surface (Table 4.1 and Figure 4.5). The differences in the FOS between the LEM and the SRM reported by Saeterbo Glamen et al. (2004) are greater than those found in the present study. Cheng et al. (2007a) suspect that this is due to the manual location of the critical failure surfaces by Saeterbo

Table 4.1 FOSs by LEM and SRM

Case	c' (kPa)	ϕ' (°)	FOS (LEM)	FOS (SRM1, non-associated)	FOS (SRM2, associated)	FOS difference from LEM (SRM1, %)	FOS difference from LEM (SRM2, %)	FOS difference between SRM1 and SRM2
1	2	5	0.25	0.25	0.26	0	4.0	4.0
2	2	15	0.50	0.51	0.52	2	4.0	2.0
3	2	25	0.74	0.77	0.78	4.0	5.4	1.3
4	2	35	1.01	1.07	1.07	5.9	5.9	0
5	2	45	1.35	1.42	1.44	5.2	6.7	1.4
6	5	5	0.41	0.43	0.43	4.9	4.9	0
7	5	15	0.70	0.73	0.73	4.3	4.3	0
8	5	25	0.98	1.03	1.03	5.1	5.1	0
9	5	35	1.28	1.34	1.35	4.7	5.5	0.7
10	5	45	1.65	1.68	1.74	1.8	5.5	3.6
11	10	5	0.65	0.69	0.69	6.2	6.2	0
12	10	15	0.98	1.04	1.04	6.1	6.1	0
13	10	25	1.30	1.36	1.37	4.6	5.4	0.7
14	10	35	1.63	1.69	1.71	3.7	4.9	1.2
15	10	45	2.04	2.05	2.15	0.5	5.4	4.9
16	20	5	1.06	1.20	1.20	13.2	13.2	0
17	20	15	1.48	1.59	1.59	7.4	7.4	0
18	20	25	1.85	1.95	1.96	5.4	5.9	0.5
19	20	35	2.24	2.28	2.35	1.8	4.9	3.1
20	20	45	2.69	2.67	2.83	0.7	5.2	6.0
21	5	0	0.20		0.23		15	
22	10	0	0.40		0.45		12.5	
23	20	0	0.80		0.91		13.8	

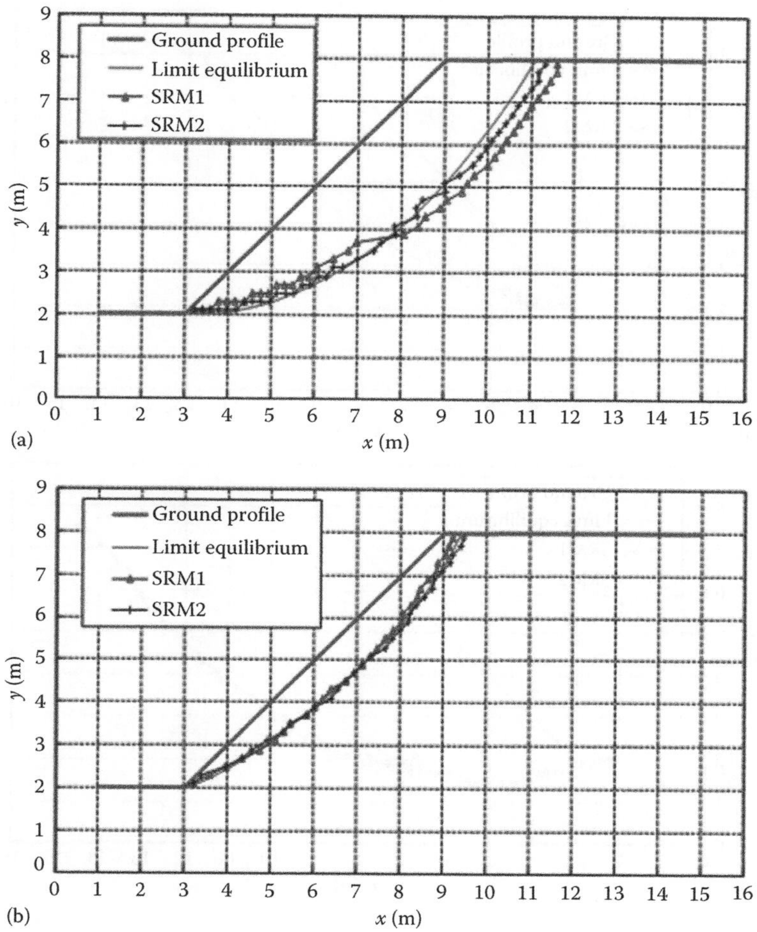

Figure 4.2 Slip surface comparison with increasing friction angle ($c'=2$ kPa). (a) $c'=2$ kPa, $\phi=5°$. (b) $c'=2$ kPa, $\phi=45°$.

Glamen et al. (2004), as opposed to the global optimization method used here. Based on Table 4.1 and Figures 4.2 and 4.3, some conclusions can be made as follows:

1. Most of the FOSs obtained by the SRM are slightly larger than those obtained by the LEM, with only few exceptions.
2. The FOSs from an associated flow rule (SRM2) are slightly greater than those from a non-associated flow (SRM1), and this difference increases with increasing friction angle. These results are reasonable and are expected. The differences between the two sets of

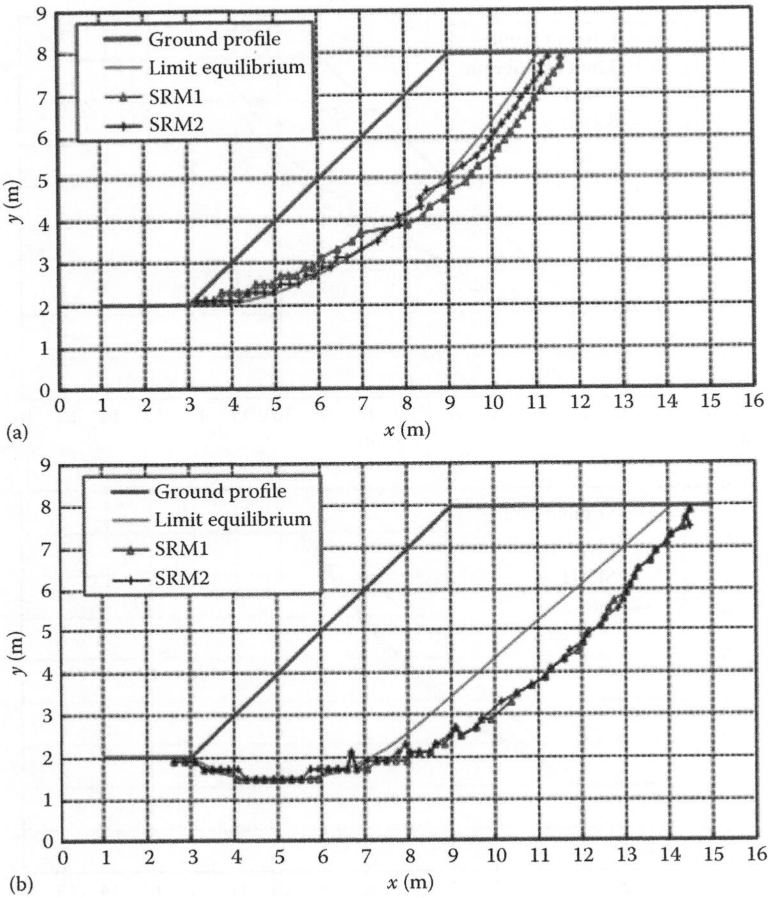

Figure 4.3 Slip surface comparison with increasing cohesion ($\phi = 5°$). (a) $c' = 2$ kPa, $\phi = 5°$. (b) $c' = 20$ kPa, $\phi = 5°$.

results are, however, small because the problem has a low level of 'kinematic constraint'.

3. When the cohesive strength of the soil is small, the differences in the FOS between the LEM and obtained from SRM (SRM1 and SRM2) are greatest for higher friction angles. When the cohesion of the soil is large, the differences in the FOS are greatest for lower friction angles. This result is somewhat different from that of Dawson et al. (2000) who concluded that the differences are greatest for higher friction angles when the results between the SRM and limit analysis are compared.

4. The failure surfaces from the LEM, SRM1 and SRM2 are similar in most cases. In particular, the critical failure surfaces obtained by SRM2 appear to be closer to those from the LEM than those obtained by SRM1. The critical failure surfaces from SRM1, SRM2 and the LEM are practically the same when the cohesive strength is small (it is difficult to differentiate clearly in Figures 4.2a, b, 4.3a and 4.4a), but noticeable differences in the critical failure surfaces are found when the cohesive strength is high (Figures 4.3b, 4.4b, 4.5a and b).
5. The right end of the failure surface moves closer to the crest of the slope as the friction angle of the soil is increased (which is a well-known result). This behaviour is more obvious for the failure surfaces

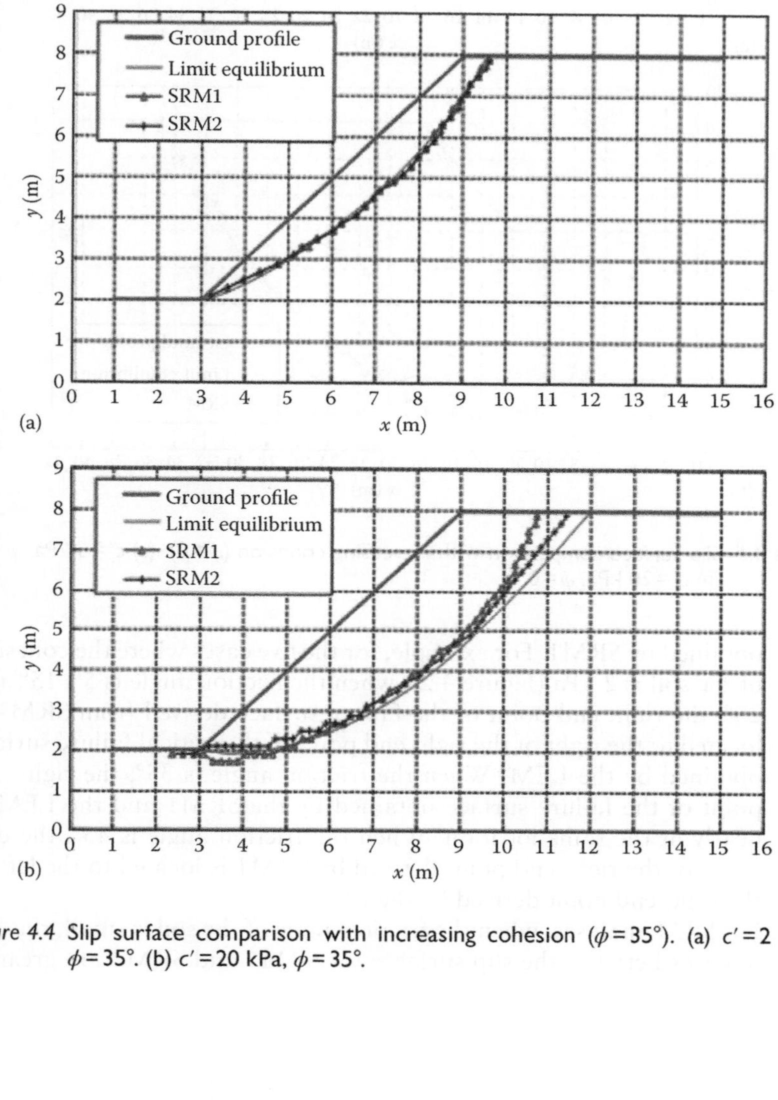

Figure 4.4 Slip surface comparison with increasing cohesion ($\phi = 35°$). (a) $c' = 2$ kPa, $\phi = 35°$. (b) $c' = 20$ kPa, $\phi = 35°$.

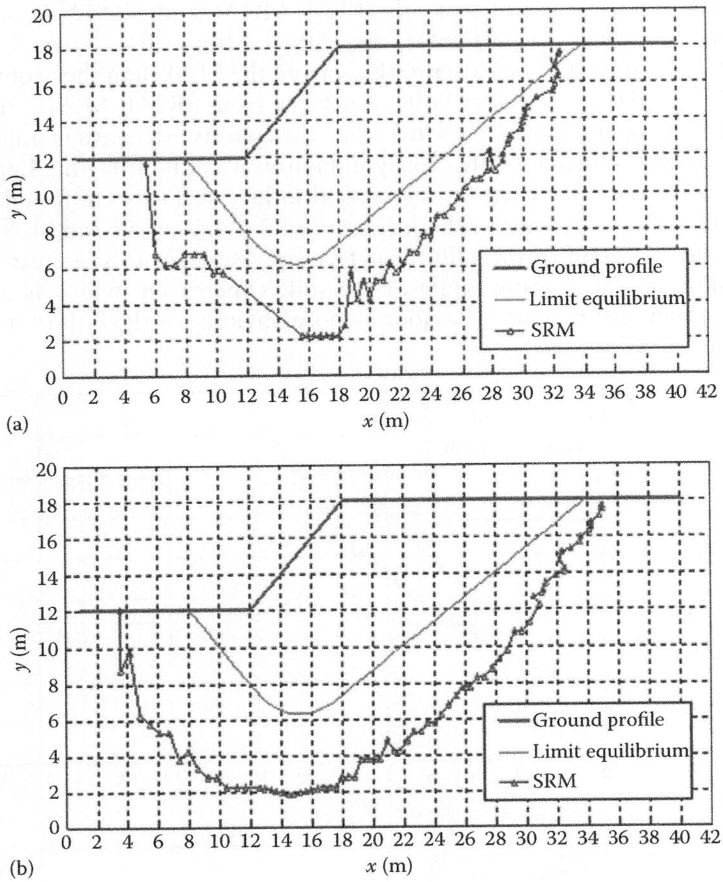

Figure 4.5 Slip surface comparison with increasing cohesion ($\phi = 0$). (a) $c' = 5$ kPa, $\phi = 0$. (b) $c' = 20$ kPa, $\phi = 0$.

obtained by SRM1. For example, for the five cases where the cohesion of the soil is 2 kPa (Figure 4.2), when the friction angle is 5°, 15° and 25°, the right end point of the failure surface derived from SRM1 is located to the right of the right end point of the critical failure surface obtained by the LEM. When the friction angle is 35°, the right end point of the failure surface obtained by the SRM1 and the LEM is nearly at the same location. When the friction angle is 45°, the distance of the right end point derived by SRM1 is located to the left of the right end point derived by the LEM.

6. For SRM analyses, when the friction angle of the soil is small, the differences between the slip surfaces for SRM1 and SRM2 are greatest

for smaller cohesion (Figure 4.3). When the friction angle is large, the differences between the slip surface for SRM1 and SRM2 are greatest for higher cohesion (Figure 4.4).

7. It can also be deduced from Figures 4.2 through 4.5 that the potential failure volume of the slope becomes smaller with increasing friction angle, but increases with increasing cohesion. This is also a well-known behaviour, as when the cohesive strength is high, the critical failure surface will be deeper.

Although there are some minor differences in the results between the SRM and the LEM in this example, the results from these two methods are generally in good agreement, which suggests that the use of either the LEM or the SRM is satisfactory in general. Cheng et al. (2007a) have however constructed an interesting case where the limitations of the SRM are demonstrated. There is an important difference between SRM and LEM, which is the inter-slice force function definition. In the SRM, the soil parameters are reduced by the FOS within the whole solution domain. When the internal stresses by the SRM are used to re-construct the inter-slice force function, Cheng found that the shape of the inter-slice force function is similar to that from the extremum principle in Section 3.12. The compliance of the Mohr–Coulomb relation along the vertical boundary between slices must however be based on the reduced soil parameters. On the other hand, for the LEM, the FOS is typically not applied along this vertical boundary in the definition. Since the FOS is usually not sensitive to the inter-slice force function, such an important difference appears to be not important for practical purpose.

4.3 STABILITY ANALYSIS OF A SLOPE WITH A SOFT BAND

A special problem with a soft band has been constructed by Cheng et al. (2007a) as it appears that similar problems have not been considered previously. The geometry of the slope is shown in Figure 4.6 and the soil properties are shown in Table 4.2. It is noted that c' is 0 and ϕ' is small for soil layer 2, which has a thickness of just 0.5 m. The critical failure surface is obviously controlled by this soft band, and slope failures under similar conditions have actually occurred in Hong Kong.

In order to consider the size effect (boundary effect) in the SRM, three different numerical models are developed to perform the SRM using Mohr–Coulomb analysis, and the widths of the domains are 28 m, 20 m and 12 m, respectively (Figure 4.7). In these three SRM models, various mesh sizes were tried until the results became insensitive to the number of elements used for analysis. For example, when the domain size is 28 m, the FOS

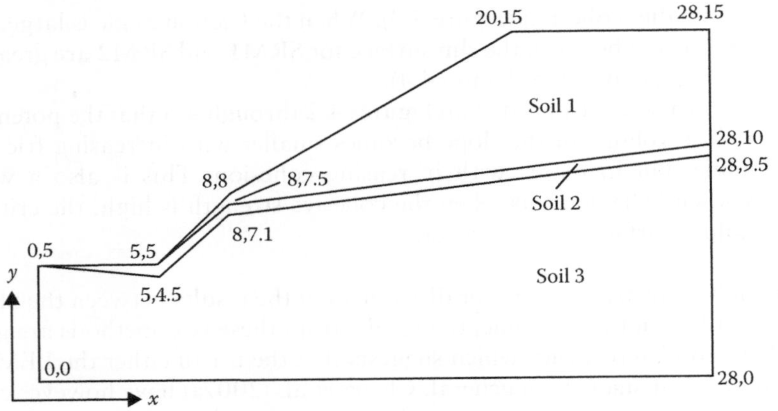

Figure 4.6 Slope with a thin soft band.

Table 4.2 Soil properties for Figure 4.6

Soil name	*Cohesion (kPa)*	*Friction angle (°)*	*Density (kN/m³)*	*Elastic modulus (MPa)*	*Poisson's ratio*
Soil 1	20	35	19	14	0.3
Soil 2	0	25	19	14	0.3
Soil 3	10	35	19	14	0.3

was found to be 1.37 (Tables 4.3 and 4.4) with 12,000 elements, 1.61 with 6,000 elements and 1.77 with 3,000 elements using SRM1 analysis and the programme Phase.

Since the FOSs for this special problem have great differences from those found using the LEM, Cheng et al. (2007a) tried several well-known commercial programmes and obtained very surprising results. The locations of the critical failure surfaces by the SRM for solution domain widths of 12, 20 and 28 m are virtually the same. The local failures from the SRM, shown in Figure 4.8b, range from $x=5$ m to $x=8$ m, and the failure surfaces are virtually the same for the three different solution domains. A major part of the critical failure surface lies within layer 2, which has a low shear strength, and is far from the right boundary. It is surprising to find that different programmes produce drastically different results (Table 4.3) for the FOS even though the locations of the critical failure surface from these programmes are very similar. For the cases shown in Figure 4.1, and other cases in a latter part of this chapter, the results are practically insensitive to the domain size, while the cases shown in Figure 4.6 are very sensitive to the size of the domain for the programmes Flac3D (SRM1 and SRM2) and Phase (SRM2). Results from the Plaxis programme appear to not be

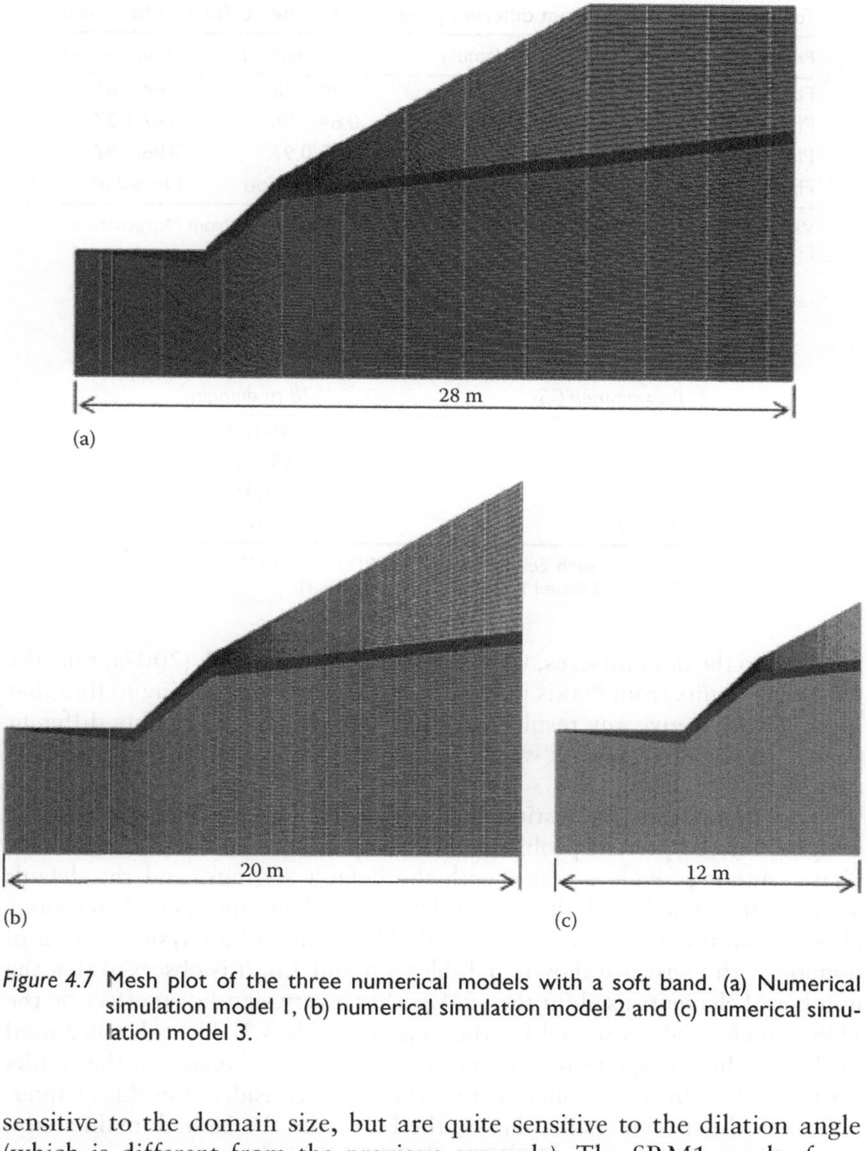

Figure 4.7 Mesh plot of the three numerical models with a soft band. (a) Numerical simulation model 1, (b) numerical simulation model 2 and (c) numerical simulation model 3.

sensitive to the domain size, but are quite sensitive to the dilation angle (which is different from the previous example). The SRM1 results from programme Phase are also not sensitive to the domain size for SRM1, but results from SRM2 behave differently. The FOSs from Flac3D appear to be overestimated when the soil parameters for the soft band are low, but the results from this programme are not sensitive to the dilation angle, which is similar to all the other examples in the present study. For SRM1, the results from Phase and Plaxis appear to be more reasonable as the results are not

Table 4.3 FOS by SRM from different programmes when c' for soft band is 0

Programme/FOS	*12 m domain*	*20 m domain*	*28 m domain*
Flac3D	1.03/1.03	1.30/1.28	1.64/1.61
Phase	0.77/0.85	0.84/1.06	0.87/1.37
Plaxis	0.82/0.94	0.85/0.97	0.86/0.97
Flac2D	No solution	No solution	No solution

Values in each cell are based on SRM1 and SRM2 (min. FOS=0.927 from Morgenstern–Price analysis).

Table 4.4 FOS by SRM from different programmes when $\phi'=0$ and $c'=10$ kPa for soft band

Programme/FOS	*28 m domain*
Flac3D	1.06/1.06
Phase	0.99/1.0
Plaxis	1.0/1.03
Flac2D	No solution

Values in each cell are based on SRM1 and SRM2 (min. FOS = 1.03 from Morgenstern–Price analysis).

sensitive to the domain sizes, while for SRM2, Cheng et al. (2007a) take the view that results from Plaxis may be better. It is also surprising to find that Flac2D cannot give any result for this problem, even after many different trials, but the programmes worked properly for all the other examples in this chapter.

There is another interesting and important issue when the SRM is adopted for the present problems. For the problem with a 12 m domain, Phase cannot provide a result with the default settings, and the default settings are varied (including the tolerance and the number of iterations allowed) until convergence is achieved. The results of analysis for a 12 m domain with Phase are shown in Tables 4.5 and 4.6. It is observed that the number of elements used for the analysis has a very significant effect on the FOS, which is not observed for the cases in Table 4.1. The tolerance used in the non-linear equation solution also has a major impact on the results for this case. This is less obvious for other cases considered in this chapter.

Besides the special results shown, the FOS from the 28 m domain analysis appears to be large for Flac3D and Phase when the strength parameters for soil layer 2 are low. In fact, it is not easy to define an appropriate FOS from SRM analysis for this problem. If the cohesive strength of the top soil is reduced to 0, the FOS can be estimated as 0.57 from the relation $\tan\phi/\tan\theta$, where θ is the slope angle. It can be seen that for the LEM, the cohesive strength 20 kPa for soil 1 helps to bring the FOS to 0.927, and a high FOS for this problem is not reasonable. Without the results from the

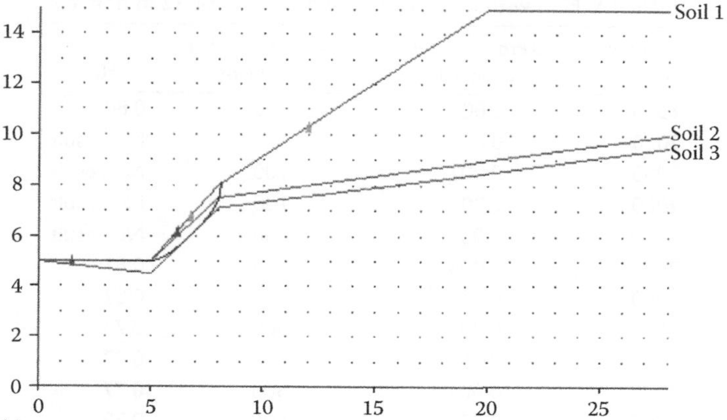

(a)

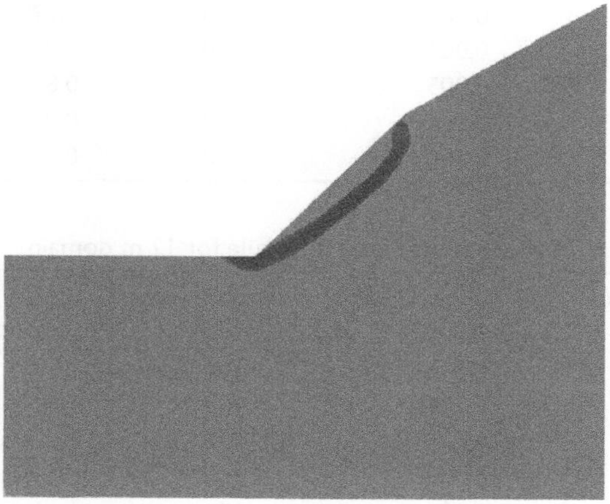
(b)

Figure 4.8 Locations of critical failure surfaces from the LEM and the SRM for a frictional soft band problem. (a) Critical solution from the LEM when the soft band is frictional material (FOS = 0.927). (b) Critical solution from the SRM for a 12 m width domain.

LEM for comparison, it may be not conservative to adopt the values of 1.64 (1.61) from the SRM based on Flac3D.

When the soil properties of the soft band are changed to $c' = 10$ kPa and $\phi' = 0$; the results of analyses are shown in Table 4.4. It is found that the critical failure will extend to a much greater distance so that a 28 m wide domain is necessary. The FOSs from the different programmes are virtually the same, which is drastically different from the results in Table 4.3 (the same meshes were used for Tables 4.3 and 4.4).

Table 4.5 FOS with non-associated flow rule for 12 m domain

Element number	*Tolerance (stress analysis)*	*Maximum number of iterations*	*FOS*
1500	0.001	100	0.80
2000	0.001	100	No result
2000	0.003	100	No result
2000	0.004	100	No result
2000	0.005	100	No result
2000	0.008	100	0.81
2000	0.01	100	0.82
2000	0.001	500	0.74
2000	0.003	500	0.77
2000	0.004	500	0.77
2000	0.005	500	0.79
3000	0.001	100	No result
3000	0.003	100	0.79
3000	0.004	100	0.8
3000	0.005	100	0.8
3000	0.01	100	0.84
3000	0.001	500	0.77

Table 4.6 FOS with associated flow rule for 12 m domain

Element number	*Tolerance (stress analysis)*	*Maximum number of iterations*	*FOS*
1000	0.001	100	1.03
1200	0.001	100	1.0
1500	0.001	100	No result
1500	0.003	100	No result
1500	0.004	100	1
1500	0.005	100	1.39
1500	0.01	100	2.09
1500	0.001	500	0.86
1500	0.003	500	0.98
3000	0.001	100	No result
3000	0.003	100	No result
3000	0.004	100	No result
3000	0.005	100	No result
3000	0.01	100	No result
3000	0.001	500	0.85
3000	0.003	500	0.89
3000	0.004	500	0.9
3000	0.005	500	1.5
3000	0.01	500	2.09

If the soil properties of soils 2 and 3 are interchanged so that the third layer of soil is the weak soil, the FOSs from SRM2 are 1.33 (with all programmes) for all three different domain sizes. The corresponding FOS from the LEM is 1.29 from Spencer's analysis. The locations of the critical failure surface from the SRM and the LEM for this case are also very close, except for the initial portion shown in Figure 4.9a and b. It appears that the presence of a soft band with frictional material, instead of major differences in the soil parameters, is the actual cause for the difficulties in the SRM analysis. Great care is required in the implementation of a robust non-linear equation solver for the SRM.

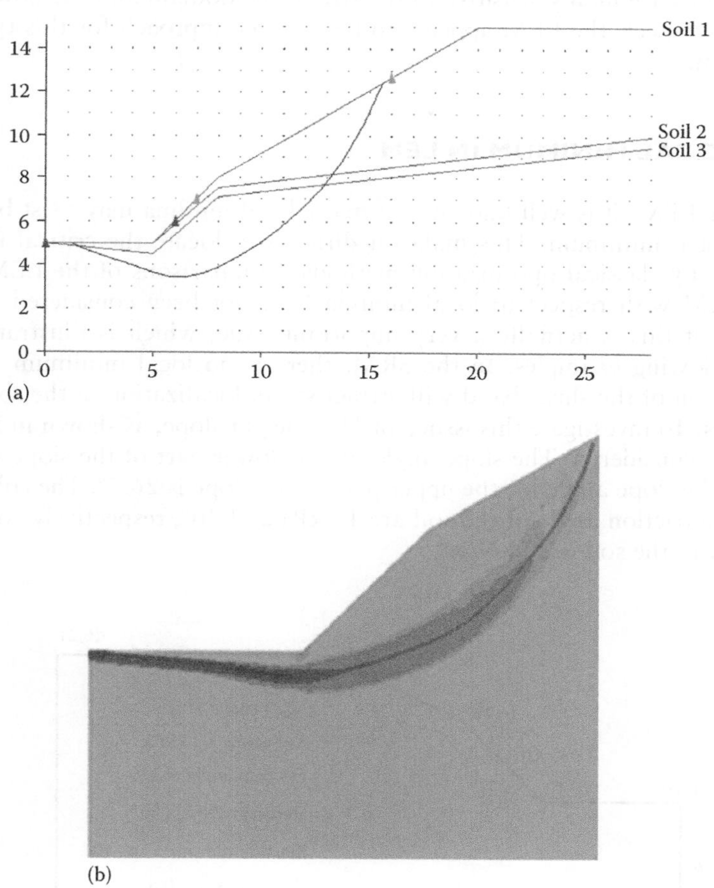

Figure 4.9 Critical solutions from the LEM and the SRM when the bottom soil layer is weak. (a) Critical failure surface from the LEM when the bottom soil layer is weak (FOS = 1.29). (b) Critical failure surface from SRM2 and a 12 m domain (FOS = 1.33).

The problems shown in Table 4.3 may reflect the limitations of commercial programmes rather than the limitations of the SRM, but they illustrate that it is not easy to compute a reliable FOS for this type of problem using the SRM. For the updated versions of these commercial programmes, it appears that the problems mentioned earlier have been solved. The results are highly sensitive to different non-linear solution algorithms, which are not clearly explained in the commercial programmes. Great care, effort and time are required to achieve a reasonable result from the SRM for this special problem and comparisons with the LEM are necessary. It is not easy to define a proper FOS from the SRM alone for the present problem as the results are highly sensitive to the size of the domain and the flow rule. In this respect, the LEM appears to be a better approach for this type of problem.

4.4 LOCAL MINIMUM IN LEM

For the LEM, it is well known that many local minima may exist besides the global minimum. This makes it difficult to locate the critical failure surface by classical optimization methods. Comparisons of the LEM and the SRM with respect to local minima have not been considered in the past, but this is actually a very important issue, which is illustrated by the following examples. In the SRM, there is no local minimum as the formation of the shear band will attract strain localization in the solution process. To investigate this issue, an 11 m height slope, as shown in Figure 4.10, is considered. The slope angle for the lower part of the slope is 45°, while the slope angle for the upper part of the slope is 26.7°. The cohesion and the friction angle of the soil are 10 kPa and 30°, respectively, and the density of the soil is 20 kN/m^3.

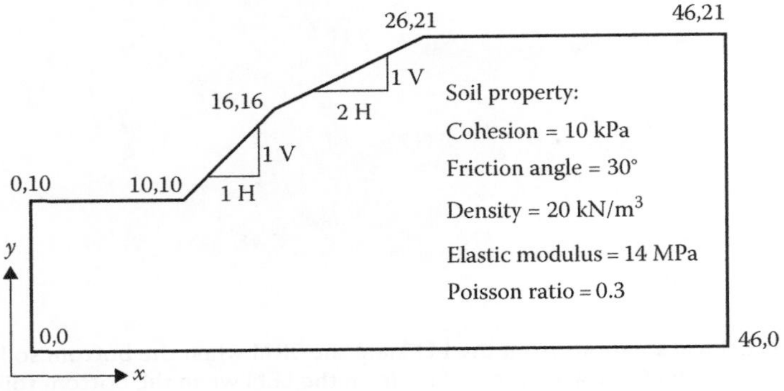

Figure 4.10 Slope geometry and soil property.

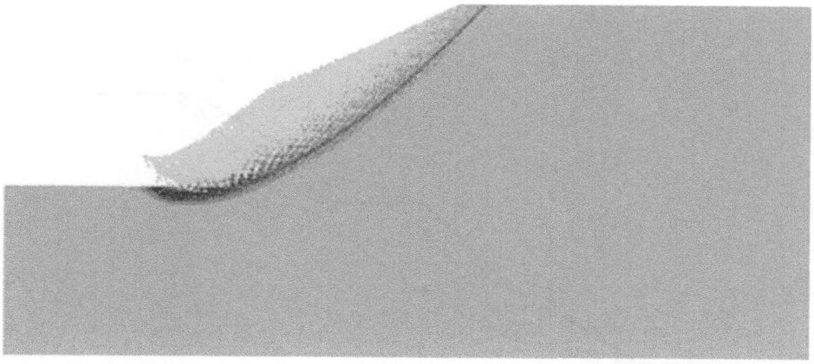

Figure 4.11 Result derived by the SRM.

The failure mechanism by the SRM is shown in Figure 4.11, and the FOS is 1.47 for both non-associated flow and associated flow. The right end point of the failure surface is located to the right of the crest of the slope. The results derived from the LEM are presented in Figure 4.12 (number of slices is 50). The global minimum FOS is 1.383, but a local minimum FOS of 1.3848 is also found. The location of the failure surface for the local minimum 1.3848 is very close to that from the SRM, and the failure surface for the global minimum from the LEM is not the critical failure surface from the SRM. Since the FOSs for the two critical failure surfaces from the LEM are so close, both failure surfaces should be considered to be probable failure surfaces in slope stabilization. For the SRM, there is only one unique failure surface from the analysis, and another possible failure mechanism cannot be easily determined. Thus, the SRM analysis may yield a local failure surface of less importance while a more severe global failure surface remains undetected, as illustrated in the next example. This is clearly a major drawback of the SRM as compared with the LEM.

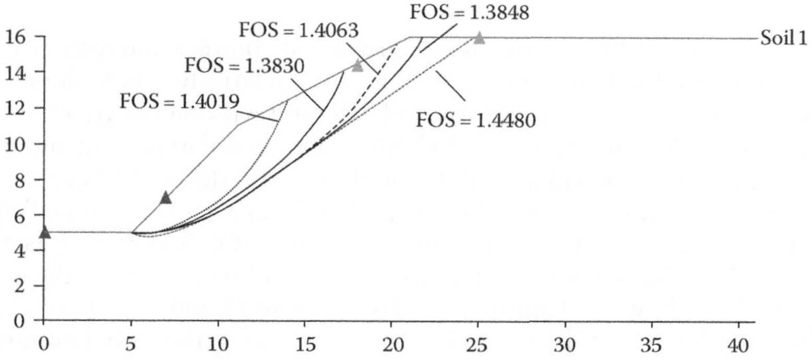

Figure 4.12 Global and local minima by the LEM.

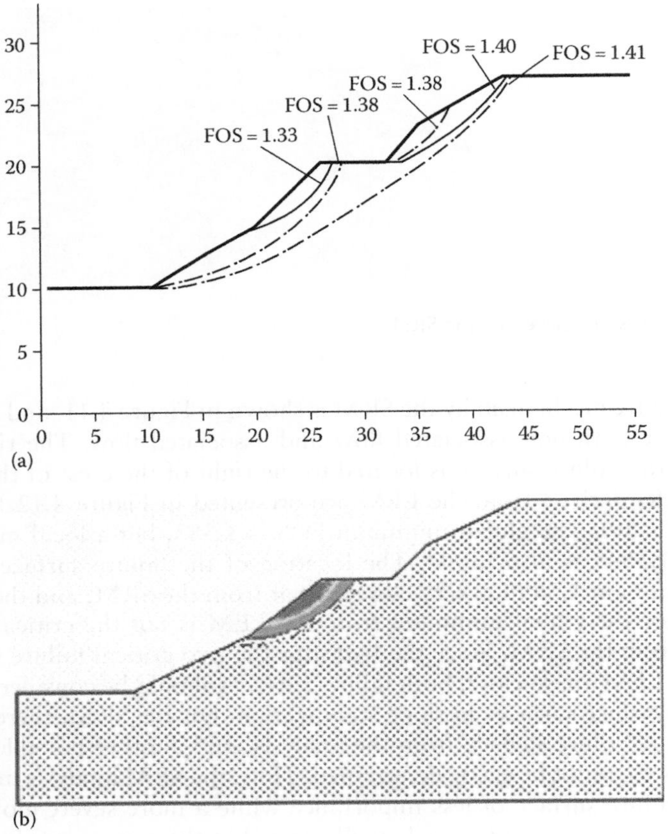

Figure 4.13 (a) Global and local minimum FOSs are very close for a slope (b) with FOS = 1.327 from the SRM.

Cheng et al. (2007a) have also constructed another interesting case, which is worth discussion. Figure 4.13 shows a relatively simple slope with a total height of 55 m in a uniform soil. The soil parameters are $c' = 5$ kPa and $\phi' = 30°$, while unit weight is 20 kN/m^3. The global minimum and local minima are determined in accordance with the procedures of Cheng (2003) and different boundaries for the left and right exit ends are specified in the study. Using the LEM, the global minimum FOS is obtained as 1.33 (Figure 4.13a), but several local minima are found with FOSs in the range 1.38–1.42, as shown in Figure 4.13a. From the SRM, only a FOS of 1.327, as shown in Figure 4.13b, is found, which is similar to the global minimum shown in Figure 4.13a. If slope stabilization is only carried out for this failure surface, the possible failure surfaces given by the local minimum in

Figure 4.13a will not be considered. Baker and Leshchinsky (2001) have proposed the concept of the 'safety map', which enables the global minimum and local minima from the LEM to be visualized easily, but the construction of such a map using the SRM is tedious. In this respect, the LEM is a better tool for slope stability analysis. It is possible that the use of the SRM may miss the location of the next critical failure surface (with a very small difference in the FOS, but a major difference in the location of the critical failure surface) so that the slope stabilization measures may not be adequate. This interesting case has illustrated a major limitation of the SRM for the design of slope stabilization works. It is true that the use of the safety map by the SRM can also overcome the limitation of the local minimum, but the evaluation of all local minima using the LEM and the modern optimization method requires only 10 min for the complete analysis, which is much faster than the use of the safety map. The assessment of the local minima and the global minimum can also give a picture similar to that by the safety map. In the authors' view, the use of the safety map concept is not compulsory.

4.5 EFFECT OF WATER ON SLOPE STABILITY ANALYSIS

For slope stability analysis with water flow, different ways to consider the seepage forces in the LEM have caused various confusions. In the traditional LEM, the boundary water forces with total weights are usually used and the water pressure enters the base force calculation but not the inter-slice forces (or consequently, the equilibrium of slice). Turnbull and Hvorslev (1967) viewed that the traditional method may yield unreasonable results for high pore pressure and suggested that the effective stress should be resolved in a direction normal to the failure surface. Greenwood (1987) and King (1989) introduced some effective-stress methods of slices, which include the inter-slice water forces. These approaches, however, further complicate the analysis and are not adopted in commercial programmes. Greenwood (1987), King (1989) and Duncan and Wright (2005) have shown that there are cases where the refined approaches may have noticeable differences from the classical methods. If the SRM is used, pore pressure will affect the effective stress on which the stability analysis is based. The confusions about the effect of water in the LEM will not appear in the SRM.

In this section, a two-dimensional 6 m height slope with 45° slope angle is analysed. A 10 m height model is developed in which water is 4 m high on the left and 10 m high on the right side. The pore pressure and the flow vector distribution for the free-surface seepage flow analysis are shown in Figure 4.14. In Figure 4.14, the total head is 10 m for the top flow line and the total head difference for each flow line is 1 m. The density, elastic

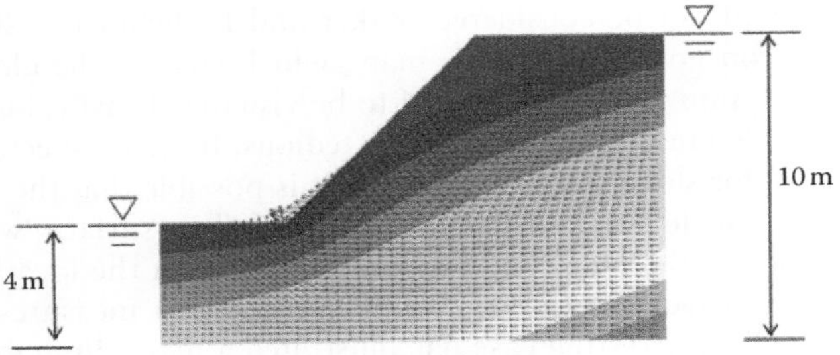

Figure 4.14 Pore water pressure and flow vector of a simple slope from a free-surface seepage analysis (total head is 10 m for the top flow line and total head difference for each flow line is 1 m).

modulus and Poisson's ratio of the soil are maintained at 20 kN/m^3, 14 MPa and 0.3, respectively, in the analysis. Cheng et al. (2007a) have demonstrated that the uses of these elastic properties in SRM are not important, and the FOSs using different elastic properties are virtually not affected by these parameters.

It is found that the location of the slip surface for sandy soil is greatly influenced by the water flow from various parametric studies, and the failure surface becomes shallower and closer to the slope toe under the influence of water seepage flow. These results are also similar to the observations of the slope failures in Hong Kong over the last 30 years (GEO, 1996a,b), where many slopes failures are initiated from toe failures under heavy rain. With reference to Figure 4.15a, there is a relatively high hydraulic gradient around the toe of the slope; hence slope failure will be limited to the region close to the toe of the slope, while a typical slope failure, which extends to the top of the slope, is obtained in Figure 4.15b where there is no water. Such results are further supported by the tremendous slope failures observed during rainy season in Hong Kong, when most of the

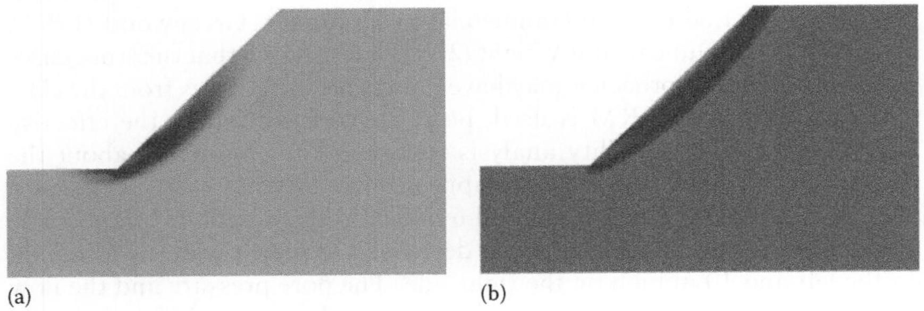

Figure 4.15 Slip surface for a slope with 1 kPa cohesion and 45° friction angle (a) with water flow (FOS = 0.82) and (b) no water (FOS = 1.30).

Table 4.7 FOS for different situations

Cohesion and friction angle	*1 kPa, 45°*	*2 kPa, 45°*	*5 kPa, 35°*	*10 kPa, 25°*
FOS with water (pore pressure generated by seepage analysis)	0.82	0.96	0.98	1.07
FOS with water (pore pressure generated by water table)	0.66	0.82	0.92	1.03
FOS for no water	1.3	1.44	1.35	1.37

slope failures initiate around the toe of slopes where there is a rapid change of the total head and, hence, a high hydraulic gradient, and such results are also predicted by the numerical analysis shown in Figure 4.15.

In the aforementioned analysis, the pore water pressure is generated by a seepage flow analysis, which is a reasonable way to obtain the pore pressure distribution. It is however possible to define the pore water pressure by a water table in the SRM similar to that in the LEM analysis. In order to investigate the difference between the two approaches, another model is developed in which the pore pressure is generated by the water table. The FOSs for the two cases are compared in Table 4.7. The FOS for the case where the pore pressure is generated by the water table is smaller than in the case where it is generated by the seepage flow analysis, as the pore water pressure calculated by the water table (free surface) is larger. It means that use of the water table (or piezometric line) is a conservative method of analysis, and the difference is very small for a clayey slope but is much larger for sandy soil. In this respect, the commonly adopted engineering design method using a phreatic surface to evaluate the pore water pressure appears to be a conservative approach, which is more important for sandy soil slopes.

4.6 SOIL NAILED SLOPES BY SRM AND LEM

Soil nailing is a simple and economic slope stabilization technique and is particularly useful for strengthening existing slopes. The fundamental concept of soil nailing is to reinforce the soil with closely spaced passive inclusions to create a coherent gravity structure and, thereby, increase the overall shear strength of the in situ soil, restraining its displacements. The basic design consists of transferring the resisting tensile forces generated in the inclusions into the ground through the friction mobilized at the nail–soil interfaces. There are many different design methods for soil nailing, including LEMs; several working stress design methods; and also, Davis's method (Shen et al., 1981), the German method (Stocker et al., 1979) and the French method (Schlosser, 1982; Schlosser and Unterreiner, 1991). The US Federal Highway Administration has also published a series of design guidelines (1996, 1997, 1999, 2001, 2003) currently used by many engineers for nailed soil and reinforced earth structures based on the LEM.

If the failure of a soil nailed slope is internal, the failure mode can be classified into the three different types: face failure, pullout failure and nail tensile. One of the important factors for proper analysis of soil nailed slopes is the determination of the nail load. According to Hong Kong's practice (which is a special case of Davis's method by Shen et al., 1981), the ultimate bond strength of a soil nail τ_f is expressed as $\tau_f = \pi Dc' + 2D\sigma_v' \tan\phi'$, where D is the hole diameter, σ_v' is the effective vertical stress on the nail, $\tan\phi'$ is the frictional coefficient between the soil and the nail, and c' is the cohesion of the soil. Since adhesion and friction between the soil and the nail will be less than c' and $\tan\phi'$, respectively, a FOS of 2.0 is given to the ultimate bond strength for design purposes. Another soil nail design practice exists in which the bond stress is assumed to be independent of the confining/overburden stress. Laboratory and field tests in Hong Kong have suggested that both design methods may be appropriate under different cases, and the actual nail load appears to be dependent on the soil type, time, the grouting pressure and the topography. In the following studies, the two methods of bond load determination are considered and the SRM and the LEM are compared on an equal basis.

In this section, a parametric study for a homogeneous soil slope with a slope height of 6 m and a slope angle of 45° is conducted (Wei and Cheng, 2010), and the slope geometry and soil nail distribution are the same as the example in Figure 4.16. In the parametric study, different shear strength

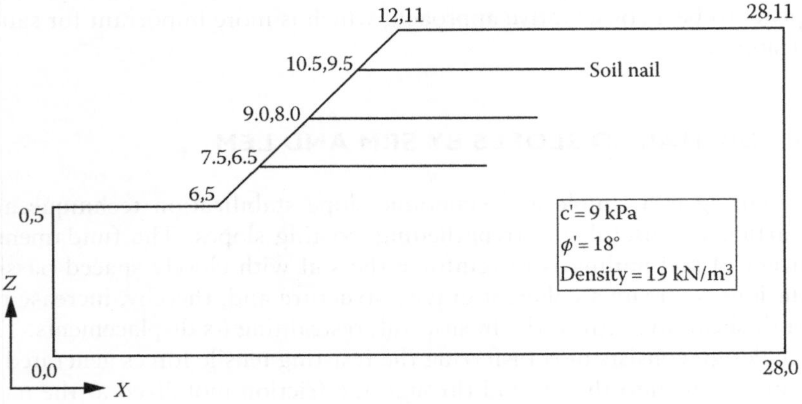

Figure 4.16 Soil nailed slope model.

properties are used and both the LEM and SRM analyses are carried out. The cohesion of the soil varies from 2, 5, 10 to 20 kPa, while the friction angle varies from 5°, 15°, 25°, 35° to 45°. The density, elastic modulus and Poisson's ratio of the soil are kept constant at 20 kN/m^3, 15 MPa and 0.35, respectively, in all the analyses, and the results of analyses are shown in Table 4.8. For the LEM, Spencer's method is adopted and the tolerance for locating the critical failure surface is 0.0001, which is good enough for the present study.

From Table 4.8, it is found that the FOSs by the SRM and the LEM are very similar under different combinations of soil parameters, and all the FOSs obtained from the SRM (associated flow rule) are slightly larger than those obtained by the LEM, with a maximum difference of 12.4%. This phenomenon is also similar to that obtained by Cheng et al. (2007a) and Wei and Cheng (2009a) for un-reinforced slopes. When the friction angle of the soil is small, the slip surfaces by the SRM and the LEM are in good agreement. When the friction angle of the soil becomes very large (e.g. $\phi=45°$), the slip surfaces by the SRM are deeper than those by the

Table 4.8 FOSs by LEM and SRM

Case	*c′ (kPa)*	*ϕ′ (°)*	*FOS (LEM)*	*FOS (SRM)*	*FOS difference between LEM and SRM (%)*
1	2	5	0.27	0.28	3.70
2	2	15	0.59	0.63	6.78
3	2	25	0.99	1.09	10.10
4	2	35	1.55	1.72	10.97
5	2	45	2.34	2.63	12.39
6	5	5	0.44	0.46	4.55
7	5	15	0.80	0.86	7.50
8	5	25	1.23	1.33	8.13
9	5	35	1.80	1.96	8.89
10	5	45	2.59	2.9	11.97
11	10	5	0.70	0.74	5.71
12	10	15	1.14	1.19	4.39
13	10	25	1.63	1.71	4.91
14	10	35	2.17	2.37	9.22
15	10	45	2.97	3.3	11.11
16	20	5	1.15	1.26	9.57
17	20	15	1.67	1.79	7.19
18	20	25	2.20	2.34	6.36
19	20	35	2.82	3.04	7.80
20	20	45	3.67	3.95	7.63

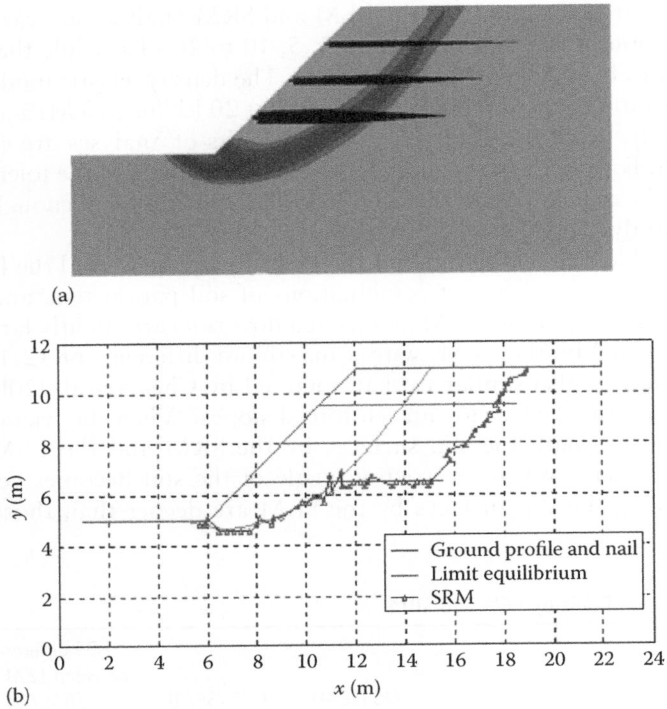

Figure 4.17 Slip surface obtained for different element size ($c'=5$ kPa, $\phi=35°$). (a) FOS = 1.96 for slightly increased element size. (b) FOS = 1.94 for very fine mesh.

LEM, but the differences between the slip surfaces from the SRM and the LEM are relatively small.

Wei and Cheng (2009a) have observed several special cases where a combined failure surface may occur. For example, when $c'=5$ kPa and $\phi=35°$ (Figure 4.17), a combined slip surface given by Figure 4.17b will be detected by the use of a very fine mesh, and such a special slip surface will not occur when the mesh size is increased slightly (Figure 4.17a). Even though there are major differences in the two slip surfaces, the differences between the FOSs are very small and such a phenomenon does not appear for the LEM. It appears that determination of a combined failure surface is controlled by the size of the mesh in the computation. Cheng however views that such a combined failure surface is not a true phenomenon as such a failure surface possesses a high FOS when the LEM is used. Wei and Cheng (2009a) suspect that due to the use of an extremely fine mesh, stress concentration occurs around the grout–nail interfaces so that the combined failure surface will come out. Actually, Cheng et al. (2007a) and Wei and Cheng (2009a) have demonstrated the limitations of the non-linear solution algorithms in evaluating the ultimate limit states of a slope, and Wei and Cheng (2009a) also sometimes experienced strange results for unreinforced slopes if an extremely fine mesh was used in the analysis. In this respect, the engineers need to be very careful in assessing the results from the SRM, particularly for a nailed slope using three-dimensional analysis.

The nail axial force and bending moment from this problem are given in Table 4.9. The maximum mobilized nail axial force increases initially and then decreases with nail inclination, and this also explains why the FOS first increases and then decreases finally. It is noticed that some bending effect is mobilized in the nail, especially when the nail inclination is large (30°–60°), but the effect of the bending moment on the nail force is very small compared with the axial nail force (Table 4.9) so that the contribution of the bending effect to the FOS is also very small. This result also supports the current method of design where the bending effect of soil nails is neglected in the analysis.

Table 4.9 Results for different nail inclinations

Inclination angle (°)	*0*	*10*	*20*	*30*	*40*	*60*
FOS by LEM	1.22	1.26	1.24	1.24	1.22	1.13
Maximum force of top nail by LEM (kN)	3.71	6.54	11.26	14.17	15.63	19.00
Maximum force of middle nail by LEM (kN)	5.9	9.21	13.85	17.06	18.80	21.54
Maximum force of bottom nail by LEM (kN)	8.74	12.39	17.28	20.47	21.72	23.10
FOS by SRM1	1.28	1.30	1.33	1.33	1.31	1.17
Maximum force of top nail by SRM1 (kN)	13.85	13.22	14.78	15.27	15.94	3.79
Maximum force of middle nail by SRM1 (kN)	17.22	21.55	26.92	30.25	33.00	23.33
Maximum force of bottom nail by SRM1 (kN)	28.15	31.41	35.80	39.57	41.69	24.65
FOS by SRM2	1.28	1.30	1.33	1.36	1.36	1.20
Maximum force of top nail by SRM2 (kN)	18.52	19.56	21.54	23.79	26.22	4.99
Maximum force of middle nail by SRM2 (kN)	21.35	26.49	28.77	33.46	36.15	22.13
Maximum force of bottom nail by SRM2 (kN)	30.32	34.98	38.84	41.59	44.76	23.77
Maximum moment (kN·m)	1.193	1.215	1.079	1.744	1.820	1.819

Table 4.10 FOS with 200 kPa top pressure (bond load controlled by overburden stress)

Nail length (m)	*8*	*12*	*16*
FOS by SRM	0.66	0.76	0.84
FOS by LEM	0.56	0.59	0.61

Although the results from the SRM and the LEM are generally similar, under the action of external pressure, more noticeable and important differences between the SRM and the LEM are found, which are discussed in this section. For a 200 kPa pressure applied on the top of the slope, both the SRM and the LEM results for different nail lengths are in Tables 4.10 and 4.11. Two different SRM models have been developed for this study. In the first model, the bond stress is assumed to be controlled by the confining pressure. The results for the first model are shown in Table 4.10 and are compared with the results by the LEM. There are major differences between the results from the SRM and the LEM, and Wei and Cheng (2009a) view that such major differences arise from the increased confining pressure and hence the bond stress in the SRM analysis. On the other hand, the overburden pressure on the nail from the external load is usually not considered in the bond stress calculation (which appears to be the practice for all commercial programmes), so a lower bond stress is determined from the LEM. In this respect, there is a major difference in the soil nail design by the SRM and the LEM under the action of the external loads if the bond load is the function of the confining stresses.

In order to eliminate the influence of the confining stresses on the bond stress, the second model is developed in which the bond stress is assumed to be independent of the confining pressure (bond stress = 5.6 kN/m), and the results of analyses are shown in Table 4.11. Since the basic assumption for the bond stress is the same for the SRM and the LEM, the differences between the results from the LEM and the SRM are smaller than in the case in Table 4.10. It can be concluded that the SRM and the LEM may give greatly different results depending on the bond load determination method, and engineers should be aware of this difference.

Table 4.11 FOS with 200 kPa top pressure (constant pullout resistance)

Nail length (m)	*8*	*12*	*16*
FOS by SRM	0.60	0.67	0.73
FOS by LEM	0.58	0.61	0.65

4.6.1 Distribution of the nail tension force and the critical slip surface by SRM

The line of maximum tension within the nail is often considered as the failure surface, which divides the soil mass into two separate zones: (1) an 'active zone' close to the ground surface where the shear stresses exerted by the soil on the reinforcement are directed outwards and tend to pull the reinforcement out of the ground and (2) a 'resistant zone' where the shear stresses are directed inwards and tend to restrain the reinforcements from being pulled out. This concept is shown in Figure 4.18 (after FHWA, 1996), and many reinforced earth structures are designed in this way. It should be noted that the line of maximum tension does not correspond to the conventional critical slip surface as defined in limiting equilibrium stability analyses, but reflects the results of the soil structure interactions between the soil and the nail/facing reinforcement system (FHWA, 1996, 1999). Classically, many researchers assume that the maximum tensile force line coincides with the potential sliding surface, and this view is supported by some model tests (Juran et al., 1984) and full-scale tests (Clouterre, 1991). In this section, Wei and Cheng (2009a) will discuss this issue by conducting strength reduction analysis of several soil nailed slopes.

First, the influence of the failure modes is investigated. Soil nailed slope failure can be broadly classified into external failure modes and internal failure modes. For external failure modes, the slip surface does not intersect with the nails, so the maximum tensile force line will not coincide with the potential sliding surface. Three different external failure modes are shown in Figure 4.19. As shown in Figure 4.19a, if the soil nail is very short, it will be a global failure where the nails are totally within the failure mass. As in Figure 4.19b, the soil nail is long and the soil mass/nail becomes an integral

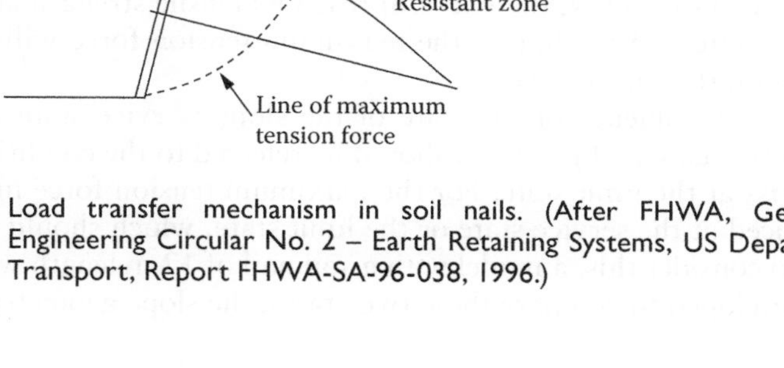

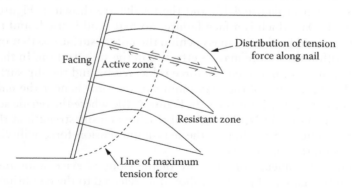

Figure 4.18 Load transfer mechanism in soil nails. (After FHWA, Geotechnical Engineering Circular No. 2 – Earth Retaining Systems, US Department of Transport, Report FHWA-SA-96-038, 1996.)

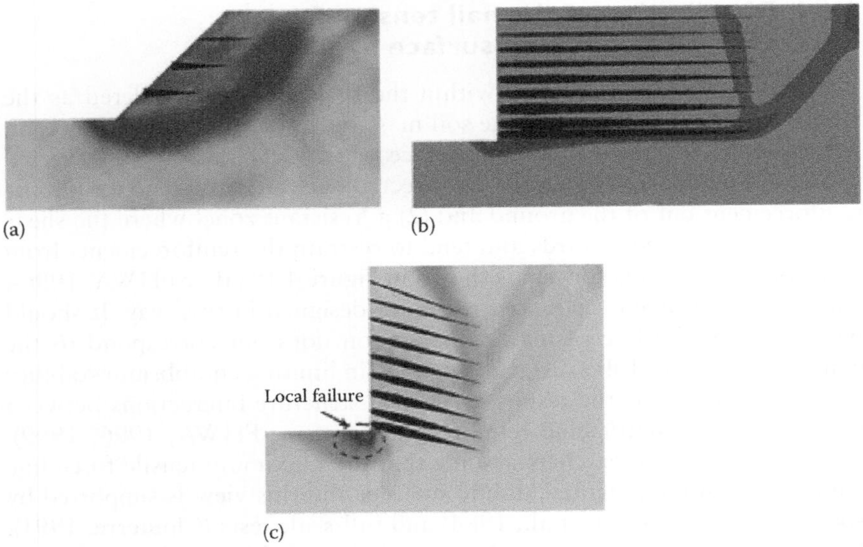

Figure 4.19 Axial tensile force distribution of soil nails in different external failure modes. (a) Global failure, (b) sliding failure and (c) bearing capacity failure.

body and fails by a sliding mode. In Figure 4.19c, which is one of Seattle's first soil nailed walls as discussed in the previous section, the reinforced soil nailed wall is more stable than the toe of the slope so that it is a local bearing capacity failure. For internal failure modes, they are usually classified into three different types: face failure, pullout failure and nail tensile failure. These three different internal failure modes are modelled for the slope as shown in Figure 4.16, and the results are shown in Figure 4.20. In Figure 4.20a, which is a face failure, no nail head is used, and the soil nail is restrained by the soil mass behind the failure surface so that only the friction within the failure mass is effective in the stabilization. In this case, the line of the maximum tension force is located behind the slip surface. In Figure 4.20b, the line of the maximum tension force is near the nail head and is in front of the slip surface. In Figure 4.20c where the tensile strength of the nail is only 10 kN, the nail will reach its tensile strength at the ultimate limit state and the line of the maximum tension force will virtually coincide with the slip surface.

Second, the influence of the state of the slope (service state and limit state) is investigated. Slip surface should be referred to the condition of the failure mass at the limit state. For the maximum tension force line, it can be referenced at the service state or the limit state, which should be stated clearly. To consider this, a model with a soil nail of 12 m length with a nail head is developed to compare these two states; the slope geometry and the

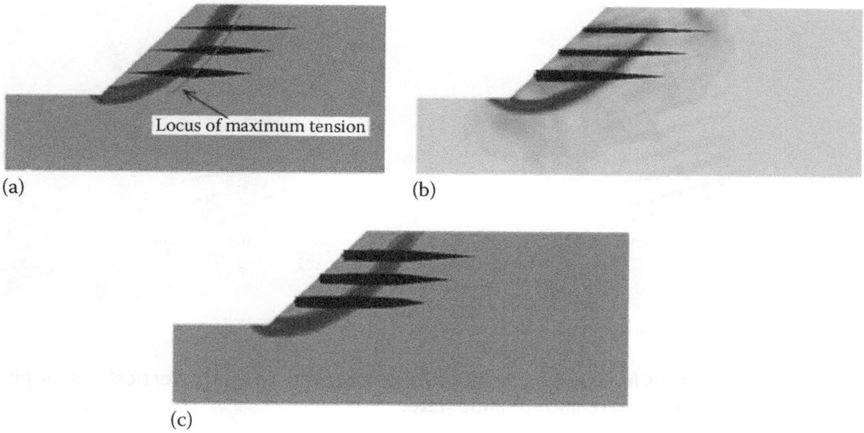

Figure 4.20 Slip surface and nail tension stress distribution in different internal failure modes. (a) No nail head (FOS = 1.20), face failure; (b) nail head is strong (FOS = 1.28), pullout failure; (c) nail head is strong, nail tensile strength = 10 kN (FOS = 1.22 with nail tensile failure).

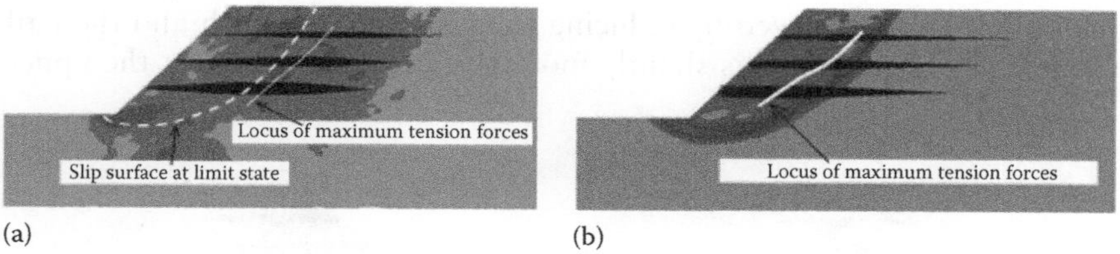

Figure 4.21 Axial tensile force distribution of soil nails in different states. (a) Service state and (b) limit state (FOS = 1.41).

soil nail distribution are the same as in the example in Figure 4.16. The results of analysis are shown in Figure 4.21. In Figure 4.22a (service state), the slip surface at the limit state is shown by the dashed line for comparison. It can be seen that the line of the maximum tension force at the service state is behind the slip surface. The force at the nail head connection at the service state is very small. In the limit state (Figure 4.22b), the line of the maximum tension force is in front of the slip surface, and a relatively large force at the nail head connection is mobilized at the limit state. With reduction of the soil shear strength, the soil nailed slope gradually transforms from the service state to the limit state, and during this process, the nail force is gradually mobilized while the line of maximum nail force will gradually move towards the nail head. In this example, which is an internal pullout failure, the slip surface appears to be located between the lines of maximum nail force at the service state and the limit state.

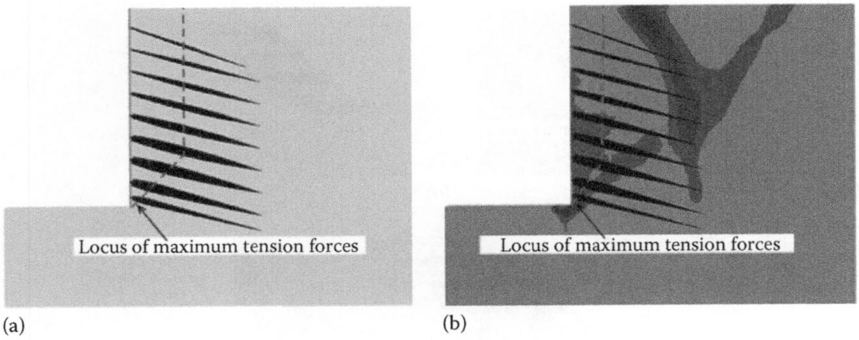

Figure 4.22 Axial tensile stress distribution in different states for a vertical cut slope. (a) Service state and (b) limit state.

In this analysis, the limit state is achieved by reducing the shear strength, which is easy to be conducted in numerical simulations but not for model tests or full-scale tests. The limit state can also be achieved by applying pressure on the top of a slope; an example is shown in Figure 4.23 where the results at the limit state are achieved by different ways. In Figure 4.23a, the limit state is achieved by reducing the soil shear strength and the nail force at the bottom row is slightly more mobilized than that at the upper

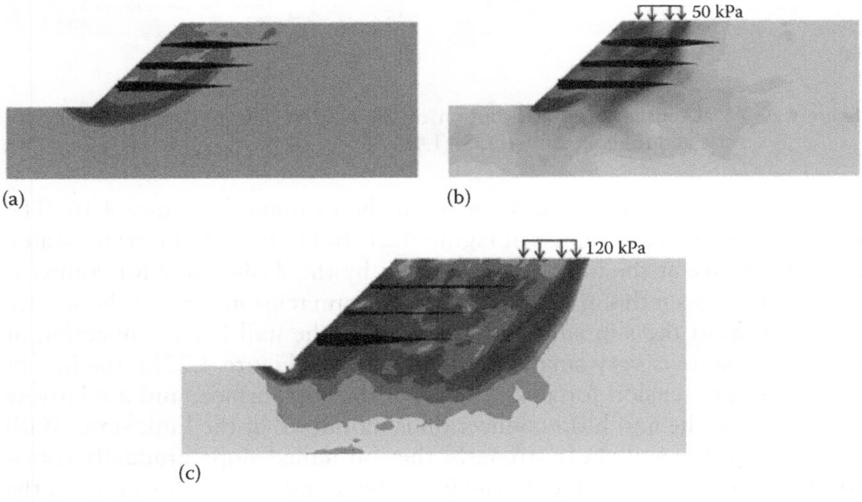

Figure 4.23 Axial tensile stress distribution in different limit states: (a) limit state by reducing strength ($f = 1.31$), (b) limit state by applying pressure near the crest and (c) limit state by applying 120 kPa pressure at the slope top near nail end.

two rows. In Figure 4.23b, the limit state is achieved by applying a pressure near the crest of the slope with an internal failure, and the nail force at the top row is more mobilized. In Figure 4.23c, the limit state, which is an external failure, is achieved by applying a pressure on the top of the slope near the end of the soil nails; the nail force at the bottom row is very large, while the nail forces at the upper two rows are much smaller.

Gassler (1993) and Stocker et al. (1979) have carried out tests by applying pressure on the top of slopes. Clouterre (1991) achieved the limit state by gradually saturating the soil, and this can be viewed as approximately equivalent to reducing the soil strength. The location of the slip surface is controlled mainly by the location of the loading as the slip surfaces starts from the edge of the loading. Since the location of the slip surface is controlled mainly by the location of the loading and the slip surface only partly intersects the soil nails, the line of the maximum tension force will not coincide with the slip surface. From the test by Shen et al. (1981), the maximum tension force measured is near to nail head. Based on this analysis, it can be concluded that the line of the maximum tension does not necessarily correspond to the conventional critical slip surface in general, but it reflects the results of the soil structure interactions between the soil and the nail/facing reinforcement system. First, the tension stress distribution is influenced by the state of the slope, and it is obviously different between the service state and the limit state. When the soil nailed slope gradually transforms from the service state to the limit state by reducing the shear strength, if face failure is prevented by the use of a nail head, the maximum tension line will move towards the slope surface (such as in Figures 4.21 and 4.22) as the nail force is gradually mobilized and the resistant zone becomes larger to maintain the slope in a stable condition. Second, the tension stress distribution is influenced by the slope failure modes at the limit state. For slopes with internal failure modes, if it is face failure, the line of maximum tension is usually located behind the slip surface (Figure 4.20a). For pullout failure, the line of the maximum tension is usually located in front of the slip surface (Figures 4.20b and 4.21b). For nail tensile failure, the line of the maximum tension will coincide well with the slip surface, which is demonstrated by both numerical simulations and full-scale tests. For slopes with external failure modes, the line of the maximum tension force will not coincide with the slip surface. Besides, both the nail tension force distribution and the slip surface are greatly controlled by the location of the external loading.

4.7 STABILIZATION OF SLOPE WITH PILES USING SRM

The use of piles as a stabilizing element to increase slope stability has been proved to be an effective solution in recent decades (D'Appolonia et al., 1967; De Beer and Wallays, 1970; Ito and Matsui, 1975; Fukuoka, 1977;

Wang et al., 1979; Ito et al., 1981, 1982; Reese et al., 1992). The piles used in slope stabilization are considered as passive piles because they are usually subjected to lateral force arising from horizontal movements of the surrounding soil. For passive piles, the problem is complicated because the lateral forces acting on the piles are dependent on the soil movements, which are in turn affected by the presence of the piles. A pile will function by its shear strength, which is different from an anchor or a soil nail.

Different methods have been used to evaluate the performance and design of piles that are used as reinforcement in slopes. Ito and Matsui (1975) have analysed the development of lateral forces on stabilizing piles when the soil is forced to squeeze between the piles. Ito et al. (1981, 1982) subsequently developed a limit equilibrium design method for pile-reinforced slopes based on this approach. Hassiotis et al. (1997) also proposed a method for design of slopes reinforced with a single row of piles in which the plastic state theory of Ito and Matsui (1975) was used to estimate the pressure acting on the piles. The simplified form of the boundary element method (Poulos, 1973; Poulos and Davis, 1980; Lee et al., 1991; Chen and Poulos, 1997) has also been used to evaluate the passive pile response around the piles. Based on this method, Lee et al. (1995) and Poulos (1995) have developed a limit equilibrium approach for design of piles. The design procedure of Poulos (1995) is as follows: (1) evaluating the total shear force required to increase the safety factor of the slope to the desired value; (2) evaluating the maximum shear force that each pile can provide to resist sliding of the potentially unstable portion of the slope by pile–soil interaction analysis using the boundary element method and (3) selecting the type and the number of piles, and the most suitable location in the slope. Chow (1996) presented a numerical method for the analysis of pile-stabilized slopes. In this method, piles are modelled using beam elements and the soil is modelled using a hybrid method that simulates the soil response at individual piles using the sub-grade reaction modulus and pile–soil–pile interaction using the theory of elasticity. Yamagami et al. (2000) presented another limit equilibrium design method for slopes with one row of piles. The basic idea of this method is to allow two different critical slip surfaces for the two sides of the piles, and the forces acting on the stabilizing piles are estimated based on a given critical slip surface where the FOS is prescribed to ensure the slope stability.

The kinematic approach of limit analysis is used by Ausilio et al. (2001) to analyse the stability of earth slopes reinforced with piles; analytical expressions for the force required to increase the safety factor to a desired value are proposed, and the most suitable location of the piles within the slope is then evaluated. For pile-stabilized slopes, Cai and Ugai (2000) have considered the effects of stabilizing piles on the stability of a slope by a three-dimensional finite-element analysis using the SRM, and the effects of pile spacing, pile head conditions, bending stiffness and pile positions on the safety factor have been analysed. Won et al. (2005) have also analysed the same slope as by Cai and Ugai (2000) using the three-dimensional finite-difference code Flac3d by the SRM. In the piled slope analysis by Won et al. (2005) and Cai and Ugai (2000), the location of the critical slip surface was determined by the maximum shear force in the pile so that a very deep critical slip surface was determined, while the maximum shear strain in the soil was not considered. Wei a The author views that the method of Cai and Ugai (2000) and Won et al. (2005) for determining the critical slip surface is unreasonable and not realistic. A study has been carried out by the authors (Wei and Cheng, 2009b), and it has been found that pile spacing has a major influence on the failure mode.

The slope considered by Won et al. (2005) and Cai and Ugai (2000) is 10 m high with a gradient of 1 V:1.5 H (Figure 4.24). Two symmetric extreme boundaries are used so that the problem consists of a row of piles with a plane of symmetry. A steel tube pile with an outer diameter of 0.8 m is used in this study. The piles are treated as a linear elastic solid material and are installed in the middle of the slope with $L_x = 7.5$ m, with centre-to-centre spacing (s) varying from 2D to 8D. The piles are embedded and fixed into the bedrock or a stable layer (infinite pile length assumption). In this model, the pile head is free. The cohesive strength, friction angle,

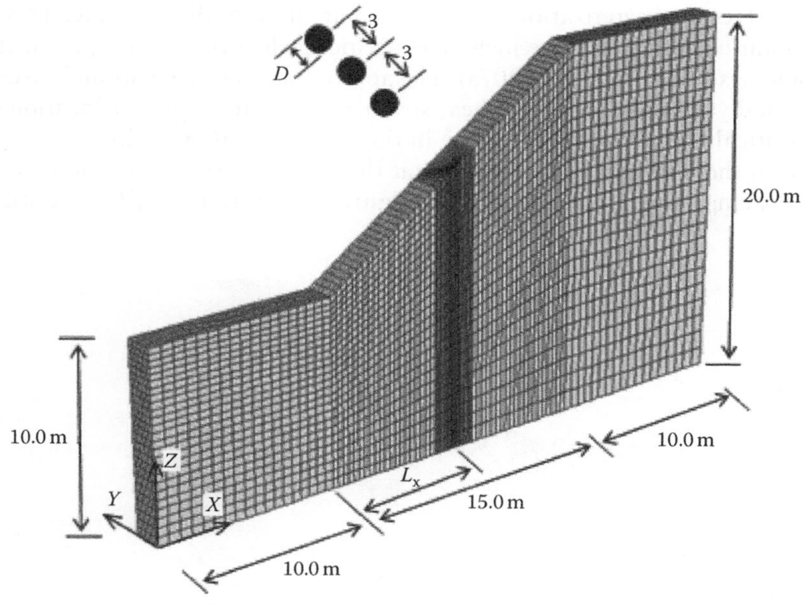

Figure 4.24 Slope model and finite-difference mesh.

elastic modulus, Poisson's ratio and unit weight of the soil are 10 kPa, 20°, 200 MPa, 0.25, and 20 kN/m^3, respectively. The elastic modulus and Poisson's ratio of the piles are 60,000 MPa and 0.2, respectively. The soil and the grout material are modelled by a Mohr–Coulomb model, while the pile is taken as an elastic material.

When the slope is not reinforced with piles, the FOS by the SRM is 1.20 using an associated flow rule with the programme Flac3D (1.13–1.15 as reported by Cai and Ugai, 2000 and Won et al., 2005). Since the FOSs are affected by the size of the mesh, the numerical iteration algorithm and the termination criterion, some noticeable differences (though still small) between the results by the authors and those by Cai and Ugai (2000) and Won et al. (2005) are noted. As a check to the result, the authors have performed a three-dimensional Spencer's analysis as outlined by Cheng and Yip (2007) and have obtained a FOS of 1.18. Since a finer mesh has been used in both the three-dimensional SRM and limit equilibrium analysis in the present study (Wei and Cheng, 2009b), the authors view that the factor of 1.18–1.2 is a slightly better result for the present problem. The critical slip surfaces and the FOSs for different pile spacing are shown in Figures 4.26 through 4.31. In most of the results in Figures 4.26 through 4.31, the differences in strain within the highly stressed regions are small so that the construction of a precise critical slip surface appears to be not meaningful. The small differences in strains at different locations within those highly stressed regions can be attributed to the location of the critical slip surface or discretization errors, which are hard to differentiate. This is a limitation of the SRM, which is less critical when there is no pile in the problem (see Cheng et al., 2007a). The approximate critical slip surface can be visually observed from the shear strain contour, and a precise location of the critical slip surface is not given in these figures (Figure 4.25).

From these figures, it is noticed that the FOS decreases with increase in pile spacing, which is as expected. In Figures 4.26 through 4.31, the critical

Figure 4.25 Slip surface of slope with no pile for Figure 1 (FOS = 1.20).

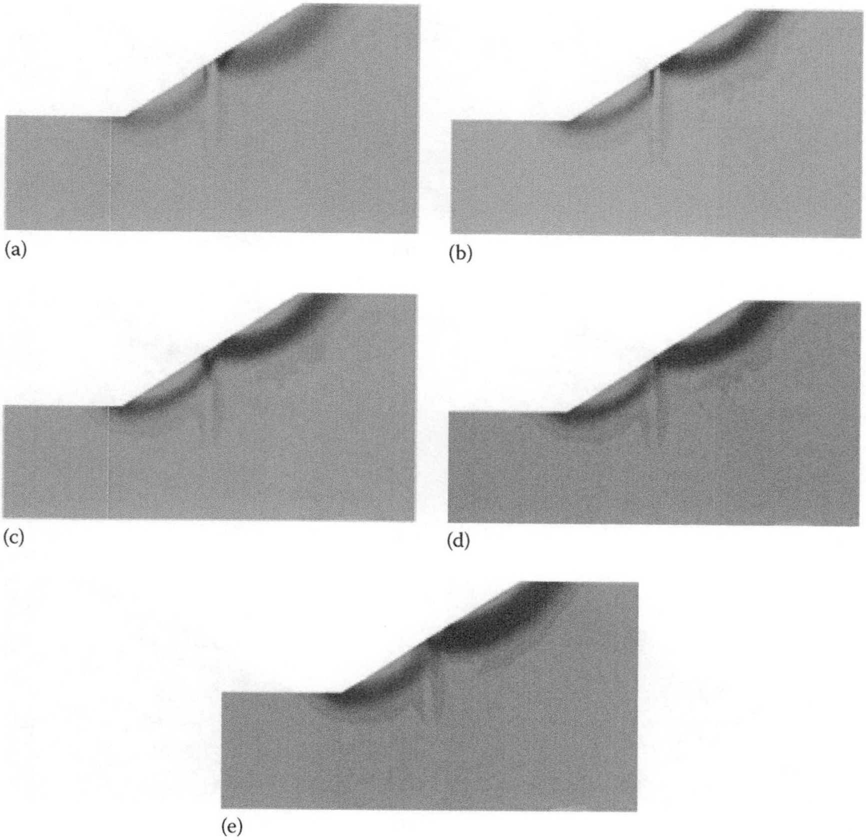

Figure 4.26 3 Slip surface at different sections for $s=2D$ (FOS = 1.78) (note: see the coordinate system in Figure 1 for the definition of y). (a) $y=0$, through pile centre line. (b) $y=0.2$ m. (c) $y=0.4$ m. (d) $y=0.6$ m. (e) $y=0.8$ m, through soil midway between piles.

slip surfaces through different sections are shown (refer to Figure 4.26; section $y=0$ is the section through the pile centre line; section $y=0.4$ m is the plane tangent to the pile outer surface; and for the section through soil midway between piles, y varies from 0.8 to 3.2 m in accordance with the pile spacing, which varies from 2D to 8D). When $s=2D$ (Figure 4.28), the critical slip surface is clearly divided into two parts even at the section through the soil midway between the piles. When the section varies from the section through the pile centre line to the section through the soil midway between the piles, the shear strain in the soil becomes more mobilized. When $s=3D$ (Figure 4.26), the critical slip surface near the pile is still divided clearly into two parts. The critical slip surfaces at the section

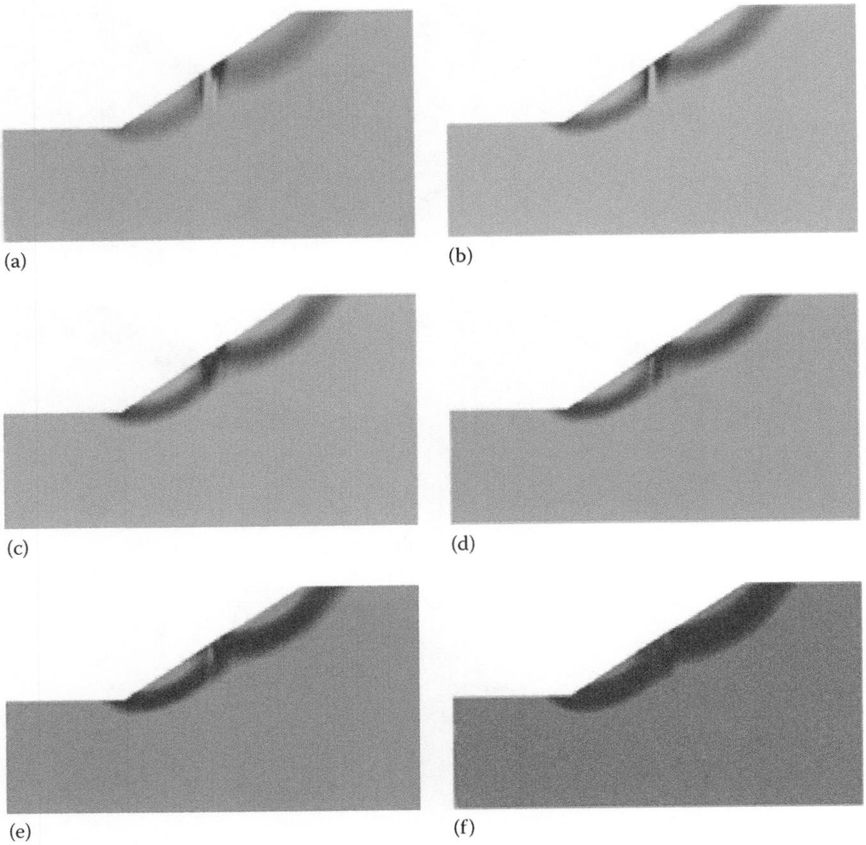

Figure 4.27 Slip surface at different sections for s = 3D (FOS = 1.72). (a) $y = 0$, through pile centre line. (b) $y = 0.2$ m. (c) $y = 0.4$ m. (d) $y = 0.6$ m. (e) $y = 0.8$ m. (f) $y = 1.2$ m, midway through soil.

through the soil midway between the piles are slightly connected, but the overall critical slip surface is still nearly divided into two parts. When pile spacing increases from 4D to 6D (Figures 4.28 through 4.30), the critical slip surface at the sections near the pile is also divided into two parts due to the presence of the piles, but the two parts of the critical slip surfaces become deeper due to shear strain mobilization along the vertical direction at the interface between the pile and the soil. The critical slip surface at the section through the soil midway between the piles becomes more connected and deeper, and at the connection of the two parts of the critical slip surfaces, there is clear shear strain mobilization in the vertical direction. When pile spacing increases to 8D (Figure 4.31), the critical slip surface at the section through the soil midway between the piles is clearly one single

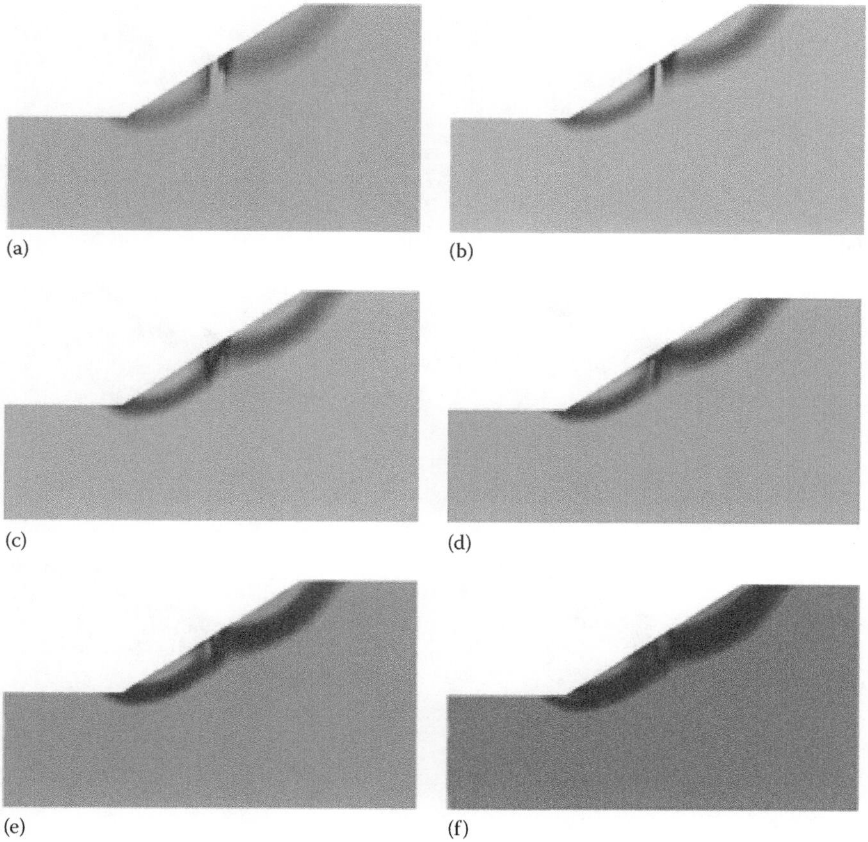

Figure 4.28 Slip surface at different sections for s=4D (FOS=1.61). (a) $y=0$, through pile centre line. (b) $y=0.2$ m. (c) $y=0.4$ m. (d) $y=0.8$ m. (e) $y=1.2$ m. (f) $y=1.6$ m, midway through soil.

critical slip surface, although the critical slip surface at the sections near the pile is still divided into two parts due to the presence of the piles. For a section that is not far from the pile, for example, $y=1.0$ m (Figure 4.31d), clear shear strain mobilization in the vertical direction is also found at the pile location. When the section is far enough from the pile, for example, $y=1.8$ m (Figure 4.31e), the vertical shear strain mobilization at the critical slip surface centre disappears.

The critical slip surfaces at the section through the soil midway between the piles are compared in Figure 4.32 for different pile spacing (the critical slip surface can be seen easily from shear strain mobilization and are also shown by dashed lines). The critical slip surface for the slope with

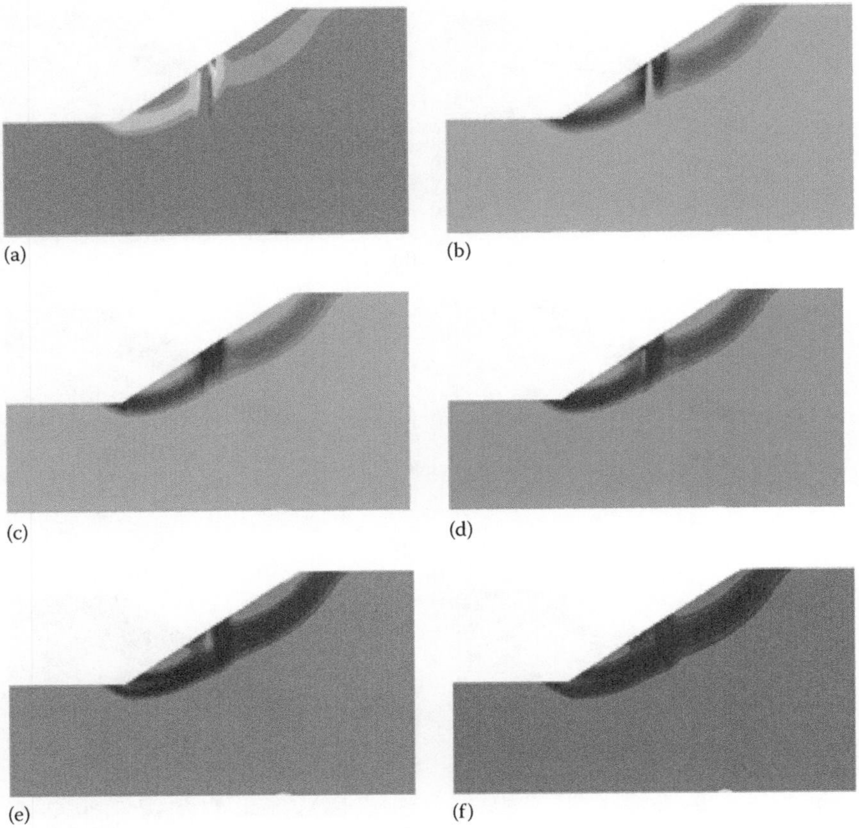

Figure 4.29 Slip surface at different sections for s = 5D (FOS = 1.55). (a) y = 0, through pile centre line. (b) y = 0.2 m. (c) y = 0.4 m. (d) y = 0.8 m. (e) y = 1.4 m. (f) y = 2.0 m, midway through soil.

no pile is shown by the solid line. When pile spacing is small, the critical slip surface is shallow and is nearly divided into two parts. With increase in pile spacing, the critical slip surface becomes deeper and the two parts of the critical slip surface become more connected. When pile spacing is large enough, the two parts of the critical slip surface gradually merge to a clear single critical slip surface, which is nearly the same as the critical slip surface for un-reinforced slopes. This means the critical slip surface of a piled slope is usually shallower than that of a slope with no pile, and this result is totally different from that of Won et al. (2005) and Cai and Ugai (2000) who obtained a very deep critical slip surface for piled slopes based on the maximum shear force in the pile, which is an analogy of the reinforced earth structure failure mechanism based on the FHWA

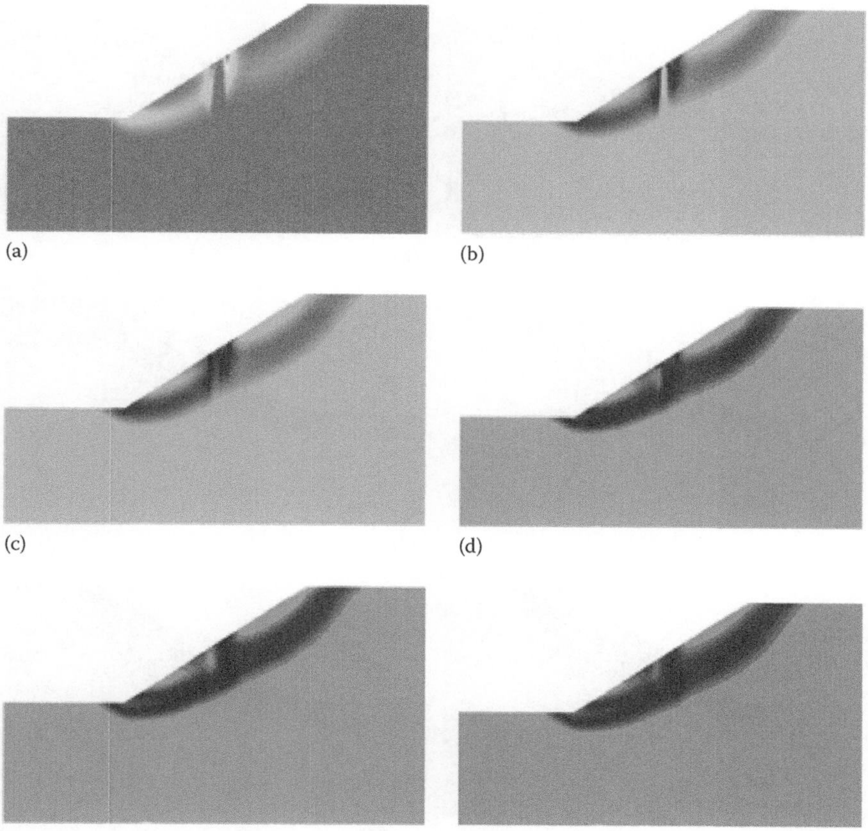

(a) (b) (c) (d) (e) (f)

Figure 4.30 Slip surface at different sections for $s = 6D$ (FOS = 1.52). (a) $y = 0$, through pile centre line. (b) $y = 0.2$ m. (c) $y = 0.4$ m. (d) $y = 1.0$ m. (e) $y = 1.6$ m. (f) $y = 2.4$ m, midway through soil.

recommendation (see Byrne et al., 1996). The failure mode for a soil nailed based on the 'internal failure mode' can be approximated by the location of the maximum nail force as the shear stresses will reverse the sign at this location. There is no such analogy to determine the critical slip surface for a piled slope that is controlled by the shear strength of the pile instead of the pullout resistance of a nail that is closely related to the mobilized shear strength of the soil. The shear force distribution for pile spacing equal to 3D is shown in Figure 4.33. The critical slip surface should always be assessed from the mobilized shear strain of the soil, and in the case of a reinforced earth structure, the mobilization of the nail force and the shear strain of the soil are closely related for a nailed

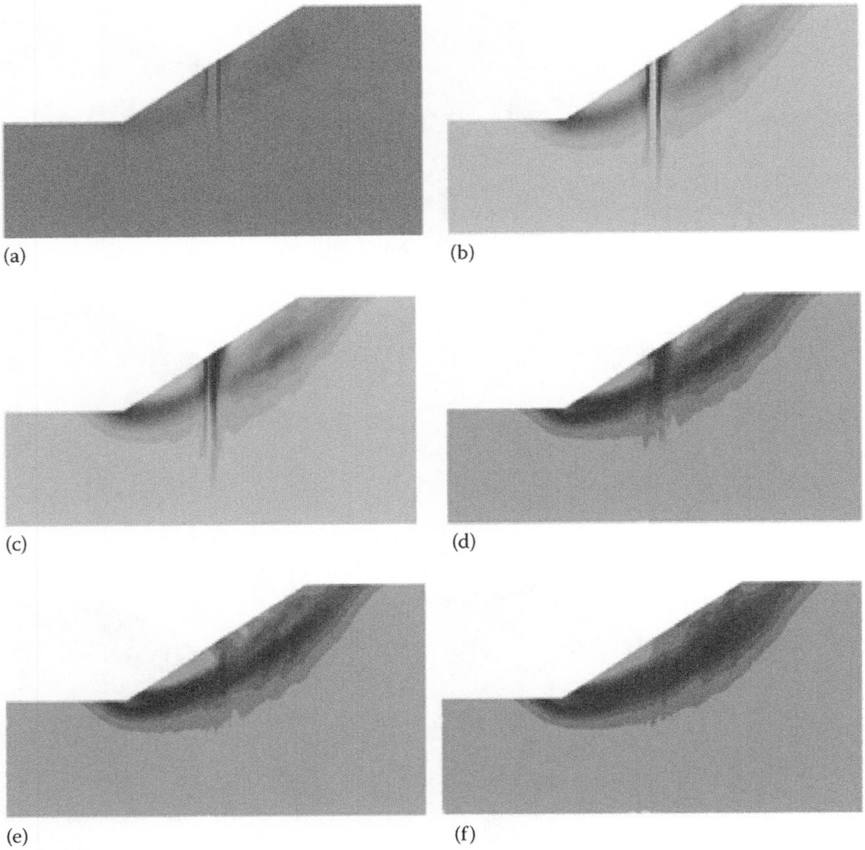

Figure 4.31 Slip surface at different sections for s = 8D (FOS = 1.42). (a) y = 0, through pile centre line. (b) y = 0.2 m. (c) y = 0.4 m. (d) y = 1.0 m. (e) y = 1.8 m. (f) y = 3.2 m, midway through soil.

slope but not for a piled slope. It is clearly seen that the location of the maximum shear force is very deep and is far away from the real critical slip surface; therefore, the maximum shear force location is not necessarily the location of the critical slip surface for a piled slope. It is also observed that there are noticeable differences in the FOS for piled slopes as obtained by the author and those by Cai and Ugai (2000) (although the trend is similar and the differences are not great); the reasons for such differences are probably similar to the previous discussion.

For slopes reinforced with one row of piles, when pile spacing is small, the critical slip surface is shallow and is nearly divided into two parts. With increase in pile spacing, the critical slip surface becomes deeper and the

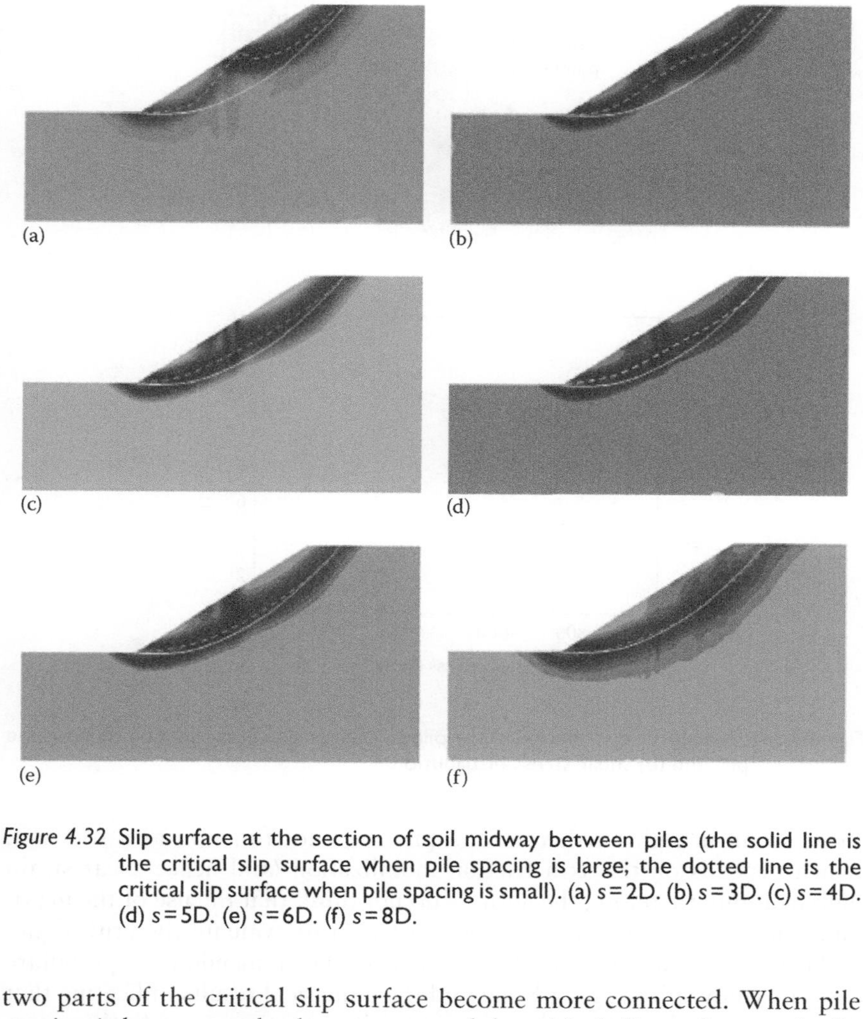

Figure 4.32 Slip surface at the section of soil midway between piles (the solid line is the critical slip surface when pile spacing is large; the dotted line is the critical slip surface when pile spacing is small). (a) s = 2D. (b) s = 3D. (c) s = 4D. (d) s = 5D. (e) s = 6D. (f) s = 8D.

two parts of the critical slip surface become more connected. When pile spacing is large enough, the two parts of the critical slip surface gradually turn into a clear single critical slip surface, which is nearly the same as the critical slip surface with no pile. This means the critical slip surface of a piled slope is usually shallower than that of a slope with no pile. With the placement of a pile, the two smaller slip surfaces will control the slope failure, while the original overall critical slip surface will no longer control the failure as there is obstruction to the failure by the pile. This present result is based on the use of maximum shear strain in soil, and the results are different from that of Cai and Ugai (2000) based on the maximum point of shear force. With reference to Figure 4.33, the location of the maximum shear force in the pile does not correspond to the location of the maximum

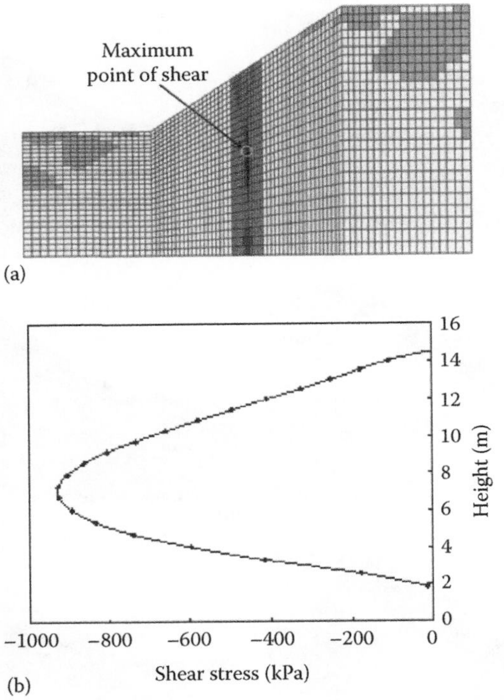

Figure 4.33 Shear force distribution of the piled slope for s = 3D. (a) Section through the pile and (b) shear stress distribution.

shear strain in soil. Furthermore, a pile does not function as a soil nail, which depends on the skin friction mobilization (and hence shear strain mobilization); therefore, the authors do not think that the use of the maximum shear force location is a good criterion to evaluate the critical slip surface of a piled slope problem. Based on the tremendous slope failure cases in sandy soil in Hong Kong and the numerical results indicating that the maximum shear force location in a pile is not necessarily the maximum shear strain location, the authors view that the critical slip surface should be a shallower failure mode for a piled slope. The present study has revealed that the location of the maximum point of shear force is very deep and is far away from the real critical slip surface, as observed from the slope failures in Hong Kong; therefore, the pile maximum shear force location is not necessarily the location of the critical slip surface of a piled slope. The present result is based on the use of normal pile stiffness and has been found to be valid even when pile stiffness is increased or decreased by 100 times. Under extreme conditions, the stiffness of a pile may be important, which requires a detailed three-dimensional analysis.

4.8 DISCUSSION AND CONCLUSION

In this chapter, a number of interesting applications of the SRM were highlighted, which are important for the proper analysis of a slope. While most research has concentrated on the FOS between the LEM and the SRM, the present work has compared the locations of critical failure surfaces by these two methods. In a simple and homogenous soil slope, the differences between the FOS and the locations of critical failure surfaces obtained from the SRM and the LEM are small and both methods are satisfactory for engineering use. It is found that when the cohesion of soil is small, the difference between the FOS obtained from the two methods is greatest for higher friction angles. When the cohesion of soil is large, the difference between FOS is greatest for lower friction angles. With regard to the flow rule, the FOS and the locations of critical failure surfaces are not greatly affected by the choice of the dilation angle (which is important for the adoption of the SRM in slope stability analysis). When an associated flow rule is assumed, the critical slip surfaces from SRM2 appear to be closer to those from the LEM than those from SRM1. The use of the SRM requires the Young's modulus, Poisson's ratio and the flow rule to be defined. The importance of the flow rule has been discussed in the previous section. Cheng et al. (2007a) have also tried different combinations of Young's modulus and Poisson's ratio and have found these two parameters to be insensitive to the results of analysis.

For the SRM, the effects of dilation angle, tolerance for non-linear equation analysis, soil moduli and domain size (boundary effects) are usually small but still noticeable. In most cases, these factors cause differences of just a few percent and are not critical for engineering use of the SRM. Since use of different LEMs will also produce differences of several percent in the FOS, the LEM and the SRM can be viewed as similar in performance for normal cases.

Drastically different results are obtained from different computer programmes for the problem with a soft band. For this special case, the FOS is very sensitive to the size of the elements, the tolerance of the analysis and the number of iterations allowed. It is strongly suggested that the LEM be used to check the results from the SRM. This is because the SRM is highly sensitive to the non-linear solution algorithms and the flow rule for this special type of problem. The SRM has to be used with great care for problems with a soft band of this nature.

The two examples with local minima for the LEM illustrate another limitation of the SRM in engineering use. With the SRM, there is strain localization during the solution and the formation of local minima is unlikely. In the LEM, the presence of local minima is a common phenomenon, and this is a major difference between the two methods. Thus, it is suggested that the LEM should be performed in conjunction with the SRM as a routine check.

Through the present study, two major limitations of the SRM have been established: (1) it is sensitive to non-linear solution algorithms/flow rule for some special cases and (2) it is unable to determine other failure surfaces that may be only slightly less critical than the SRM solution, but still require treatment for good engineering practice. If the SRM is used for routine analysis and design of slope stabilization measures, these two major limitations have to be overcome and it is suggested that the LEM should be carried out as a cross reference. If there are great differences between the results from the SRM and the LEM, great care and engineering judgment should be exercised in assessing a proper solution. There is one practical problem in applying the SRM to a slope with a soft band. When the soft band is very thin, the number of elements required to achieve a good solution is extremely large so that very significant computer memory and time would be required. Cheng (2003) has tried a slope with a 1 mm soft band and has effectively obtained the global minimum FOS by the simulated annealing method. If the SRM is used for a problem with a 1 mm thick soft band, it would be extremely difficult to define a mesh with a good aspect ratio unless the number of elements is huge. For the SRM with a 500 mm thick soft band, about 1 h of CPU time for a small problem (several thousand elements) and several hours for a large problem (over 10,000 elements) would be required for the Phase programme, whereas the programme Flac3D would require 1–3 days (for small to large meshes). If a problem with a 1 mm thick soft band is to be modelled with the SRM, the computer time and memory required will be huge and the method would not be applicable for this special case. The LEM is perhaps better than the SRM for these cases.

For the SRM, there are further limitations that are worth addressing. Shukha and Baker (2003) have found that there are minor but noticeable differences in the FOSs from Flac using square elements and distorted elements. The use of distorted elements is however unavoidable in many cases. Furthermore, when both the soil parameters c' and ϕ' are very small, it is well known that there are numerical problems with the SRM. The failure surface in this case will be deep and wide, and a large domain would be required for analysis. It has been found that the solution time is extremely long and a well-defined critical failure surface is not well established from the SRM. For the LEM, there is no major difficulty in estimating a FOS and the critical failure surface under these circumstances.

The advantage of the SRM is the automatic location of the critical failure surface without the need for a trial-and-error search. With the use of modern global optimization techniques, the location of critical failure surfaces by a simulated annealing method, a genetic algorithm or other methods as discussed in Chapter 3, is now possible, and a trial-and-error search with the LEM is no longer required. While the LEM suffers from the limitation of an inter-slice shear force assumption, the SRM requires a flow rule and suffers from being sensitive to non-linear solution algorithms/flow rule for some special cases. Zheng et al. (2005, 2006, 2008) have also suggested precaution in adopting a suitable value of Poisson's ratio in SRM analysis, which is also used by some commercial programmes.

Griffith and Lane (1999) have suggested that a non-associated flow rule should be adopted for slope stability analysis. As the effect of flow rule on the SRM is not negligible in some cases, such as those involving a soft band, the flow rule is indeed an issue for proper slope stability analysis. It can be concluded that both the LEM and the SRM have their own merits and limitations, and use of the SRM is not really superior to the use of the LEM in routine analysis and design. Both methods should be viewed as providing an estimation of the FOS and the probable failure mechanism, but engineers should also appreciate the limitations of each method when assessing the results of their analyses.

While the 2D SRM is available in several commercial programmes, there still are various difficulties with the 3D SRM; the authors have tested two commercial programmes. For simple and normal problems, there are no major issues with the 3D SRM, and the results are also close to those by the 3D LEM. There are however various difficulties with the 3D SRM for complicated non-homogeneous problems with contrasting soil parameters. More importantly, many strange results may appear when soil nails are present, and there is lack of a good termination criterion for the determination of the FOS under this case. The authors have also found that reliance on the default settings for 3D SRM programmes may not be adequate for many cases, and there is a lack of a clear and robust method for the FOS determination when a soil nail is present. The authors are still working on this issue from various aspects, and in general, it is the authors' view that the 3D SRM is far from being mature for ordinary engineering use.

Chapter 5

Three-dimensional slope stability analysis

5.1 LIMITATIONS OF THE CLASSICAL THREE-DIMENSIONAL LIMIT EQUILIBRIUM METHODS

All slope failures are three-dimensional (3D) in nature, but two-dimensional (2D) modelling is usually adopted as this greatly simplifies the analysis. At present, there are many drawbacks in most of the existing 3D slope stability methods, which include the following:

1. Direction of slide is not considered in most existing slope stability formulations so that the problems under consideration are symmetrical in geometry and loading.
2. Location of the critical non-spherical 3D failure surface under general conditions is a difficult N–P hard type of global optimization problem, which has not been solved effectively and efficiently.
3. Existing methods of analyses are numerically unstable under transverse horizontal forces.

Because of the limitations mentioned earlier, 3D analysis based on limit equilibrium is seldom adopted in practice. Cavounidis (1987) has demonstrated that the factor of safety (FOS) for a normal slope under 3D analysis is greater than that under 2D analysis, and this can be important for some cases.

Baligh and Azzouz (1975) and Azzouz and Baligh (1983) presented a method that extended the concepts of the 2D circular arc shear failure method to 3D slope stability problems. The method was appropriate for a slope in cohesive soil. The results obtained showed that 3D effects could lead to 4%–40% increase in the FOS. Hovland (1977) proposed a general 3D method for cohesive-frictional soils. The method was an extension of the 2D ordinary method of slices (Fellenius, 1927). The intercolumn forces and pore water pressure were not taken into consideration. Two special cases have been analysed: (1) a cone-shaped slip surface on a vertical slope and (2) a wedge-shaped slip surface. It was shown that the 3D FOS were generally higher than for the 2D ones, and the ratio of FOS in 3D to that in 2D was quite sensitive to the magnitudes of cohesion and friction angles and to the shape of the slip surface in 3D.

Chen and Chameau (1982) extended the Spencer 2D method to 3D. The sliding mass was assumed to be symmetrical and divided into several vertical columns. The intercolumn forces had the same inclination throughout the mass, and the shear forces were parallel to the base of the column. It was shown that: (1) the configuration of a sliding mass in 3D had significant effects on the FOS when the length of the sliding mass was small; (2) for gentle slopes, the dimensional effects were significant for soils with high cohesion and low friction angles; (3) in certain circumstances the 3D FOS for cohesionless soils may be slightly less than the 2D one.

Hungr (1987) directly extended the Bishop simplified 2D method of slices (1955) to analyse the slope stability in 3D. The method was derived based on two key assumptions: (1) the vertical forces of each column were neglected and (2) both the lateral and the longitudinal horizontal force equilibrium conditions were neglected. Hungr et al. (1989) presented a comparison of 3D Bishop and Janbu simplified methods (1957) with other published limit equilibrium solutions. It was concluded that the Bishop simplified method might be conservative for some slopes with non-rotational and asymmetric slip surfaces. The method appeared reasonably accurate in the important class of problems involving composite surfaces with weak basal planes.

Zhang (1988) proposed a simple and practical method of 3D stability analysis for concave slopes in plane view using equilibrium concepts. The sliding mass was symmetrical and divided into many vertical columns. The slip surface was approximately considered as the surface of an elliptic revolution. To render the problem statically determinate, the forces acting on the sides and ends of each column which was perpendicular to the potential direction of movement of sliding mass were neglected in the equilibrium conditions. The investigations using the method showed that: (1) the stability of concave slopes in the plane view increased with the decreases in their relative curvature; (2) the effect of a plane curvature on the stability of concave slopes in the plane view increased with the increase in the lateral pressure coefficient. However, the lateral pressure coefficient had only a small effect on the stability of the straight plane.

By using the method of columns, Lam and Fredlund (1993) extended the 2D general limit equilibrium formulation (Fredlund and Krahn, 1977) to analyse a 3D slope stability problem. The intercolumn force functions of an arbitrary shape to simulate various directions for the intercolumn resultant forces were proposed. All the intercolumn shear forces acting on the various faces of the column were assumed to be related to their respective normal forces by the intercolumn force functions. A geostatistical procedure (i.e., the Kriging technique) was used to model the geometry of a slope, the stratigraphy, the potential slip surface and the pore water pressure conditions. It was found that the 3D FOSs determined by the method (Lam and Fredlund, 1993) were relatively insensitive to the form of the intercolumn force functions used. Lam and Fredlund (1993), however, have not given a clear and systematic way for solving a general 3D problem.

Chang (2002) developed a 3D method for analysing the slope stability based on the sliding mechanism observed in the 1988 failure of the Kettleman Hills Landfill slope and the associated model studies. Using a limit equilibrium concept, the method assumed the sliding mass to be a block system in which the contacts between blocks were inclined. The lines of intersection of the block contacts were assumed to be parallel, which enabled the sliding kinematics. In consideration of the differential straining between blocks, the shear stresses on the slip surface and the block contacts were evaluated based on the degree of shear strength mobilization on those contacts. The overall FOS was calculated based on the force equilibrium of the individual block and the entire block system as well. Due to the assumed interblock boundary pattern, the method was not fully applicable for dense sands or overly consolidated materials under drained conditions. Zheng (2012) has developed a mesh-/column-free method where the stability of the global mass is considered without the use of intercolumn force functions. Under such formulation, the internal acceptability of the system is not checked. This method gives FOSs similar to those obtained from other methods which consider the internal force distribution for a normal problem with smooth slip surface, and this is a reflection that the internal forces will not have a great effect on the analysis for smooth slip surfaces. In the method developed by Zheng (2012), the thrust line is outside the soil mass, and the Mohr–Coulomb criterion can be violated within the soil mass; this method appears to be applicable for normal slip surfaces, while the results may need to be analysed further for irregular slip surfaces.

In addition, 3D stability formulations based on the limit equilibrium method and variational calculus have been proposed by Leshchinsky et al. (1985), Ugai (1985) and Leshchinsky and Baker (1986). The functionals are the force and/or moment equations where the FOS can be minimized while satisfying several other conditions. The shape of a slip surface can be determined analytically. In such approaches, the minimum FOS and the associated failure surface can be obtained at the same time. These methods were, however, limited to homogeneous and symmetrical problems only. In the follow-up studies, Leshchinsky and Huang (1992) developed a generalized approach which is appropriate for symmetrical slope stability problems only. The analytical solutions approach based on variational analysis are difficult to obtain for practical problems with complicated geometric forms and loading conditions. Cheng is currently working on the modern optimization method to replace the tedious variational principle.

Table 5.1 Summary of some 3D limit equilibrium methods

Method	*Related 2D method*	*Assumptions*	*Equilibrium*
Hovland (1977)	Ordinary method of slice	No intercolumn force	Overall moment equilibrium
Chen and Chameau (1982)	Spencer method	Constant inclination	Overall moment equilibrium Overall force equilibrium
Hungr (1987)	Bishop simplified method	Vertical equilibrium	Overall moment equilibrium Vertical force equilibrium
Lam and Fredlund (1993)	General limit equilibrium	Intercolumn force function	Overall moment equilibrium Overall force equilibrium
Huang and Tsai (2000)	Bishop simplified method	Consider direction of slide	Overall moment equilibrium Vertical force equilibrium
Cheng and Yip (2007)	Bishop and Janbu simplified, MP	Consider direction of slide	Overall and local force equilibrium and overall moment equilibrium for MP

Most of the existing 3D methods rely on an assumption of a plane of symmetry in the analysis, and the commonly used methods are summarized in Table 5.1. For complicated ground conditions, this assumption is no longer valid, and the failure mass fails along a direction with *least resistance* so that the sliding direction also controls the FOS of a slope. Stark and Eid (1998) have also demonstrated that the FOS of a 3D slope is controlled by the direction of slide and a symmetric failure may not be suitable for a general slope. Yamagami and Jiang (1996, 1997) and Jiang and Yamagami (1999) have developed the first method for asymmetric problems where the classical stability equations (without direction of slide/direction of slide is zero) are used and the direction of slide is considered by minimizing the FOS with respect to the rotation of axes. The Yamagami and Jiang formulation (1996, 1997) can be very time consuming even for a single failure surface as the formation of columns and the determination of geometry information with respect to the rotation of axes is the most time-consuming computation in stability analysis. Huang and Tsai (2000) have proposed the first 3D asymmetrical Bishop method where sliding direction is taken into consideration for determining the safety factor. The generalized 3D slope stability method by Huang et al. (2002) is practically equivalent to the Janbu rigorous method with some simplifications on the transverse shear forces. It is difficult to completely satisfy the line of thrust constraints in the Janbu rigorous method, which is well known in 2D analysis; the generalized 3D method by Huang et al. (2002) also faces converge problems, making it less useful for practical problems.

At the verge of failure, the soil mass can be considered a rigid body. There can be three possibilities for the direction of slide:

1. Soil columns move in the same direction with a unique sliding direction – adopted by Cheng and Yip (2007) and many other researchers in the present formulation.
2. Soil columns move towards each other – this violates the assumption of rigid failure mass and is not considered.
3. Soil columns move away from each other – adopted by Huang and Tsai (2000) and Huang et al. (2002).

Since the sliding directions of soil columns are not unique in Huang and Tsai's formulation (2000) and Huang et al. (2002) and some columns move apart, the summation process for determining the FOS may not be applicable as some of the columns separate from the others. Cheng and Yip (2007) have demonstrated in a later section that under a transverse load, the requirement of different sliding directions for different soil columns may lead to failure to converge. For soil columns moving away from each other, the distinct element method is the recommended method of analysis and a simple illustration is given in Section 2.9. Since the parameters required for distinct element analysis are different from the classical soil strength parameters, it is not easy to adopt the results from distinct element analysis directly, and the results should be considered as the qualitative analysis of the slope stability problem.

The assumption of a unique sliding direction may be an acceptable formulation for the analysis of the ultimate limit state, and the present formulation is based on this assumption. It is not wrong to assume that all soil columns slide in one unique direction at the verge of failure. After failure is initiated, the soil columns may separate from each other and sliding directions can be different among different columns.

5.2 NEW FORMULATION FOR 3D SLOPE STABILITY ANALYSIS IN BISHOP, JANBU AND MORGENSTERN–PRICE METHODS BY CHENG AND YIP

5.2.1 Basic formulation with consideration of sliding direction

For 3D analysis, the potential failure mass of a slope is divided into a number of columns. At the ultimate equilibrium condition, the internal and external forces acting on each soil column are as shown in Figure 5.1. For the sake of simplicity, the weight of soil and the vertical load are assumed to act at the centre of each column. This assumption is not exact but is

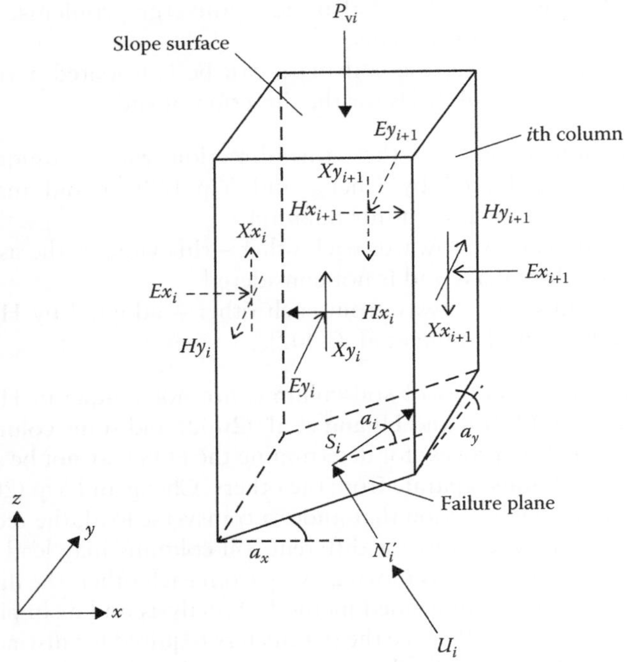

Figure 5.1 External and internal forces acting on a typical soil column.

applicable if the width of each column is small enough, and the resulting equations are highly simplified and sufficiently good for practical purposes. The assumptions required in the present 3D formulation are as follows:

1. The Mohr–Coulomb failure criterion is valid.
2. For the Morgenstern–Price (MP) method, the FOS is determined based on the sliding angle where FOS with respect to force and moment are equal.
3. The sliding angle is the same for all soil columns (Figure 5.2).

By the Mohr–Coulomb criterion, the global FOS, F, is defined as

$$F = \frac{S_{fi}}{S_i} = \frac{C_i + N_i' \cdot \tan\phi_i}{S_i} \tag{5.1}$$

where

F is the FOS
S_{fi} is the ultimate resultant shear force available at the base of column i
N_i' is the effective base normal force
C_i is $c'A_i$ and A_i is the base area of the column

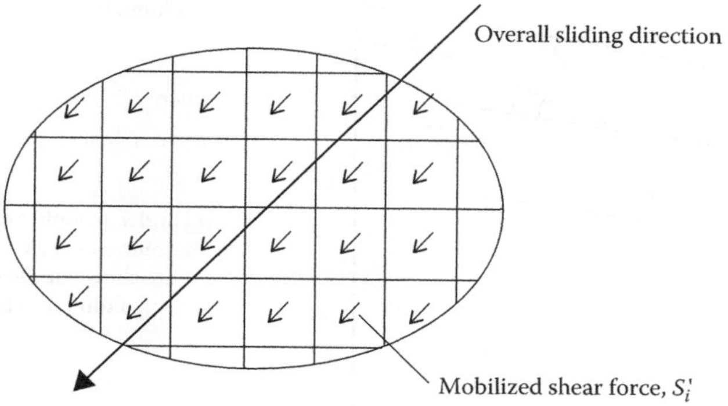

Figure 5.2 Unique sliding direction for all columns (on plan view).

The base shear force S and base normal force N with respect to x, y and z directions for column i are expressed as the components of forces by Huang and Tsai (2000) and Huang et al. (2002):

$$S_{xi} = f_1 \cdot S_i; \quad S_{yi} = f_2 \cdot S_i; \quad \text{and} \quad S_{zi} = f_3 \cdot S_i$$

$$N_{xi} = g_1 \cdot N_i; \quad N_{yi} = g_2 \cdot N_i; \quad \text{and} \quad N_{zi} = g_3 \cdot N_i \tag{5.2}$$

in which $\{f_1 \cdot f_2 \cdot f_3\}$ and $\{g_1 \cdot g_2 \cdot g_3\}$ are unit vectors for S_i and N_i (see Figure 5.1). The projected shear angle a' (individual sliding direction) is the same for all the columns in the x–y plane in the present formulation, and by using this angle, the space shear angle a_i (see Figure 5.3) can be found for each column and is given by Huang and Tsai (2000) as Equation 5.3:

$$a_i = \tan^{-1}\left\{\frac{\sin\theta_i}{\left[\cos\theta_i + \left(\cos a_{yi}/\tan a' \cdot \cos a_{xi}\right)\right]}\right\} \tag{5.3}$$

where

a_i is the space sliding angle for the sliding direction with respect to the direction of slide projected to the x–y plane (see also a' in Figure 5.2 and Equation 5.3)

a_x a_y base inclination along the x and y directions measure at centre of each column (shown at the edge of the column for clarity)

Ex_i and Ey_i are the intercolumn normal forces in the x and y directions, respectively

Hx_i and Hy_i are the lateral intercolumn shear forces in the x and y directions, respectively

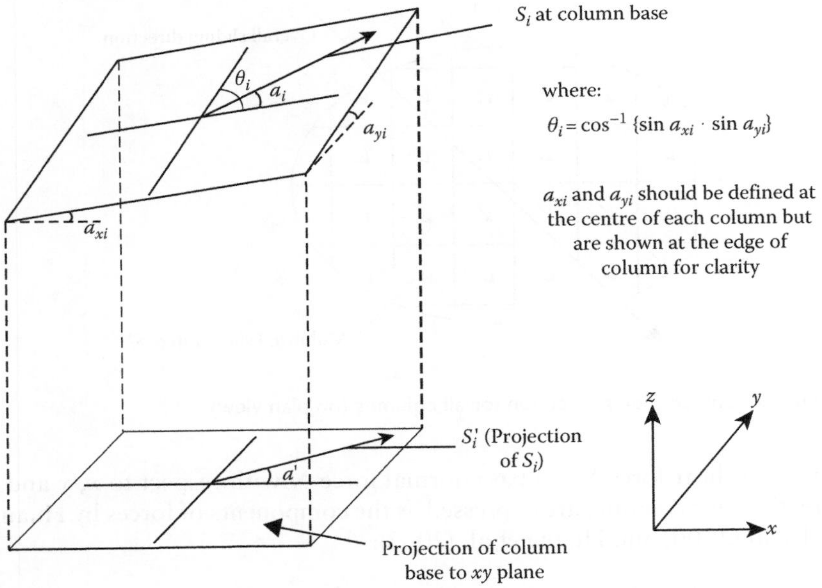

Figure 5.3 Relationship between projected and space shear angle for the base of column *i*.

N'_i and U_i are the effective normal forces and base pore water force, respectively
P_{vi} and S_i are the vertical external forces and base mobilized shear force, respectively
Xx_i and Xy_i are the vertical intercolumn shear forces in a plane perpendicular to the x and y directions

$$n_i = \left\{ \frac{\pm\tan a_{xi}}{J}, \frac{\pm\tan a_{yi}}{J}, \frac{1}{J} \right\} = \{g_1, g_2, g_3\}$$

[–ve adopted by Huang and Tsai (2000) and +ve adopted by Cheng and Yip (2007)]

$$s_i = \left\{ \frac{\sin(\theta_i - a_i)\cdot\cos a_{xi}}{\sin\theta_i}, \frac{\sin a_i \cdot \cos a_{yi}}{\sin\theta_i}, \frac{\sin(\theta_i - a_i)\cdot\sin a_{xi} + \sin a_i \cdot \sin a_{yi}}{\sin\theta_i} \right\}$$

$$= \{f_1, f_2, f_3\}$$

in which

$$J = \sqrt{\tan^2 a_{xi} + \tan^2 a_{yi} + 1}$$

An arbitrary intercolumn shear force function $f(x, y)$ is assumed in the present analysis, and the relationships between the intercolumn shear and normal forces in the x and y directions are given as

$$Xx_i = Ex_i \cdot f(x,y) \cdot \lambda_x; \quad Xy_i = Ey_i \cdot f(x,y) \cdot \lambda_y \tag{5.4}$$

$$Hx_i = Ey_i \cdot f(x,y) \cdot \lambda_{xy}; \quad Hy_i = Ex_i \cdot f(x,y) \cdot \lambda_{yx} \tag{5.5}$$

where

λ_x and λ_y are the intercolumn shear force X mobilization factors in the x and y directions, respectively

λ_{xy} and λ_{yx} are the intercolumn shear force H mobilization factors in the xy and yx planes, respectively

Taking the moment about the z axis at the centre of the ith column, the relations between lateral intercolumn shear forces can be expressed as

$$\Delta y_i \cdot \left(Hx_{i+1} + Hx_i\right) = \Delta x_i \cdot \left(Hy_{i+1} + Hy_i\right) \tag{5.6}$$

From Equation 5.6,

$$Hx_{i+1} = \frac{\Delta x_i}{\Delta y_i} \cdot \left(Hy_{i+1} + Hy_i\right) - Hx_i \tag{5.7}$$

From Equation 5.6,

$$Hy_{i+1} = \frac{\Delta y_i}{\Delta x_i} \cdot \left(Hx_{i+1} + Hx_i\right) - Hy_i \tag{5.8}$$

where Δx_i and Δy_i are the widths of the column defined in Figure 5.4. Hx_i and Hy_i for the exterior columns should be zero in most cases or equal to the applied horizontal forces if defined. By using the property of complementary shear (or moment equilibrium in the xy plane), Hy_{i+1} or Hx_{i+1} can be determined from Equations 5.5 and 5.7 or 5.8; thus only λ_{xy} or λ_{yx} needs to be determined but not both. The important concept of complementary shear force, which is similar to the complementary shear stress ($\tau_{xy} = \tau_{yx}$) in elasticity, has not been used in any 3D slope stability analysis method in the past but is crucial in the present formulation. It should be noted that Huang et al. (2002) assumed Hy_i to be 0 for asymmetric problems in order to render the problem determinate, which is valid for symmetric failure only. Although the concept of complementary shear is applicable only in an infinitesimal sense, if the size of the column is not great, this assumption greatly simplifies the equations. More importantly, as discussed in the later sections of this chapter, Cheng and Yip (2007) have demonstrated that the effect of λ_{xy} or λ_{yx} is small and the error in this assumption is actually not important.

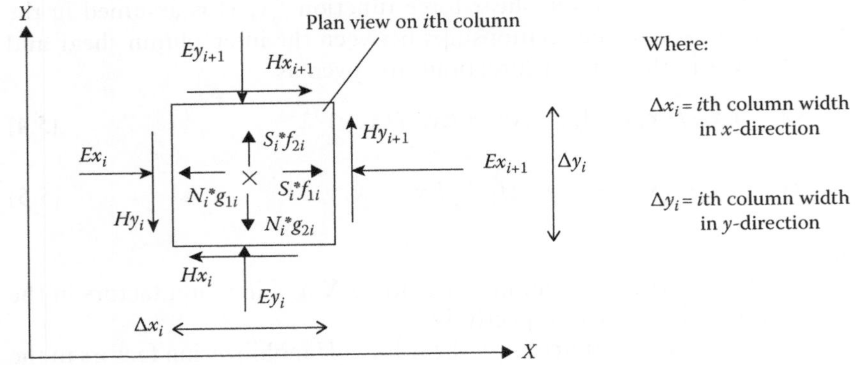

Figure 5.4 Force equilibrium in the *xy* plane.

Force equilibrium in x, y and z directions: Considering the vertical and horizontal forces equilibrium for the *i*th column (Figures 5.5 and 5.6) in the *z*, *x* and *y* directions gives:

$$\Sigma F_z = 0 \rightarrow N_i \cdot g_{3i} + S_i \cdot f_{3i} - \left(W_i + P_{vi}\right) = \left(Xx_{i+1} - Xx_i\right) + \left(Xy_{i+1} - Xy_i\right) \tag{5.9}$$

$$\Sigma F_x = 0 \rightarrow S_i \cdot f_{1i} - N_i \cdot g_{1i} + P_{hxi} - Hx_i + Hx_{i+1} = Ex_{i+1} - Ex_i \tag{5.10}$$

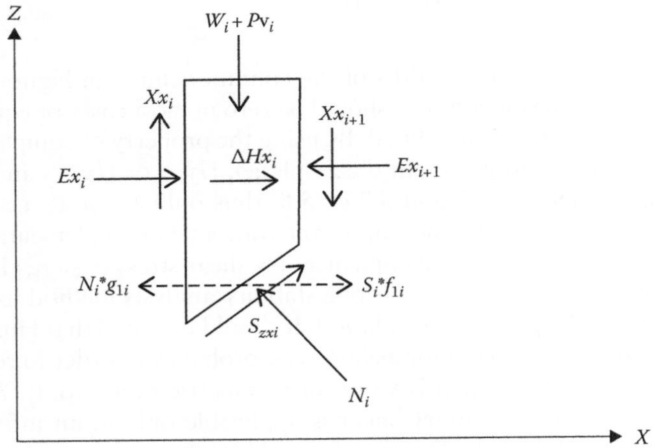

Figure 5.5 Horizontal force equilibrium in the *x* direction for a typical column (ΔHx_i = net lateral intercolumn shear force).

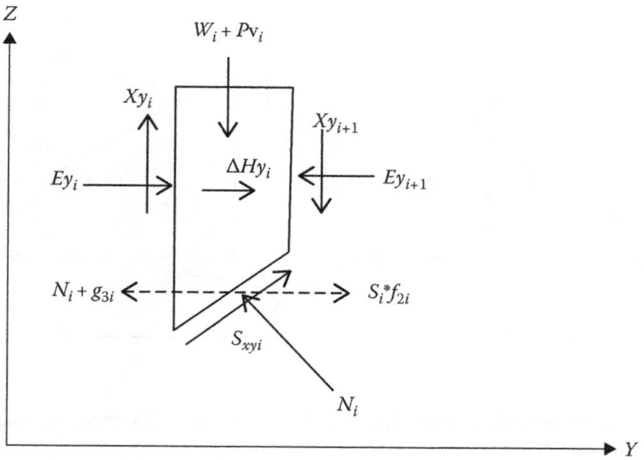

Figure 5.6 Horizontal force equilibrium in the y direction for a typical column (ΔHy_i = net lateral intercolumn shear force).

$$\Sigma F_y = 0 \rightarrow S_i \cdot f_{2i} - N_i \cdot g_{2i} + P_{hyi} - Hy_i + Hy_{i+1} = Ey_{i+1} - Ey_i \tag{5.11}$$

Solving Equations 5.1, 5.4 and 5.9, the base normal and shear forces can be expressed as

$$N_i = A_i + B_i \cdot S_i; \quad S_i = \frac{C_i + (A_i - U_i) \cdot \tan\phi_i}{F\left(1 - \frac{B_i \cdot \tan\phi_i}{F}\right)} \tag{5.12}$$

$$A_i = \frac{W_i + P_{vi} + \Delta Ex_i \cdot \lambda_x + \Delta Ey_i \cdot \lambda_y}{g_{3i}}; \quad B_i = -\frac{f_{3i}}{g_{3i}};$$

$$U_i = u_i A_i \left(u_i = \text{average pore pressure at } i\text{th column}\right)$$

Overall force and moment equilibrium in the x and y directions: Considering the overall force equilibrium in the x direction:

$$-\sum Hx_i + \sum N_i \cdot g_{1i} - \sum S_i \cdot f_{1i} = 0 \tag{5.13}$$

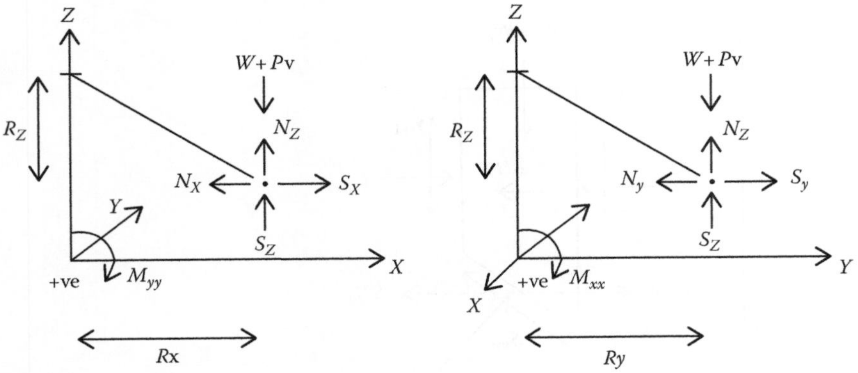

Figure 5.7 Moment equilibrium in the *x* and *y* directions. (Earthquake loads and net external moments are not shown for clarity.)

Let $F_x = F$ in Equation 5.1; using Equation 5.12 and rearranging Equation 5.13, the directional safety factor F_x can be determined as:

$$F_x = \frac{\Sigma\left[\left(N_i - U_i\right)\cdot \tan\phi_i + C_i\right]\cdot f_{1i}}{\Sigma N_i \cdot g_{1i} - \Sigma Hx_i}, \quad 0 < F_x < \infty \tag{5.14}$$

From the overall moment equilibrium in the *x* direction (Figure 5.7):

$$\sum\left(W_i + P_{vi} - N_i \cdot g_{3i} - S_i \cdot f_{3i}\right)\cdot RX + \sum\left(N_i \cdot g_{1i} - S_i \cdot f_{1i}\right)\cdot RZ = 0 \tag{5.15}$$

RX, RY and RZ are the lever arms to the moment point. Similarly, considering the overall force equilibrium in the *y* direction:

$$-\sum Hy_i + \sum N_i \cdot g_{2i} - \sum S_i \cdot f_{2i} = 0 \tag{5.16}$$

Let $F_y = F$ in Equation 5.1; using Equation 5.12 and rearranging Equation 5.16, the directional safety factor (F_y) can be determined as

$$F_y = \frac{\Sigma\left[\left(N_i - U_i\right)\cdot \tan\phi_i + C_i\right]\cdot f_{2i}}{\Sigma N_i \cdot g_{2i} - \Sigma Hy_i}, \quad 0 < F_y < \infty \tag{5.17}$$

The overall moment equilibrium in the *y* direction (Figure 5.7) is as follows:

$$\sum\left(W_i + P_{vi} - N_i \cdot g_{3i} - S_i \cdot f_{3i}\right)\cdot RY + \sum\left(N_i + g_{2i} - S_i \cdot f_{2i}\right)\cdot RZ = 0 \tag{5.18}$$

Based on a trial sliding angle, λ_x is changed at specified intervals in Equation 5.14, until the calculated F_x satisfies the overall moment equilibrium Equation 5.15 in the x direction. A similar procedure is applied to λ_y until the calculated F_y also satisfies the overall moment equilibrium equation (5.18) in the y direction. If F_x is not equal to F_y, the sliding angle is varied until $F_x = F_y$ and then force as well as moment equilibrium is achieved. Since all the equilibrium equations have been used in the formulation, there is no equation to determine λ_{xy} unless additional assumptions are specified. In the present formulation, Cheng and Yip (2007) suggest that λ_{xy} can be specified by the user or can be determined from the minimization of the FOS with respect to λ_{xy}. The problem associated with λ_{xy} and the importance of this parameter will be further discussed in Section 5.2.8.

5.2.2 Reduction to 3D Bishop and Janbu simplified methods

Solving the 3D asymmetric MP method takes a relatively long time, and the convergence is less satisfactory as compared with the simplified method. The initial solutions from 3D Janbu or Bishop analysis can be adopted to accelerate the MP solution, and many engineers may still prefer the simplified method for routine design. The proposed MP formulation can be simplified by considering only the force or moment equilibrium equations and neglecting all the intercolumn vertical and horizontal shear forces. Consider the *overall moment equilibrium* in the x direction and about an axis passing through (x_0, y_0, z_0) (centre of rotation of the spherical failure surface) and parallel to the y axis. Let $F_{my} = F$ in Equation 5.1; then rearranging Equation 5.15 gives

$$F_{my} = \frac{\sum \{K_{yi} \cdot [f_{1i} RZ_i + f_{3i} RX_i]\}}{\sum (W_i + P_{vi}) \cdot RX_i + \sum N_i \cdot (g_{1i} \cdot RZ_i - g_{3i} \cdot RX_i)} \tag{5.19}$$

The corresponding F_{mx} is obtained from Equation 5.18 as

$$F_{mx} = \frac{\sum \{K_{xi} \cdot [f_{2i} RZ_i + f_{3i} RY_i]\}}{\sum (W_i + P_{vi}) \cdot RY_i + \sum N_i \cdot (g_{2i} \cdot RZ_i - g_{3i} \cdot RY_i)} \tag{5.20}$$

in which

$$K_{yi} = \frac{\left\{C_i + \left[\dfrac{(W_i + P_{vi})}{g_{3i}} - U_i\right] \tan\phi_i\right\}}{1 + \dfrac{f_{3i} \cdot \tan\phi_i}{g_{3i} \cdot F_{my}}}; \quad K_{xi} = \frac{\left\{C_i + \left[\dfrac{(W_i + P_{vi})}{g_{3i}} - U_i\right] \tan\phi_i\right\}}{1 + \dfrac{f_{3i} \cdot \tan\phi_i}{g_{3i} \cdot F_{mx}}}$$

Considering the *overall moment equilibrium* about an axis passing through (x_0, y_0, z_0) and parallel to the z axis gives

$$\sum(-N_i \cdot g_{1i} + S_i \cdot f_{1i}) \cdot RY + \sum(N_i \cdot g_{2i} - S_i \cdot f_{2i}) \cdot RX = 0 \tag{5.21}$$

Let $F_{mz} = F$ in Equation 5.1; then rearranging Equation 5.21 gives

$$F_{mz} = \frac{\sum\left[K_{zi} \cdot (f_{2i} \cdot RX_i - f_{3i} \cdot RY)\right]}{\sum N(g_{2i} \cdot RX_i - g_{1i} \cdot RY)}; \quad K_{zi} = \frac{\left\{C_i + \left[\dfrac{(W_i + P_{vi})}{g_{3i}} - U_i\right]\tan\phi_i\right\}}{1 + \dfrac{f_{3i} \cdot \tan\phi_i}{g_{3i} \cdot F_{mz}}} \tag{5.22}$$

For the 3D asymmetric Bishop method, at moment equilibrium point, the directional FOSs, F_{mx}, F_{my} and F_{mz}, are equal to each other. Under this condition, the global FOS, F_m, based on moment can be determined as:

$$F_m = F_{mx} = F_{my} = F_{mz} \tag{5.23}$$

The sliding direction can be found by changing the *projected shear direction* at a *specified angular interval*, until F_{mx}, F_{my} and F_{mz} are equal to each other. In reality, there is no way to ensure complete 3D moment equilibrium in the Bishop method as Equation 5.21 is redundant and is not used in the present method or the method by Huang and Tsai (2000) and Huang et al. (2002) as Equations 5.19 and 5.20 are sufficient for solving the FOS. The left-hand side (LHS) of Equation 5.21 can hence be viewed as an *unbalanced moment term*. For a completely symmetric slope, this term is exactly zero and the 3D moment equilibrium is automatically achieved. In general, this term is usually small if the asymmetrical loading or sliding direction is not great. Cheng and Yip (2007) therefore adopt Equations 5.19 and 5.20 in the formulation, which is equivalent to assigning $F_{mx} = F_{my}$. This is a limitation of the present 3D asymmetric Bishop simplified method as well as all the other existing 3D Bishop methods for general asymmetric problems as Equation 5.21 is a redundant equation.

By neglecting the intercolumn shear forces for Janbu analysis, Equations 5.14 and 5.17 simplify to

$$F_{sx} = \frac{\sum A_{xi}\left(f_{1i} + \dfrac{f_{3i} \cdot g_{1i}}{g_{3i}}\right)}{\sum \dfrac{g_{1i}}{g_{3i}} \cdot (W_i + P_{vi})}; \quad A_{xi} = \frac{\left\{C_i + \left[\dfrac{(W_i + P_{vi})}{g_{3i}} - U_i\right]\tan\phi_i\right\}}{1 + \dfrac{f_{3i} \cdot \tan\phi_i}{g_{3i} \cdot F_{sx}}} \tag{5.24}$$

$$F_{sy} = \frac{\sum A_{yi}\left(f_{2i} + \frac{f_{3i} \cdot g_{2i}}{g_{3i}}\right)}{\sum \frac{g_{2i}}{g_{3i}} \cdot \left(W_i + P_{vi}\right)}; \quad A_{yi} = \frac{\left\{C_i + \left[\frac{\left(W_i + P_{vi}\right)}{g_{3i}} - U_i\right]\tan\phi_i\right\}}{1 + \frac{f_{3i} \cdot \tan\phi_i}{g_{3i} \cdot F_{sy}}} \tag{5.25}$$

For the 3D asymmetric Janbu method, at force equilibrium point, the directional FOSs, F_{sx} and F_{sy}, are equal to each other. Under this condition, the global FOS, F_f, based on force can be determined as

$$F_f = F_{sx} = F_{sy} \tag{5.26}$$

Since the FOS is also used in vertical force equilibrium, 3D force equilibrium is completely achieved in the 3D Janbu simplified method.

5.2.3 Numerical implementation of Bishop, Janbu and MP methods

To determine the FOS, the domain under consideration is divided into a regular grid. An initial value for the projected shear angle a' is chosen for analysis and a_i is then computed from Equation 5.3. In the program SLOPE3D developed by Cheng, an initial value of 2° is chosen for a' and an increment of a' is chosen to be 1° in the analysis. Once a_i is defined, the unit vectors n_i and s_i, as given in Figure 5.1, can be determined. Equations 5.19 and 5.20 are used to compute F_{my} and F_{mx} for the Bishop method until convergence is achieved. For the Janbu method, Equations 5.24 and 5.25 are used to compute F_{sx} and F_{sy} until convergence is achieved. If $F_{my} \neq F_{mx}$ or $F_{sx} \neq F_{sy}$, a' increases by 1° for the next loop. From the difference between two consecutive directional safety factors during the sliding angle determination, the bound between a' can then be determined. Suppose a' is bound between 10° and 11°, the directional safety factors are computed again for a' based on 10.5°; a' is then bound within a 0.5° range and a simple interpolation is used to compute a refined value for a'. This formulation is relatively simple to operate and is good enough for analysis.

There are four major parameters to be determined for 3D MP analysis: F, sliding angles, λ_x and λ_y and λ_{xy}, which are prescribed by the engineer or determined from a minimization of the FOS. To accelerate the MP solution, 3D Janbu simplified analysis or Bishop analysis is performed in the first stage. The sliding angle and the intercolumn normal forces Ex_i and Ey_i from the simplified analysis are taken as the trial initial solution for the calculation of the intercolumn shear forces Xx_i and Xy_i in the first step.

To solve for the FOS, Cheng and Yip (2007) have tried two methods, which are as follows:

1. Simple triple looping technique
   To solve for a', λ_x and λ_y, a triple looping technique can be adopted; a', λ_x and λ_y are varied sequentially until all the previous equations are satisfied. This formulation is simple to be programmed, but convergence is extremely slow even for a modern CPU.
2. Double looping Brent method
   Cheng and Yip (2007) have found that the non-linear equations solver by Brent (1973) is suitable for the 3D Janbu and Bishop methods. If the directional safety factors from simplified methods are used as the initial values in the Brent method, convergence with the Brent method is good. Using the Brent method, one level of looping is removed and is replaced by the solution for a system of non-linear equations $F_{sx} - F_{sy} = 0$ and $F_{mx} - F_{my} = 0$. Unlike the simplified 3D method of analysis, if arbitrary values of the directional safety factors (other than those from simplified 3D analysis) are used as the initial values in the Brent method, failure to convergence can happen easily. Such behaviour is possibly induced by the effect of intercolumn shear forces on the analysis, which is neglected in the corresponding simplified 3D stability analysis.

5.2.4 Numerical examples and verification

Based on the present formulation and the formulation by Huang and Tsai (2000), Cheng et al. (2005) and Cheng and Yip (2007) have developed a program SLOPE3D, and several examples are used for the study of the proposed formulations. In this chapter, function $f(x, y)$ is taken to be 1.0 for MP analysis, and the method is hence actually Spencer analysis. Cheng and Yip (2007) have tried $f(x, y)$ for limited cases and the results from the use of $f(x) = 1.0$ and $f(x, y) = \sin(x, y)$ are virtually the same, which is similar to that obtained from the corresponding 2D analysis. It is expected that for a highly irregular failure surface, the results may be sensitive to the choice of $f(x, y)$. The first example is a laterally symmetric slope (Figure 5.8) considered by Baligh and Azzouz (1975) with an assumed spherical sliding surface, and the results of analysis are given in Table 5.2. For Example 5.1, SLOPE3D gives a sliding angle exactly equal to 0°, and the results are also very close to those obtained by other researchers except the one obtained by Hungr (1987). As per Cheng and Yip (2007) the result 1.422 obtained by Hungr (1987) is actually not correct as Hungr (1987) obtained this based on the following equation:

$$F = \frac{\sum \left(W_i \tan\phi_i + C_i A_i \cos\gamma_z \right) / m_\alpha}{\sum W_i \sin\alpha_y} \tag{5.27}$$

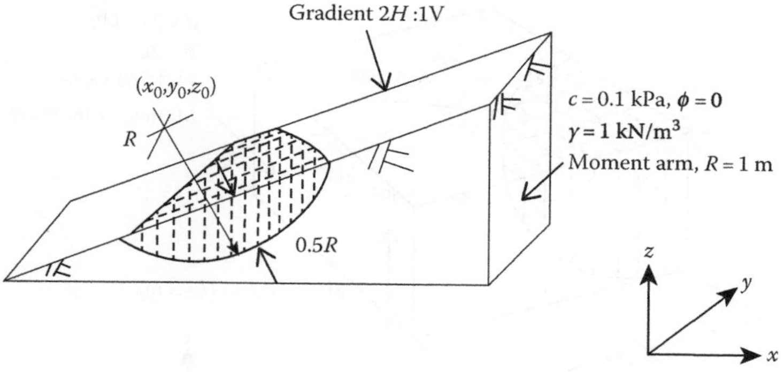

Figure 5.8 Slope geometry for Example 5.1.

Table 5.2 Comparison of F_s for Example 5.1

Method	*Baligh and Azzouz (1975)*	*Hungr et al. (1989)*	*Lam and Fredlund (1993)*	*Huang and Tsai (2000)*	*SLOPE 3D (Huang's approach)*	*SLOPE 3D (Cheng's approach)*
Bishop simplified	1.402	1.422 (1.39)	1.386 (1200c)	1.399 (5300c)	1.390 (8720c)	1.390 (8720c)
Janbu simplified	—	—	—	—	1.612 (8720c)	1.612 (8720c)

Note: Number of columns in the analyses. For Hungr's result, FOS after correction is 1.39.

Equation 5.27 is not correct because the moment equilibrium is considered about the centre of rotation of the spherical failure surface. Moment is a vector and should be defined about an axis instead of a point so that the moment contribution from each section cannot be added directly as in Equation 5.27. To correct Equation 5.27, the moment equilibrium should be considered about an axis passing through the centre of rotation and the smaller radius at each section should be adopted and is given by:

$$F = \frac{\sum (W_i \tan\phi_i + C_i A_i \cos\gamma_z) r_i / m_\alpha}{\sum W_i r_i \sin\alpha_y} \tag{5.28}$$

where $r_i = \sqrt{R^2 - y_i^2}$ and R is the radius of the spherical failure mass. In Equation 5.28, the radius at each section r_i is smaller than the global radius of rotation R and cannot be cancelled out because r_i is different at different sections. By using Equation 5.27, Cheng and Yip (2007) have obtained the value 1.42 (same as Hungr) while 1.39 is obtained from Equation 5.28 for Example 5.1.

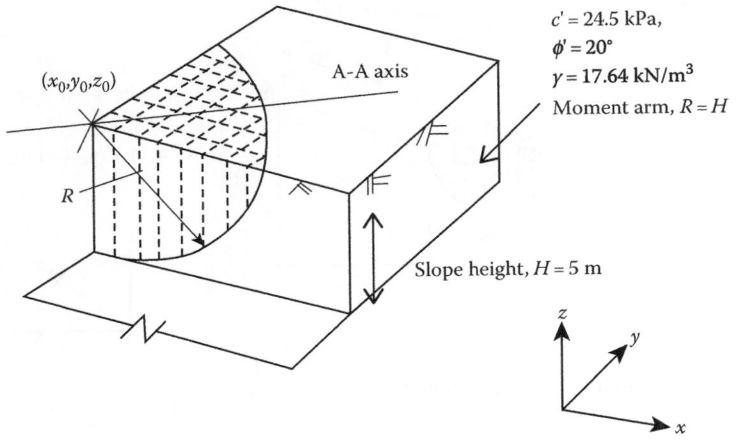

Figure 5.9 Slope geometry for Example 5.2.

Example 5.2 (Huang and Tsai, 2000) is a vertical cut slope (Figure 5.9) with an assumed spherical sliding surface. The failure mass is symmetrical about an axis inclined at 45° to the x axis, and this result is predicted with both Huang's formulation and the present formulation. The results based on the present formulations agree well with the results obtained by Huang and Tsai (2000), which have demonstrated that the new formulation will give results close to those obtained from Huang's formulation.

Example 5.3 is a vertical cut slope (Figure 5.10) in which wedge-like failure is considered for the analysis. The FOS for this rigid block failure is

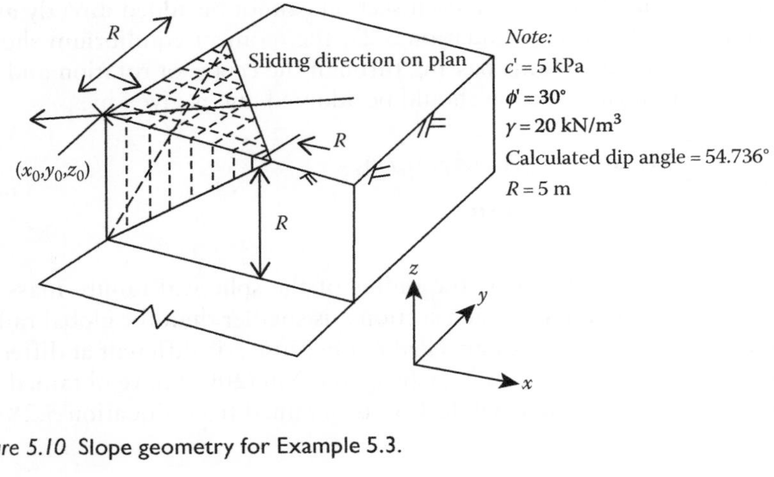

Figure 5.10 Slope geometry for Example 5.3.

determined explicitly from the simple rigid block failure as 0.726. Similar results are obtained from the present 3D Janbu and MP analyses. This has demonstrated that if the correct failure mode is adopted, the present formulation can give reasonable results for 3D analysis. For 3D Bishop analysis, the FOS based on the moment point (x_0,y_0,z_0) is not correct as the present failure mode is a sliding failure, while Bishop method does not fulfil horizontal force equilibrium. The value 0.62 obtained from the Bishop method should not be adopted because the Bishop method does not satisfy horizontal force equilibrium while the wedge actually fails by sliding. If any moment point is chosen for the Bishop analysis, the FOS will be different and this is a well-known problem for the Bishop method. For the determination of the 3D FOS, the failure mechanism should be considered in the selection of a suitable method of analysis. Example 5.4 is an asymmetric rigid block failure with a size of 2 m×4 m, as shown in Figure 5.11. The FOS and the sliding with respect to the x axis are can be determined explicitly as 0.2795° and 63.4°, respectively, which are also determined exactly from SLOPE3D with 3D Janbu and MP analyses. The FOS for Examples 5.3 and 5.4 in Table 5.3 are correctly predicted by the present formulation, which confirms its validity.

For the FOS of a simple slope where $c' = 0$, F_s is given as tan ϕ'/tan θ, where θ is the slope angle. The critical failure surface is a planar surface parallel to the ground surface. From a spherical/elliptical optimization search, the minimum FOSs for a simple slope from the present formulations are equal to tan ϕ/tan θ for Bishop, Janbu and MP analyses, and the results comply with the requirement of basic soil mechanics.

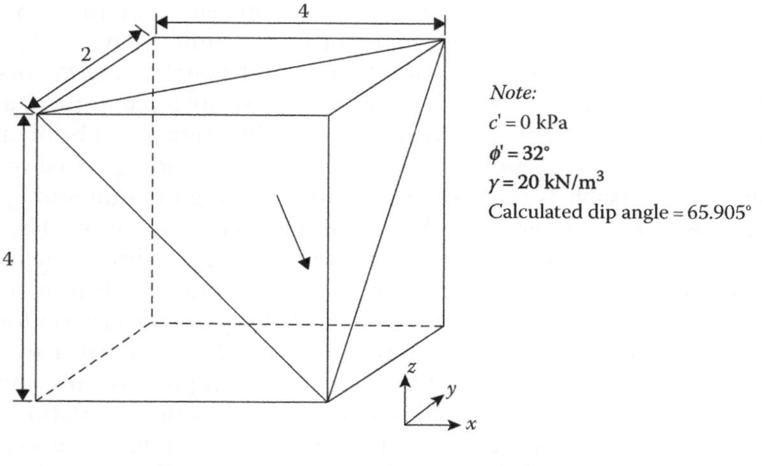

Figure 5.11 Slope geometry for Example 5.4 (F=0.2795, sliding direction=63.4° with respect to the x axis).

Table 5.3 Comparison of F_s for Examples 5.2, 5.3 and 5.4

	Example 5.2			*Example 5.3*		
Method	*Huang and Tsai (2000)*	*Slope 3D (Huang's approach)*	*Slope 3D (Cheng's approach)*	*Analytical*	*Slope 3D (Huang's approach)*	*Slope 3D (Cheng's approach)*
Bishop simplified	1.766[a]	1.781 (45°)[b,d]	1.801 (45°)[b,c,d]	0.726	0.620 (45°)[a,d]	0.620 (45°)[b,c,d]
Janbu simplified	—	2.820 (45°)[b,d]	2.782 (45°)[b,c,d]		0.722 (45°)[a,d]	0.722 (45°)[b,c,d]
MP			1.803 (45°)[b,c,d]			0.724 (45°)[b,c,d]

[a] 2039 columns are used by Huang and Tsai for Examples 5.2 and 5.3, respectively.
[b] 10,000, 23,871 and 2,500 columns are used by Cheng for Examples 5.2, 5.3 and 5.4, respectively.
[c] The unique overall sliding direction (or the equilibrium point) in degree.
[d] The average overall sliding direction in degree.

The results in Examples 5.1–5.4 show that the present theory gives sliding directions similar to those computed from Huang and Tsai's method (2000). Cheng and Yip (2007) have also worked on many other examples, and the differences between the FOS and sliding direction by Huang and Tsai's formulation (2000) and the present formulation are small in general if there is no transverse load.

5.2.5 Comparison between Huang's method and the present methods for transverse earthquake load

In Huang and Tsai's method (2000), the mobilized shear force, S_i, has two components on the bottom plane of each column, namely, S_{xzi} and S_{yzi}. Besides the global safety factor F_s, two additional safety factors are further defined as F_{sx} and F_{sy} for the mobilized shear force in the x and y directions, respectively. The individual sliding direction a_i can be obtained by using the *calculated and converged* values of F_{sx} and F_{sy} based on different methods (such as the Bishop method). By using the values of F_{sx}, F_{sy} and a_i, the corresponding F_s can be determined in each iteration. The final solution can then be obtained if the tolerance of the analysis is achieved. However, by using this method, before the final solution for F_s is obtained, *three convergent criteria* need to be satisfied. They are the criteria for F_{sx}, F_{sy} and a_i. As a_i is mainly determined by F_{sx} and F_{sy}, the correct solution cannot be obtained for F_s unless the computational process can converge for F_{sx} and F_{sy}, respectively. Furthermore, a_i appears in the solution of F_s as well as in the solution of F_{sx} and F_{sy} in each iteration; therefore, if a_i is determined incorrectly, F_{sx} and F_{sy} are incorrectly determined, which yields a wrong value for F_s. Huang and Tsai's formulation (2000) may

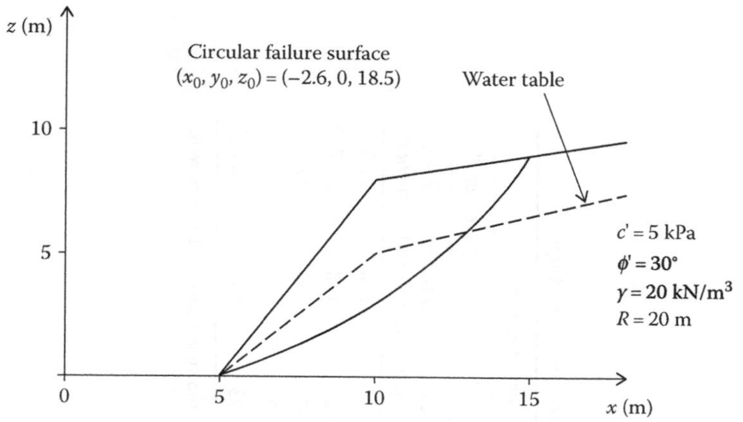

Figure 5.12 Slope geometry for Example 5.5.

therefore face difficulties with regard to convergence for more complicated problems.

By using the present method, only *two convergent criteria* need to be satisfied. They are the criteria for F_{sx} and F_{sy}; a_i is unique in the present formulation and is determined by a simple looping process instead of the three convergent criteria so that the errors associated with Huang and Tsai's method (2000) are eliminated. F_s is determined directly from the values of the directional safety factor F_{sx} or F_{sy} at the equilibrium point. To illustrate the important difference between the two methods, Example 5.5 (Figure 5.12), where transverse earthquake load Q_y is applied normal to the section as shown, is considered and the analysis results are given in Table 5.4. Huang and Tsai's method (2000) fails to converge in analysis. In examining the intermediate results, Cheng and Yip (2007) find that even though F_{sx} and F_{sy} can converge in the first and second iteration, the converged values appear to be unreasonable. This is because in Huang and Tsai's method, an initial constant value is assigned for all a_i values. As the computation process starts, F_{sy} is determined incorrectly based on these initial a_i values, which are the same among different columns in the first step. This leads to the value of F_s being determined wrongly in a consequent iteration, as given in Table 5.5. Also, based on the value of F_{sy}, a new set of a_i values are calculated with F_{sx}. Thus both F_{sx} and F_{sy} are determined incorrectly in the subsequent iteration because they are directly determined from the a_i values. In the present study, a_i has been assigned from 0.05 to 0.5 rad (2.9° to 28.6°) in an increment of 0.05 using Huang and Tsai's method, but convergence is still not achieved.

On the other hand, the present formulation can converge without any problem, as demonstrated in Figures 5.13 and 5.14. Although F_{sy} is

Table 5.4 Comparison between the present method and Huang and Tsai's method with transverse earthquake

Earthquake load[a] *(%)*		*Bishop simplified method*				*Janbu simplified method*			
		Safety factor, F_s		*Sliding direction (°)*		*Safety factor,* F_s		*Sliding direction (°)*	
Q_x	Q_y	*Huang and Tsai (2000)*	*Present method* (F_s)	*Huang and Tsai (2000)*	*Present method (°)*	*Huang and Tsai (2000)*	*Present method* (F_s)	*Huang and Tsai (2000)*	*Present method (°)*
50	50	Fail	0.3366	Fail	29.925	Fail	0.3064	Fail	31.84
30	30	Fail[b]	0.4787	Fail	22.004	Fail[b]	0.4524	Fail	23.789
30	20	Fail	0.4888	Fail	15.147	Fail	0.463	Fail	16.476
30	10	Fail	0.4953	Fail	7.73	Fail	0.4699	Fail	8.444
30	1	Fail	0.4975	Fail	0.636	Fail	0.4723	Fail	0.732

[a] Q_x and Q_y are the earthquake load (in % of soil weight) in the *x* direction and normal to the section shown in Figure 5.12, respectively.
[b] See Table 5.5 for details on failure to converge.

Table 5.5 Factors of safety during analysis based on Huang and Tsai's methods

Iteration No.	Bishop simplified method[a] Converged safety factor F_{sx}	F_{sy}	F_s	Janbu simplified method[a] Converged safety factor F_{sx}	F_{sy}	F_s
1	0.498	0.583×10^{-3}	-0.149×10^{-4}	0.472	0.592×10^{-3}	-0.151×10^{-4}
2	0.211×10^{-2}	0.276×10^{-4}	-0.165×10^{-6}	0.226×10^{-2}	0.419×10^{-4}	0.581×10^{-5}
3	Fail	Fail	Fail	Fail	Fail	Fail

[a] 30% earthquake load is set in both the *x* and *y* directions.

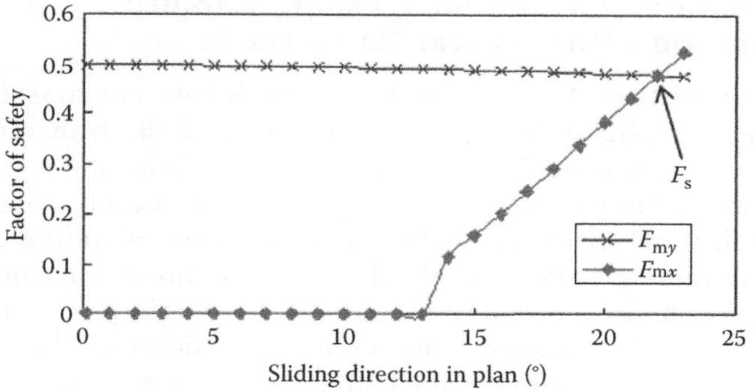

Figure 5.13 Convergent criteria based on the present method – by using the Bishop simplified method (30% earthquake load in both the *x* and *y* directions).

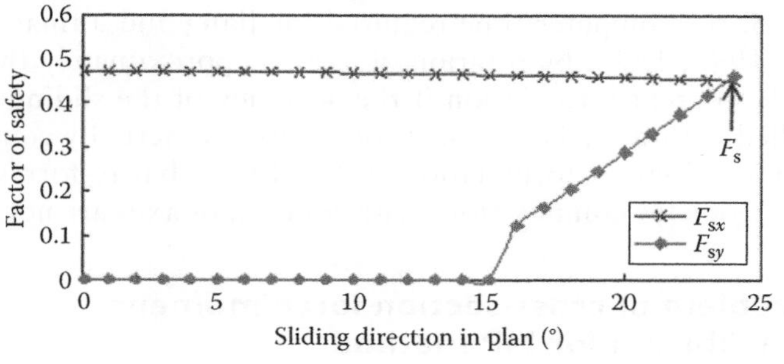

Figure 5.14 Convergent criteria based on the present method – by using the Janbu simplified method (30% earthquake load in both the *x* and *y* directions).

unreasonable when the sliding angle is small, once the sliding angle is reasonable, F_{sy} is also reasonable. As per Cheng and Yip (2007), the non-unique sliding direction in Huang and Tsai's formulation is the main cause for the failure to converge, as this method gives larger sliding angles for those soil columns near the edge of the failure mass based on the figures from Huang and Tsai (2000). Figures 5.13 and 5.14 have also demonstrated that the FOS can be very sensitive to the sliding angle, and the use of varying sliding angles between different soil columns may not be a good idea. When a transverse load is present, the sliding angles at the edge of the failure soil mass are greatly increased and loss of contact generates internal tensile forces between soil columns and unreasonable F_{sy} so that convergence is not achieved.

5.2.6 Relation of proposed 3D slope stability method with classical 3D methods

Most of the existing 3D slope stability methods have not considered sliding direction explicitly in their formulation, and the transverse direction has not been considered at all. Jiang and Yamagami (1999) and Yamagami and Jiang (1996, 1997) suggest that the axes be rotated when using classical 3D methods (without consideration of sliding direction) until the minimum FOS is obtained. The FOS and sliding direction as determined from this axes rotation is reasonable but the process is tedious, and as such this formulation is not commonly adopted. Cheng and Yip (2007) have tried the Bishop and Janbu methods for Examples 5.2 and 5.3, and the results are shown in Figure 5.15. The curves in Figure 5.15 are symmetric about a sliding direction of 45°, and the minimum FOS are as given in Table 5.3 using the present formulation where the sliding direction is considered. If a high accuracy for the determination of the sliding direction is required, Jiang and Yamagami's formulation (1996, 1997) can be very time consuming which is experienced by Cheng and Yip (2007) for the cases shown in Figure 5.15. For the cases shown in Figure 5.15, the computer time required for Jiang and Yamagami's formulation (1996, 1997) by rotation of axes is approximately three times that for the present formulation if the accuracy of the sliding direction is controlled within 1°. The present formulation is actually equivalent to Jiang and Yamagami's formulation (1996, 1997), but re-formulation of mesh and geometry computations with rotation of axes are not required.

5.2.7 Problem of cross-section force/moment equilibrium for MP method

Three-dimensional asymmetric MP formulation is highly statically indeterminate, which indicates that cross-sectional force or moment

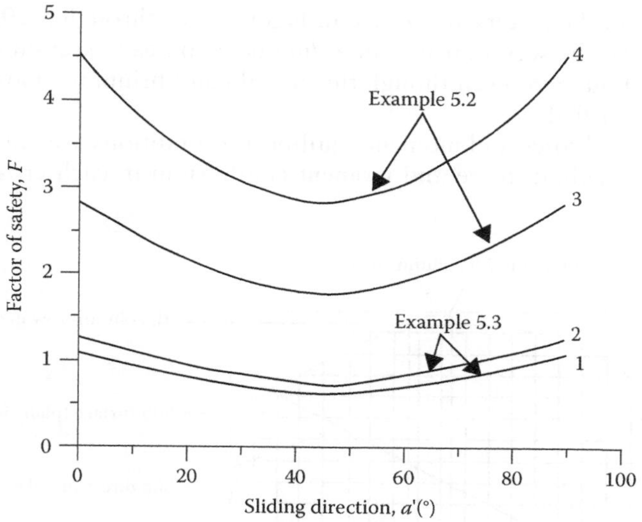

Figure 5.15 FOS against the sliding direction using classical 3D analysis methods (curves 1 and 2 are Bishop and Janbu analyses for Example 5.3 while curves 3 and 4 are Bishop and Janbu analyses for Example 5.2).

equilibrium cannot be enforced simultaneously in the analysis. In fact, it is one of the major theoretical difficulties in 3D limit equilibrium analysis as the number of redundant equations is much more than for the corresponding 2D analysis. In the 2D MP method, interslice normal (and shear) force of the last slice can be determined from the last interface (from last second slice). The equation of horizontal force equilibrium becomes redundant for the last slice. However, horizontal force equilibrium can still be maintained for all slices as overall horizontal force equilibrium is enforced. In the calculation of the interslice normal force, the calculation progresses from slice to slice, which automatically ensures that horizontal force equilibrium is satisfied even for the last slice. However, this condition cannot be enforced automatically for 3D analysis as the safety factor equations are based on the overall force equilibrium as given by Equations 5.13 and 5.15 instead of sectional horizontal force equilibrium. Therefore, horizontal force equilibrium cannot be automatically enforced in *each cross section*. Also, there is no way to enforce *cross-section moment equilibrium* in the analysis as the overall moment equilibrium instead of sectional moment equilibrium is used in the analysis. In fact, cross-section force or moment equilibrium is a common problem in all 3D analysis methods. In order to investigate the importance of cross-sectional equilibrium, Example 5.5, as shown in Figure 5.12 with one half of the soil mass loaded with a surcharge, is considered with $\lambda_{xy} = 0$ in the analysis, and the results are shown in Figures 5.16 through 5.20. In these figures, it can be seen that net forces/moments on each section exist and fluctuate about *zero* even though the global equilibrium for moment and force are satisfied.

As sectional force and moment equilibrium conditions are not enforced in 3D MP method, force and moment equilibrium in each cross section

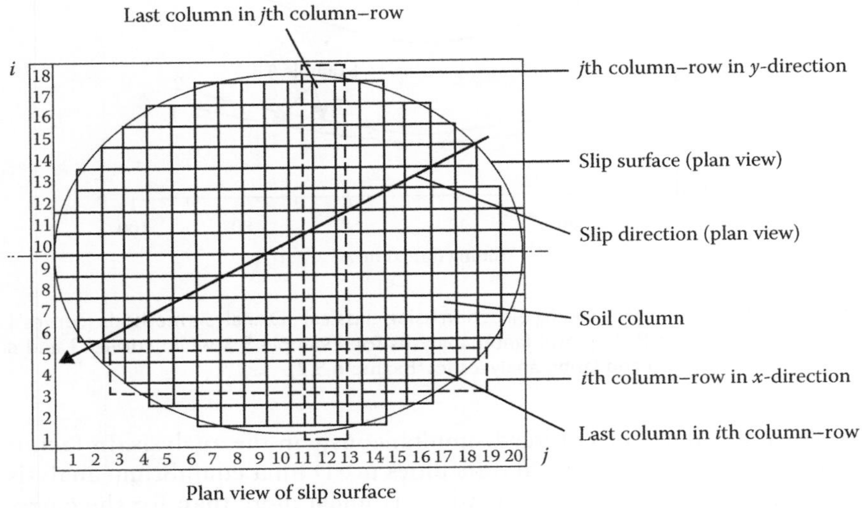

Figure 5.16 Column–row within potential failure mass of slope for Example 5.1.

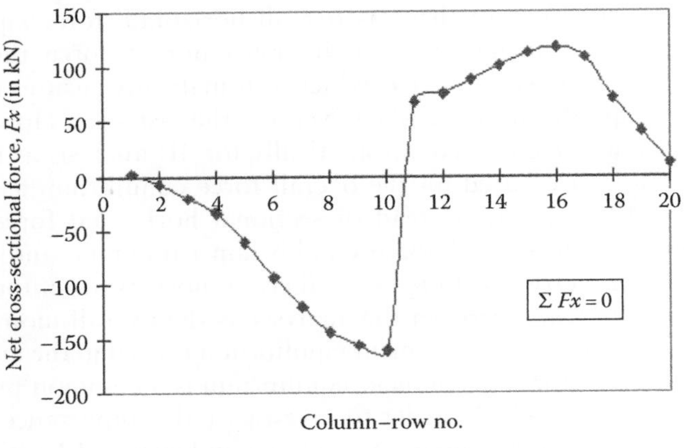

Figure 5.17 Cross-sectional force equilibrium condition in the *x* direction.

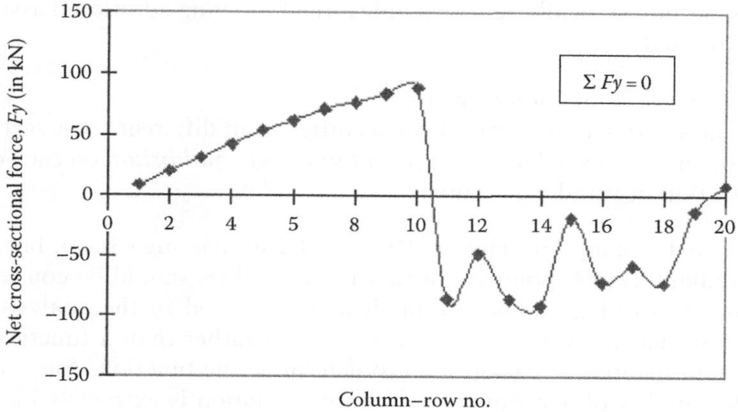

Figure 5.18 Cross-sectional force equilibrium condition in the *y* direction.

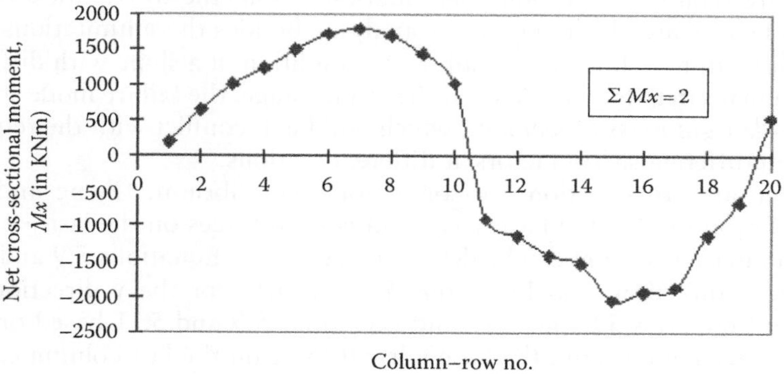

Figure 5.19 Cross-sectional moment equilibrium condition in the *x* direction.

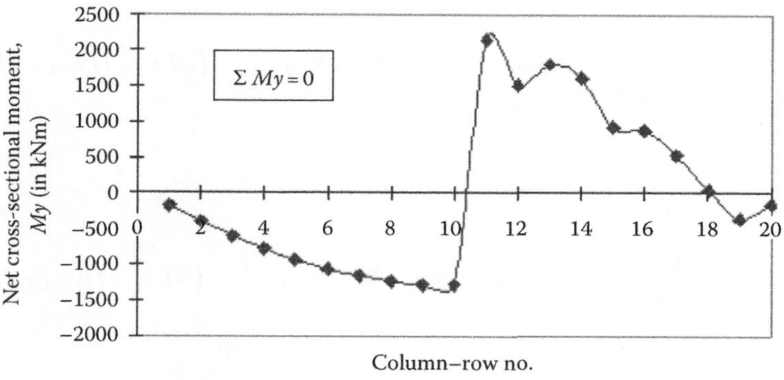

Figure 5.20 Cross-sectional moment equilibrium condition in the *y* direction.

cannot be automatically achieved unless the following additional assumptions are used:

1. λ_x or λ_y is a function of y or x.
2. The safety factor is considered as different in different cross sections, and it is subjected to the *force and moment equilibrium* on each cross section in x and y directions.

For the first assumption, Hungr (1994) had a similar suggestion, but Lam and Fredlund (1994) pointed out that these λ values should be considered as percentages of the intercolumn shear forces used in the analysis and suggested that this value should be a constant rather than a function. As there is no theoretical background to determine the functions for λ_x or λ_y, and the number of iterations required for a solution is extremely high for such formulation, this formulation has not been used in any existing 3D slope analysis model. For the second assumption, a large number of iterations are required. Additional assumptions about the distribution of the safety factors are also required for analysis. Besides these limitations, it is also difficult to define the overall safety condition of a slope with different directional safety factors. Also, at the failure stage, the failure mode should be *whole rigid mass movement*, which will be in conflict with the requirement of different safety factors at different sections.

To ensure cross-section horizontal force equilibrium, Cheng and Yip (2007) propose that the base shear and normal forces on the last (or first) column in each section can be determined by using Equations 5.9 and 5.10 for the x direction, and Equations 5.9 and 5.11 for the y direction. By solving Equations 5.9 and 5.10 and Equations 5.9 and 5.11 based on sectional force equilibrium, the base normal force on the last column can be expressed as Equation 5.29 for the x direction and Equation 5.30 for the y direction, respectively, as

$$N_i = \frac{1}{g_{3i} \cdot \left(1 + \dfrac{g_{1i} \cdot f_{3i}}{f_{1i} \cdot g_{3i}}\right)} \cdot \left[\left(-Xx_i - Xy_i + W_i + P_{vi}\right) - \frac{f_{3i}}{f_{1i}} \cdot \left(-Ex_i - Hx_{i+1} + Hx_i\right)\right] \tag{5.29}$$

$$N_i = \frac{1}{g_{3i} \cdot \left(1 + \dfrac{g_{2i} \cdot f_{3i}}{f_{2i} \cdot g_{3i}}\right)} \cdot \left[\left(-Xx_i - Xy_i + W_i + P_{vi}\right) - \frac{f_{3i}}{f_{2i}} \cdot \left(-Ey_i - Hy_{i+1} + Hy_i\right)\right] \tag{5.30}$$

In the present formulation, the intercolumn normal and shear forces are calculated based on the adjoining column of the last column by using Equations 5.9 through 5.11; the base shear and normal force can then be found in the iteration process. The use of Equations 5.29 and 5.30 is equivalent to enforcing cross-sectional horizontal force equilibrium in the last column of each section; hence cross-sectional force equilibrium can be achieved. If Equation 5.29 or 5.30 is used, sectional force equilibrium can be achieved in either direction only and equilibrium along both x and y sections cannot be maintained *simultaneously*. For 2D analysis, the base normal forces for the last one or two slices may be negative if the cohesive strength is high (Abramson et al., 2002). If the base force for the last column is not realistic, unrealistic numerical results may be introduced in the back-calculation of the base and intercolumn forces from the last column. After carrying out the computational analysis, it is found that the iteration process is more difficult to converge with the use of Equation 5.29 or 5.30. This situation is not surprising as Equations 5.29 and 5.30 become additional constraints to the convergence. The more constraints to a problem, the more difficult it will be to get the analysis converged. For those problems where cross-section force equilibrium can be achieved with converged results, the safety factors are virtually the same as those where cross-section force equilibrium is not enforced, and the results are given in Table 5.6. In general, Cheng and Yip (2007) do not suggest the enforcement of the cross-section force equilibrium in the analysis as convergence is usually more difficult to achieve. If the cross-section horizontal force equilibrium is not enforced in the analysis, there will be overall unbalanced moment about the z axis. It is not possible to achieve better convergence and eliminate unbalanced moment about the z axis unless additional assumptions are introduced in the solution.

Table 5.6 Comparison between overall equilibrium method and cross-sectional equilibrium method using 3D MP method for Example 5.5

Q (kPa)	*Method*	*FOS*	*S.D. (°)*	λ_x	λ_y
300	1	0.616	10.99	0.894	−0.1835
	2	0.619	10.93	0.9195	0.2886
200	1	0.649	9.39	0.8361	−0.1512
	2	0.651	9.33	0.8941	0.2586
100	1	0.704	6.54	0.708	−0.0857
	2	0.706	6.48	0.8451	0.1944

Note: Method 1 = overall equilibrium method; Method 2 = cross-sectional equilibrium method.

Table 5.7 Effect of λ_{xy} on safety factor and sliding direction for Example 5.5 (292 Columns)

λ_{xy}	*0*	*0.05*	*0.1*	*0.15*	*0.2*	*0.25*
F	0.6186	0.6187	0.6188	0.6188	Fail	Fail
S.D.	10.929°	10.920°	10.911°	10.902°	Fail	Fail
λ_x	0.9195	0.926	0.9325	0.9384	Fail	Fail
λ_y	0.289	0.289	0.289	0.289	Fail	Fail

5.2.8 Discussion on λ_{xy} for MP analysis

To examine the effect of λ_{xy} on the safety factor, a uniformly distributed pressure of 300 kPa is applied to half of the failure mass for the problem shown in Figure 5.12, and the results are given in Table 5.7. It can be seen that λ_{xy} is not sensitive to the analysis. This situation is not surprising as the vertical shear forces Xx_i and Xy_i (with same direction as weight of soil) are more important than the horizontal shear forces Hx_i and Hy_i in the present problem. However, if λ_{xy} becomes greater than 0.2, the iteration process fails to converge (unless a relatively large tolerance is adopted).

Huang et al. (2002) believed that the disturbing force induced by the torque due to Hx/Hy was minor and Hy is actually taken as 0 in their formulation, which conflicts with the concept of complementary shear. Actually, no additional equation is available to determine λ_{xy}. Cheng and Yip (2007) have considered the use of moment equilibrium about the z axis to determine λ_{xy}, but unbalanced moment ΣMz can actually come from the sectional force equilibrium problem as mentioned earlier. If sectional equilibrium is enforced, ΣMz will actually be 0 so that λ_{xy} is indeterminate. To avoid the introduction of additional assumption in determining λ_{xy}, it is suggested that λ_{xy} be prescribed by the engineer in the present formulation. Alternatively, Cheng and Yip (2007) suggest that λ_{xy} can be determined from the minimization of the FOS with respect to λ_{xy}. Since λ_{xy} is not a major factor in the analysis, λ_{xy} can be prescribed to be 0 for most cases without major problem.

5.2.9 Discussion on 3D limit equilibrium stability formulation

In this chapter, new 3D slope stability methods are developed which are based on force/moment equilibrium. Fundamental principles of limit equilibrium are used with an extension of the 2D Bishop simplified method, the Janbu simplified method or the MP method. The new formulations possess several important advantages, which are as follows:

1. Simple extension of the corresponding 2D formulation
2. Ability to determine unique sliding direction

3. Better convergence, which is not affected by the initial choice of a_i
4. Possibility of comparing the results from the present study to those by other researchers for some well-known cases
5. Applicability to highly non-symmetric problems with transverse load

By using this new formulation, the unique sliding direction can be determined with the corresponding safety factor. The limitations of Huang and Tsai's method (2000) are overcome by the new formulations as proposed while the assumption of a plane of symmetry can be eliminated in the analysis of 3D slopes. Cheng and Yip (2007) have tried over 100 cases for 3D Bishop and Janbu analysis using Huang's formulation and the present formulation for both symmetric and asymmetric problems. The FOSs and sliding directions from all these examples are extremely close between these two formulations, and the differences are small and negligible for all these cases. For a normal problem with no transverse load, the present formulation and Huang's formulation are practically the same.

Transverse earthquake load has not been considered in the past due to the lack of a suitable 3D analysis model. The present study on transverse earthquake load has demonstrated the limitation of Huang's method as transverse loads greatly affect the spread of the sliding directions and hence the convergence. Huang and Tsai's method (2000) faces difficulty in convergence with transverse load as the sliding direction is not unique. For the present formulation, the sliding direction is unique, and only two convergent criteria have to be met for directional safety factors determination. Convergence is hence greatly improved under the present formulation.

In the numerical examples, the present formulation is found to be reasonable in the determination of the safety factor and sliding direction of 3D slope. In particular, the analytical results for the wedge type failure in Examples 5.3 and 5.4 are exactly the same as those obtained from the present formulations. In Examples 5.1 and 5.2, Huang and Tsai's method (2000) gives results similar to those from the present formulation. However, the sliding direction from Huang and Tsai's analysis is based on the average of the sliding directions of all the columns. Conceptually, this is a major limitation as the spread of the individual sliding direction can be major if the problem is highly asymmetric. In fact, there is another fundamental problem in taking the average individual sliding direction. If there is a major variation in the sizes of columns, it is not clear whether the size of the column should be considered in the averaging process or not. Finally, if the sliding direction is not unique, some of the columns could be separating from each other, and the summation of the overturning and restoring moment/force process is strictly not applicable. In view of all these limitations, the requirement on unique sliding direction appears to be important for 3D analysis, and this has been solved effectively by the present formulation.

Cheng and Yip (2007) have demonstrated that the present formulation is equivalent to Yamagami and Jiang's formulation (1996, 1997) but is more convenient to be used for general conditions. No rotation of axes is required to determine the FOS, and there is significant reduction of works for location of critical failure surfaces where thousands of trials are required.

In general, the behaviour of the present formulation is similar to that of the corresponding 2D formulation. For example, based on limited case studies, it was found that the FOS is usually not sensitive to $f(x, y)$ ($f(x) = \sin(x, y)$ has been tried) in most cases. While 3D Janbu and Bishop methods suffer from the limitation of incomplete equilibrium condition, the 3D MP method suffers from the limitation which exists only for 3D analysis. Overall force and moment equilibrium can be maintained under the MP formulation, while cross-sectional equilibrium is violated unless Equations 5.29 and 5.30 are used, and cross-sectional equilibrium in both directions cannot be maintained simultaneously. As there are more equations than unknowns (more serious than the 2D condition), convergence with the MP method is also more difficult as compared with the corresponding 2D condition. Due to the indeterminacy of the system, Cheng and Yip (2007) believe that it is not possible to maintain overall and local equilibrium without additional assumptions in general 3D analysis. In the present formulation, enforcement of cross-sectional horizontal and moment equilibrium may affect convergence of the solution, and Cheng and Yip (2007) suggest that it is not worthwhile to impose these constraints in analysis.

In spite of the limitations in 3D MP formulation, the applicability of proposed 3D asymmetrical analysis has been demonstrated by several examples where the results from the 3D MP method are similar to those from the 3D Bishop simplified method and 3D Janbu simplified method. The proposed formulation can also predict the exact sliding directions and safety factors for the simple sliding wedge in Examples 5.3 and 5.4, which validates the applicability of the present formulation.

5.3 THREE-DIMENSIONAL LIMIT ANALYSIS

The extensions of the upper-bound technique to 3D geotechnical problems are being investigated. Michalowski (1989) presented a 3D slope stability method for drained frictional-cohesive material based on the upper-bound technique of limit analysis. The slip surface was approximated by a number of planar surfaces, whose lines of intersection were perpendicular to the plane of symmetry, in combination with so-called end surfaces that extended to the slope top or the slope surface. A typical failure mechanism used in the method consists of rigid-motion blocks separated by planar velocity discontinuity surfaces. The limit load involved in an energy balance equation was found directly. The minimum of the FOS or the limit load was obtained by searching all kinematically admissible mechanisms of failure. The simplification regarding the geometry of a slip surface and the surface of a slope made in the method limited the application to practical problems. While the approach proposed by Michalowski (1989) was limited to homogeneous slopes, more recently, Farzaneh and Askari (2003) modified and extended Michalowski's approach to deal with inhomogeneous symmetrical cases.

Chen et al. (2001a,b) presented another 3D method based on the upper-bound theorem, in which an assumption of a so-called neutral plane is needed so that the failure surface is generated by elliptical lines based on the slip surface in the neutral plane and extended in the direction perpendicular to the neutral plane. Wang (2001) demonstrated its applications to several large-scale hydropower projects.

The common features for the upper-bound methods proposed by Michalowski (1989) and Chen et al. (2001a,b) are that they both employed the column techniques in 3D limit equilibrium methods to construct the kinematically admissible velocity field and have exactly the same theoretical background and numerical algorithm which involves a process of minimizing the FOS. The only difference is that Michalowski (1989) and Farzaneh and Askari (2003) used vertical columns, while Chen et al. (2001a,b) and Wang (2001) utilized non-vertical columns, providing the flexibility to handle relatively complicated geometry and layered rocks and soils.

Lyamin and Sloan (2002b) presented a new upper-bound limit analysis using linear finite elements and non-linear programming. The formulation permitted kinematically admissible velocity discontinuities at all inter-element boundaries and furnished a kinematically admissible velocity field by solving a non-linear programming problem. The objective corresponded to the dissipated power (which was minimized), and the unknowns were subject to linear equality constraints as well as linear and non-linear inequality constraints. The optimization problem could be solved very efficiently using an interior point, two-stage, quasi-Newton algorithm.

As an illustration, a landslide that occurred in Hong Kong is considered by the 3D rigid element method (Chen, 2004). In the morning of 23 July 1994, a minor landslide occurred at a cut slope at milestone 141/2, Castel Peak Road, New Territories, Hong Kong (Figure 5.21). In the afternoon on the same day, a second landslide occurred. On 7 August 1994, a further landslide occurred at the same slope. This landslide resulted in the death of 1 man and in 17 other people being hospitalized. Due to the limited available information on the first and second landslides, only the third landslide, called the Castel Peak Road landslide herein, is analysed in this section. This landslide was typically of three dimensions and encompassed approximately 300 m^3 of soil and rock.

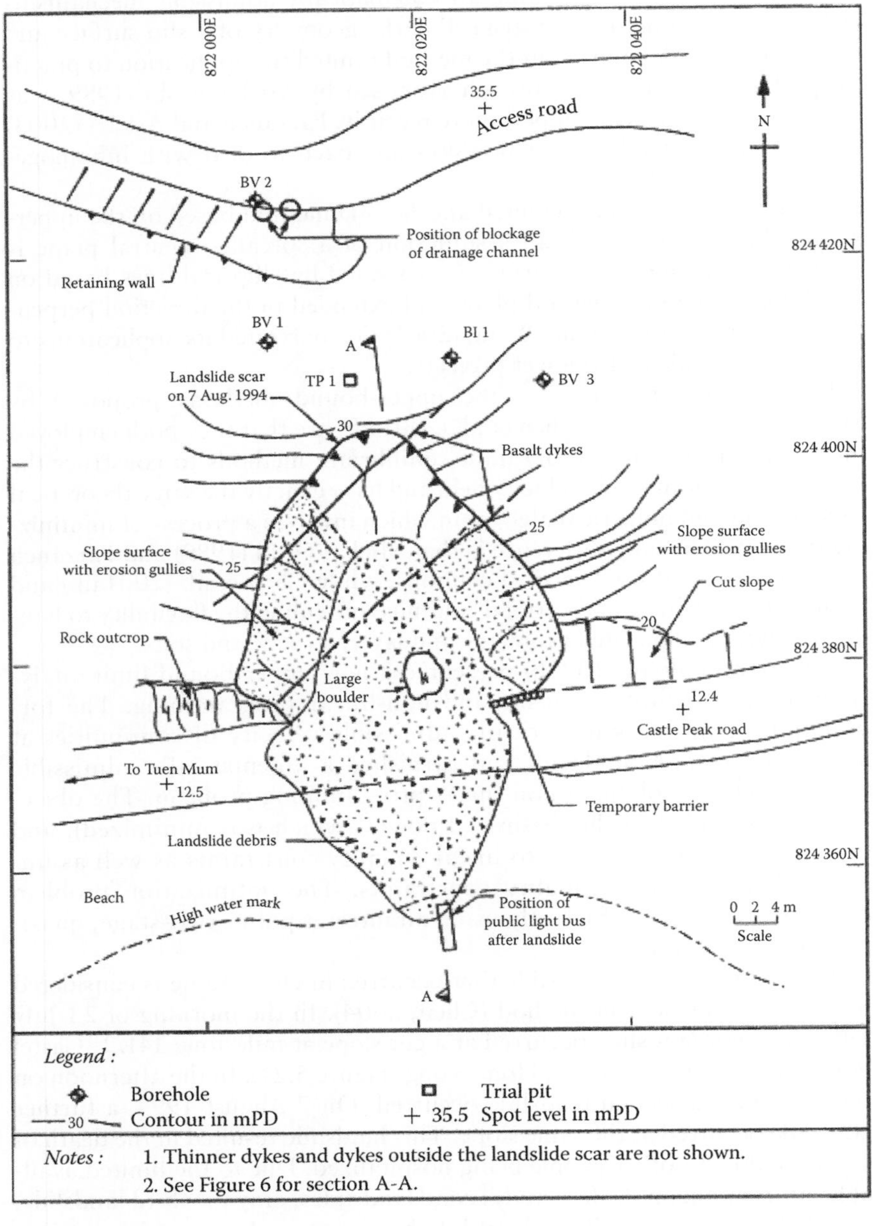

Figure 5.21 A plan view of a landslide in Hong Kong.

Site investigation showed that the ground at the location of the landslides generally comprised partially weathered fine-grained and medium-grained granite, which was a soil of silty sand. Rock of medium-grained granite (slightly to moderately decomposed) was exposed in the cut slope at the western edge of the landslide scar. The partially weathered granite exhibited a well-developed, black-stained relict joint structure. Results of laboratory tests on undisturbed samples of the weathered granite have shown that the strength of soil at this site is akin to that of similar material found in other parts of Hong Kong.

The granite at the site was intruded by a number of sub-vertical basalt dykes, which ran in the northeast direction. Two completely decomposed basalt dykes approximately 800 mm thick were exposed within the landslide scar. Field assessment and laboratory tests have revealed that the completely decomposed basalt dykes were much less permeable than the partially weathered granite. Therefore, the dykes acted as barriers to water (Pun and Yeo, 1995).

Before the landslide on 7 August, on 6 August there was heavy rain in the area along Castle Peak Road, with a total of about 287 mm of rain being recorded. Water seepage was observed in the landslide scar on the uphill side of the two decomposed basalt dykes for a period of at least 1 week after the landslide on 7 August, indicating that the groundwater level was high behind the dykes. In addition, inspection of the access road on the hillside above the landslide scar found that a drainage channel along the edge of the road was completely blocked. This indicates, to a certain degree, that the groundwater level in front of the dykes flowed along the ground surface. However, no measured data for the exact value of the groundwater level behind the basalt dykes are available. The parametric study of the different groundwater level is to be conducted to reveal the essential influence of the groundwater level in the landslide mass on the stability of the slope.

A 3D slope stability analysis was conducted for the landslide on 7 August shown in Figure 5.21. The unit weight of the landslide material was taken to be 20.6 kN/m^3. The peak strength parameters were measured to be $c' = 6.7$ kPa and $\phi' = 35.5°$. The groundwater level in the sliding mass plays an important role on the stability of landslide. Four different cases are therefore investigated in 3D slope stability analyses, that is, (1) no groundwater involved in the failure mass; (2) water level at 5.50 m; (3) water level at 6.15 m; and (4) water level at 6.80 m.

Since the slip surface of the landslide exhibits an apparent 3D characteristic, the 3D upper-bound method is suitable for the analysis. Figure 5.22a illustrates how to generate the slip body in a slope model; Figure 5.22b shows the geometry model. The mesh generated for the slip body is shown in Figure 5.22d. When the groundwater table is at the toe of the landslide (Case 1), the 3D FOS is 1.463, which indicates that the slope is stable if no water is infiltrated into the sliding mass. However, while the groundwater

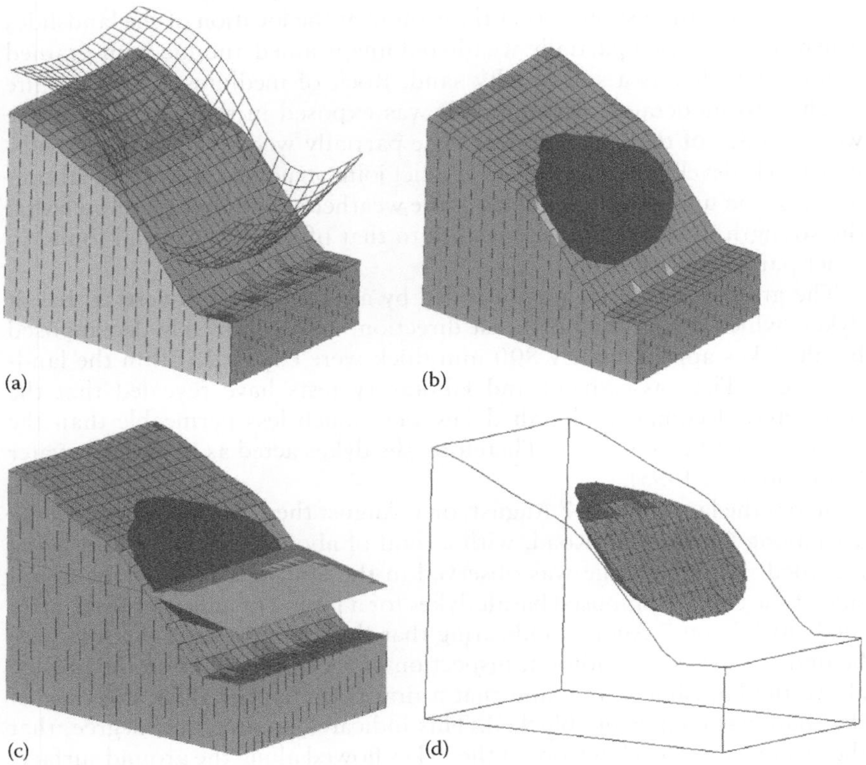

Figure 5.22 Three-dimensional slope model – (a) schematic diagram for generation of a slip body, (b) geometry model, (c) schematic diagram of groundwater and (d) mesh generation for a slip body.

level increases gradually, the 3D FOS decreases accordingly. At the level of the groundwater 5.50 m over the base line (Case 2), the 3D FOS is 1.108. While the level of groundwater is 6.15 m (Case 3), the FOS calculated is 1.002, which is close to 1.0. It indicates that the slope is in the limit state and the slope will collapse at the level of the groundwater 6.15 m. When groundwater relative to the base line is 6.92 m (Case 4), the 3D FOS is 0.922, which is less than unity and the slope has collapsed.

5.3.1 Three-dimensional bearing capacity problem with an inclined slope

There are also several 3D limit analysis models (Giger and Krizek, 1975; Michalowski, 1989; Chen et al., 2001a,b; Farzaneh and Askari, 2003) in the literature. Michalowski (1989), Chen et al. (2001a,b) and Farzaneh and

Askari (2003) have considered the 3D problem by the limit analysis and the upper-bound theorem of plasticity, which are based on the 3D models by Chen (1975). As a special case of 3D slope stability problem where a patch load is applied on top of a slope, Cheng has proposed an approximate failure mechanism and has developed the equations for the ultimate 3D condition based on limit analysis. The failure mechanism as proposed is a more reasonable mechanism based on the kinematically admissible approach, and it is a further development of the works based on some of the researches discussed earlier by using a more reasonable 3D radial shear failure zone through which other 3D failure wedges are connected together. In general, a solution for a simple problem can be obtained within a very short time as the analytical expressions are available. The present solutions have given good solutions when compared with some previous studies and a laboratory test. The laboratory test has also revealed some interesting progressive failure phenomena and deformation characteristics for a slope with a patched load on the top.

5.3.1.1 *Failure mechanism of the patched load acting on the top surface of a slope (D = 0 m)*

A simple 3D slope failure mechanism with zero embedment depth ($D = 0$ m) is shown in Figure 5.23. Figure 5.23b shows the failure mechanism at the section through the applied load, while the end effects are shown in Figure 5.23c and the bird's eye view of the 3D failure mechanism is shown in Figure 5.23d. The total work done is calculated as follows.

5.3.1.2 *Work done rate produced along load length L*

Based on Figure 5.23b, the resistance work done rate dissipated by the cohesion c along velocity discontinuity plane $ac \cdot L$ is given as

$$\dot{W}_{R1} = c \cdot ac \cdot L \cdot v_0 \cos\phi \tag{5.31}$$

where

$ac = B \sin\xi / \sin(\zeta + \xi)$ and $r_0 = bc = B \sin\zeta / \sin(\zeta + \xi)$
B is the width of the footing
L is the length of the footing normal to the section as shown in Figure 5.23b but excluding the two end effects

Resistance work done rate dissipated in the radial shear zone bcd is written as

$$\dot{W}_{R2} = c v_0 r_0 L \frac{\exp(2\Theta \tan\phi) - 1}{\tan\phi} \tag{5.32}$$

Resistance work done rate dissipated by cohesion c along velocity discontinuity plane $dg \cdot L$ is given by:

$$\dot{W}_{R3} = c \cdot dg \cdot L \cdot v_3 \cos\phi \qquad (5.33)$$

in which $v_3 = v_0 \exp(\Theta \tan\phi)$, and $bd = r_0 \exp(\Theta \tan\phi)$.

As shown in Figure 5.23b, point b is taken as the reference point (0, 0, 0) of the coordinates axes, and positive directions are towards left and downward. If the coordinate of point b is $x_b = 0$, $y_b = 0$, $z_b = 0$, the corresponding coordinate of point i is $x_i = b$, $y_i = 0$, $z_i = 0$, the coordinate of point d is

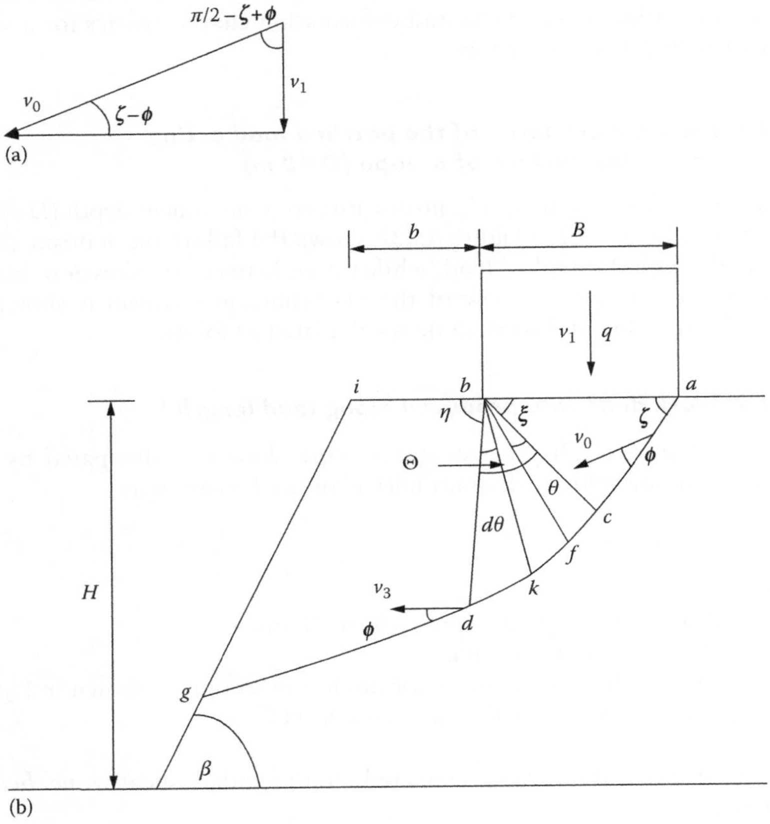

Figure 5.23 Three-dimensional failure mechanism for bearing capacity problem: (a) velocity diagram for wedge *abc*, (b) failure mechanism (excluding the two ends).

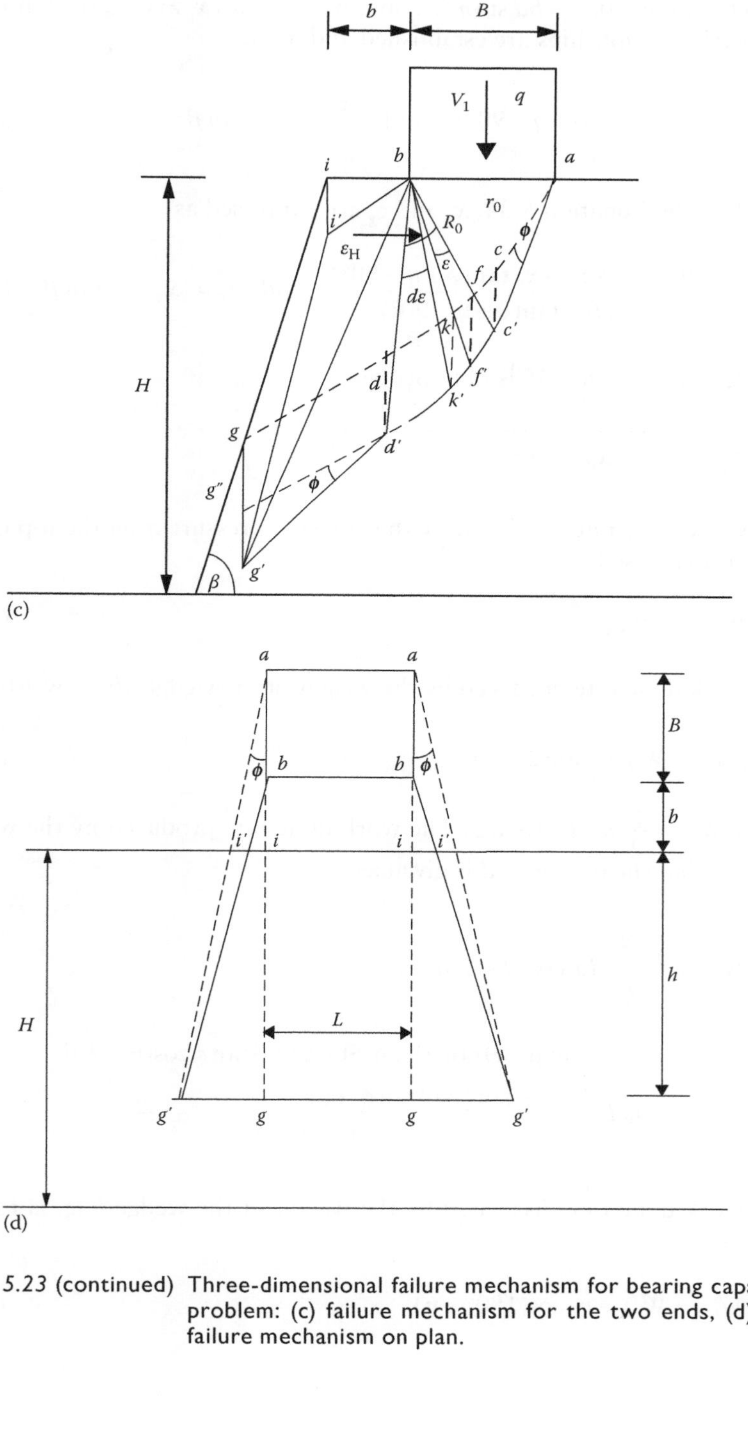

Figure 5.23 (continued) Three-dimensional failure mechanism for bearing capacity problem: (c) failure mechanism for the two ends, (d) 3D failure mechanism on plan.

$x_d = bd\cos\eta$, $y_d = 0$, $z_d = bd\sin\eta$. To obtain the values x_g and z_g, the following geometric relationships are established and used:

$$\frac{z_g - z_d}{x_g - x_d} = \tan(\phi + \eta - 90°) \quad \text{and} \quad \frac{z_g - z_i}{x_g - x_i} = \tan\beta \tag{5.34}$$

Based on the Equation 5.34, x_g and z_g are expressed as

$$x_g = \frac{b\tan\beta + z_d - x_d\tan(\phi + \eta - 90°)}{\tan\beta - \tan(\phi + \eta - 90°)} \quad \text{and} \quad z_g = (x_g - b)\tan\beta \tag{5.35}$$

and dg in Equation 5.35 is obtained as

$$dg = \sqrt{(x_g - x_d)^2 + (z_g - z_d)^2} \tag{5.36}$$

The work done rate produced by the external pressure q on the top of the slope is expressed as

$$\dot{W}_{D1} = qBLv_1 \tag{5.37}$$

The work done rate produced by the weight of the wedge *abc* is written as:

$$\dot{W}_{D2} = W_{abc} v_0 \sin(\zeta - \phi) \tag{5.38}$$

where $W_{abc} = \frac{\gamma}{2} ac \cdot B \cdot L\sin\zeta$. The work done rate produced by the weight of the radial shear zone *bcd* is given as

$$\dot{W}_{D3} = \frac{\gamma}{2}\int_0^{\Theta} r^2 Lv\cos(\theta + \xi)\,d\theta$$

$$= \frac{\gamma}{2} r_0^2 Lv_0 \frac{\exp(3\Theta\tan\phi)[\sin(\Theta + \xi) + 3\tan\phi\cos(\Theta + \xi)] - \sin\xi - 3\tan\phi\cos\xi}{1 + 9\tan^2\phi} \tag{5.39}$$

The work done rate dissipated by the weight of the wedge *bdgi* is formulated as

$$\dot{W}_{D4} = W_{bdgi} \cdot v_3 \cos(180° - \eta) \tag{5.40}$$

$$W_{bdgi} = \gamma L(S_{big} + S_{bdg}) \tag{5.41}$$

where $S_{bdg} = \dfrac{1}{2} bd \cdot dg \cos\phi$ and $S_{big} = \dfrac{1}{2} bi \cdot gi \sin\beta$, in which $bi = b$ and $gi = \sqrt{(x_g - x_i)^2 + (z_g - z_i)^2}$.

5.3.1.3 Work done rate produced at the two end-failure zones of the footing

1. End-failure zone 1
   As shown in Figure 5.23c, cc' is a horizontal line normal to the plane abc. In order to ensure that the sliding velocity of the soil mass of the end-failure zone 1 is equal to v_0, the angle between ac and should be equal to ϕ; therefore, $cc' = r_0 \tan\phi$, $r_0 = ac$. The work done rate by the velocity discontinuity plane acc' is then expressed as

   $$\dot{W}_{\mathrm{RE1}} = c \cdot S_{acc'} \cdot v_0 \cos\phi \tag{5.42}$$

   where $S_{acc'} = \dfrac{1}{2} ac \cdot cc'$. The work done rate produced by the weight of the wedge $abc - c'$ is expressed as

   $$\dot{W}_{\mathrm{DE1}} = W_{abc-c'} \cdot v_0 \sin(\zeta - \phi) \tag{5.43}$$

   in which

   $$ab = B, S_{abc} = \frac{1}{2} ab \cdot ac \sin\zeta, W_{abc-c'} = \frac{\gamma}{3} S_{abc} \cdot cc'.$$

2. End-failure zone 2
   As shown in Figure 5.23c, $b - cdd'c'$ is the 3D end radial shear failure zone 2. If we assume that $c'd'$ is a spiral and the centre of the spiral $c'd'$ is at point b, a relationship $R = R_0 \exp(\varepsilon \tan\phi)$ will exist in which $R_0 = bc'$ and $R = bf'$. For triangle bff', the velocity v is normal to both lines bf and bf', so we can deduce that the velocity v is vertical to triangle bff' and line ff'. In order to ensure a kinematically compatible velocity yield for the soil mass of the end radial shear failure zone 2, for small unit $b - fkk'f'$, the horizontal angle between $v = v_0 \exp(\theta \tan\phi)$ and line $f'k'$ should be equal to ϕ. It should be pointed out that $cc'd'd$ is normal to the plane abc, and the corresponding relationship between $Rd\varepsilon$ and $rd\theta$ is expressed as

   $$\frac{rd\theta}{\cos\phi} = Rd\varepsilon \tag{5.44}$$

in which $r = r_0 \exp(\theta \tan \phi)$, $R = R_0 \exp(\varepsilon \tan \phi)$, and $r_0 = R_0 \cos \phi$. Integrating both sides of Equation 5.44 yields

$$\theta = \varepsilon, \Theta = \varepsilon_{\mathrm{H}} \tag{5.45}$$

in which Θ is an angle between *bc* and *bd*, and ε_{H} is the angle between line *bc'* and line *bd'*.

a. Velocity discontinuity curve plane *bc'd'*
   Velocity discontinuity plane area *bfk'* is expressed as

$$S_{bfk'} = \frac{1}{2} R^2 \sin d\varepsilon = \frac{1}{2} R^2 d\varepsilon \tag{5.46}$$

   Therefore the resistance work done rate produced by *c* along the velocity discontinuity plane *bc'd'* is integrated as

$$\dot{W}_{\mathrm{RE2}} = \int_0^{\varepsilon_{\mathrm{H}}} c \cdot S_{bfk'} \cdot v \cos\phi = \frac{1}{2} c v_0 R_0^2 \cos\phi \int_0^{\Theta} \exp(3\theta \tan\phi) d\theta$$

$$= \frac{\cos\phi}{6\tan\phi} c v_0 R_0^2 \left[\exp(3\Theta \tan\phi) - \right] \tag{5.47}$$

b. Velocity discontinuity plane *cc'd'd*
   Line *ff'* is normal to line *bf*; therefore *ff'* is expressed as

$$ff' = (R^2 - r^2)^{\frac{1}{2}} = r_0 \tan\phi \exp(\theta \tan\phi) \tag{5.48}$$

   Unit area of *ff'k'k* is expressed as

$$S_{ff'k'k} = ff' \cdot R d\varepsilon = r_0 R_0 \tan\phi \exp(2\theta \tan\phi) d\theta \tag{5.49}$$

   Therefore, the resistance work done rate produced by the velocity discontinuity area *cc'd'd* is obtained as

$$\dot{W}_{\mathrm{RE3}} = \int_0^{\Theta} S_{ff'k'k} \cdot v \cdot c \cos\phi = \frac{1}{3} r_0^2 v_0 c [\exp(3\Theta \tan\phi) - 1] \tag{5.50}$$

c. Radial shear zone *b* – *cc'd'd*
   Area of triangle *bff'* is expressed as

$$S_{bff'} = \frac{1}{2} bf \cdot ff' = \frac{1}{2} r_0^2 \tan\phi \exp(2\theta \tan\phi) \tag{5.51}$$

in which $bf = r_0 \exp(\theta \tan\phi)$. Dissipated work done rate produced in the radial zone is given as

$$\dot{W}_{RE4} = \int_0^{\Theta} S_{bff'} \cdot c \cdot v d\theta = \frac{1}{2} r_0^2 c v_0 \tan\phi \int_0^{\Theta} \exp(3\theta \tan\phi) d\theta$$

$$= \frac{1}{6} r_0^2 c v_0 [\exp(3\Theta \tan\phi) - 1] \tag{5.52}$$

d. Weight of the radial zone $b - cc'd'd$ $(0 \leq \theta \leq \Theta)$
The weight of the unit wedge $b - ff'kk'$ is obtained as

$$W_{b-ff'k'k} = \frac{\gamma}{3} S_{ff'k'k} \cdot bf \cos\phi = \frac{\gamma}{3} r_0^2 R_0 \sin\phi \exp(3\theta \tan\phi) d\theta \tag{5.53}$$

Therefore the driving work done rate produced by the weight of the wedge is expressed as

$$\dot{W}_{DE2} = \int_0^{\Theta} W_{b-ff'k'k} \cdot v \cos(\xi + \theta)$$

$$= \frac{\gamma}{3} r_0^2 R_0 v_0 \sin\phi \frac{\exp[4\Theta \tan\phi] \times [\sin(\Theta + \xi) + 4\tan\phi \cos(\Theta + \xi)] - \sin\xi - 4\tan\phi \cos\xi}{1 + 16\tan^2\phi} \tag{5.54}$$

3. End-failure zone 3
As shown in Figure 5.23c, line *dg* is parallel to line *d'g"*. In order to ensure that the kinematical velocity of soil mass of the end-failure zone 3 is compatible, the angle between line *d'g'* and line *d'g"* should be equal to ϕ. It should be mentioned that straight lines *gg"g'* and *ii'* are both in the slope surface, both triangle *bd'g'* and triangle *bg'i'* are located on the same velocity discontinuity plane *bd'g'i'b* and triangle *bii'* is located in the top surface of the slope.
   a. Velocity discontinuity plane *dd'g'g*.
   The area of velocity discontinuity plane *dd'g'g* is expressed as

$$S_{dd'g'g} = dd' \cdot dg + \frac{1}{2} dg \cdot dg \cdot \tan\phi \tag{5.55}$$

in which $dd' = r_0 \tan\phi \exp(\Theta \tan\phi)$. Resistance work done rate produced by c along the velocity discontinuity plane is then obtained as

$$\dot{W}_{RE5} = c \cdot S_{dd'g'g} \cdot v_3 \cos\phi \tag{5.56}$$

b. Velocity discontinuity plane $bd'g'i'$
As shown in Figure 5.23c, the coordinate of point d' is $x_{d'} = x_d$, $y_{d'} = dd'$, $z_{d'} = z_d$, the coordinate of point g' is $x_{g'} = x_g$, $y_{g'} = dd' + dg \cdot \tan\phi$, $z_{g'} = z_g$ and the coordinate of i' is $x_{i'} = x_i$, $y_{i'} = y_i$, $z_{i'} = z_i$. The equation of the plane formed by points b, d' and g' is written as

$$\begin{vmatrix} x - x_b & y - y_b & z - z_b \\ x_{d'} - x_b & y_{d'} - y_b & z_{d'} - z_b \\ x_{g'} - x_b & y_{g'} - y_b & z_{g'} - z_b \end{vmatrix} = 0 \tag{5.57}$$

As point i' should be on the velocity discontinuity plane $bd'g'i'$, x, y, z are replaced by $x_{i'}$, $y_{i'}$, $z_{i'}$ in Equation 5.57, and $y_{i'}$ is given as

$$y_{i'} = b \frac{z_{d'} \cdot y_{g'} - y_{d'} \cdot z_{g'}}{z_{d'} \cdot x_{g'} - x_{d'} \cdot z_{g'}} \tag{5.58}$$

The area of the velocity discontinuity plane $bd'g'$ is expressed as

$$S_{bd'g'} = \sqrt{\begin{aligned} &\frac{1}{16}(bd' + d'g' + bg')(bd' + d'g' - bg') \\ &\times (bd' - d'g' + bg')(-bd' + d'g' + bg') \end{aligned}} \tag{5.59}$$

in which corresponding $bd' = R_0 \exp(\Theta \tan\phi)$, $d'g' = dg/\cos\phi$ and $bg' = \sqrt{(x_b - x_{g'})^2 + (y_b - y_{g'})^2 + (z_b - z_{g'})^2}$.
The area of the velocity discontinuity plane $bg'i'$ is written as

$$S_{bg'i'} = \sqrt{\begin{aligned} &\frac{1}{16}(bg' + bi' + g'i')(bg' + bi' - g'i') \\ &\times (bg' - bi' + g'i')(-bg' + bi' + g'i') \end{aligned}} \tag{5.60}$$

where

$$g'i' = \sqrt{(x_{i'} - x_{g'})^2 + (y_{i'} - y_{g'})^2 + (z_{i'} - z_{g'})^2}$$
$$bi' = \sqrt{b^2 + y_{i'}^2}$$

The resistance work done rate produced by c along the velocity discontinuity plane is written as

$$\dot{W}_{RE6} = c \cdot (S_{bd'g'} + S_{bg'i'}) \cdot v_3 \cos\phi \tag{5.61}$$

c. Wedge $b - dd'g'g$
Weight of the wedge $b - dd'g'g$ is expressed as

$$W_{b-dd'g'g} = \frac{\gamma}{3} S_{dd'g'g} \cdot bd \cos\phi \tag{5.62}$$

The corresponding resistance work done rate produced by the weight of wedge is obtained as

$$\dot{W}_{DE3} = W_{b-dd'g'g} \cdot v_3 \cos(180° - \eta) \tag{5.63}$$

d. Wedge $b - gg'i'i$
Weight of the wedge $b - gg'i'i$ is given by

$$W_{b-gg'i'i} = \frac{\gamma}{3} \cdot S_{gg'i'i} \cdot b \sin\beta \tag{5.64}$$

in which

$$S_{gg'i'i} = \frac{1}{2} gi \cdot (y_{i'} + y_{g'}), \quad gi = \sqrt{(x_g - x_i)^2 + (z_g - z_i)^2}$$

Then the resistance work done rate produced by the weight of the wedge is

$$\dot{W}_{DE4} = W_{b-gg'i'i} \cdot v_3 \cos(180° - \eta) \tag{5.65}$$

5.3.1.4 Determining the value of the safety factor

The total resistance work done rate of the failure mechanism shown in Figure 5.23 is expressed as

$$\dot{W}_R = \dot{W}_{R1} + \dot{W}_{R2} + \dot{W}_{R3} + 2\left(\dot{W}_{RE1} + \dot{W}_{RE2} + \dot{W}_{RE3} + \dot{W}_{RE4} + \dot{W}_{RE5} + \dot{W}_{RE6}\right) \tag{5.66}$$

The total driving work done rate is obtained as

$$\dot{W}_{D} = \dot{W}_{D1} + \dot{W}_{D2} + \dot{W}_{D3} + \dot{W}_{D4} + 2\left(\dot{W}_{DE1} + \dot{W}_{DE2} + \dot{W}_{DE3} + \dot{W}_{DE4}\right) \tag{5.67}$$

The safety factor F is obtained by means of a simple looping method until the following equation is satisfied

$$\dot{W}_{R} - \dot{W}_{D} = f(\zeta, \xi, \eta) = 0 \tag{5.68}$$

where the angles ζ, ξ and η related to the F value are the critical failure angles ζ_{cr}, ξ_{cr} and η_{cr}. While solving the non-linear Equation 5.68, it was found that the solution is very sensitive to the parameters near to the critical solution. A small change of even 0.5° can sometimes have a noticeable effect to the solution of Equation 5.68 under such condition.

5.3.1.5 *Failure mechanism of the patched load acting at an embedded depth from the top surface of slope (D>0 m)*

General 3D slope failure mechanism with $D>0$ m is shown in Figure 5.24. The work done rate for various components are formulated as follows.

5.3.1.6 *Work done rate produced along footing length L*

a. Resistance work done rate dissipated by cohesion c along velocity discontinuity plane $dg \cdot L$ is calculated by

$$\dot{W}_{R3} = c \cdot dg \cdot L \cdot v_3 \cos\phi \tag{5.69}$$

in which $v_3 = v_0 \exp(\Theta \tan\phi)$ and $bd = r_0 \exp(\Theta \tan\phi)$. As shown in Figure 5.24a, if the coordinate of point b is $x_b=0$, $y_b=0$, $z_b=0$, the corresponding coordinate of point i is $x_i=b$, $y_i=0$, $z_i=-D$, the coordinate of point d is $x_d = bd\cos\eta$, $y_d=0$, $z_d = bd\sin\eta$, the coordinate of point h is $x_h=0$, $y_h=0$, $z_h=-D$. The values of x_g and z_g are obtained through the geometric relationships:

$$\frac{z_g - z_d}{x_g - x_d} = \tan(\phi + \eta - 90°) \quad \text{and} \quad \frac{z_g - z_i}{x_g - x_i} = \tan\beta \tag{5.70}$$

Based on Equation 5.70, x_g and z_g are expressed as

$$x_g = \frac{b\tan\beta + D + z_d - x_d\tan(\phi + \eta - 90°)}{\tan\beta - \tan(\phi + \eta - 90°)}, \quad z_g = (x_g - b)\tan\beta - D \tag{5.71}$$

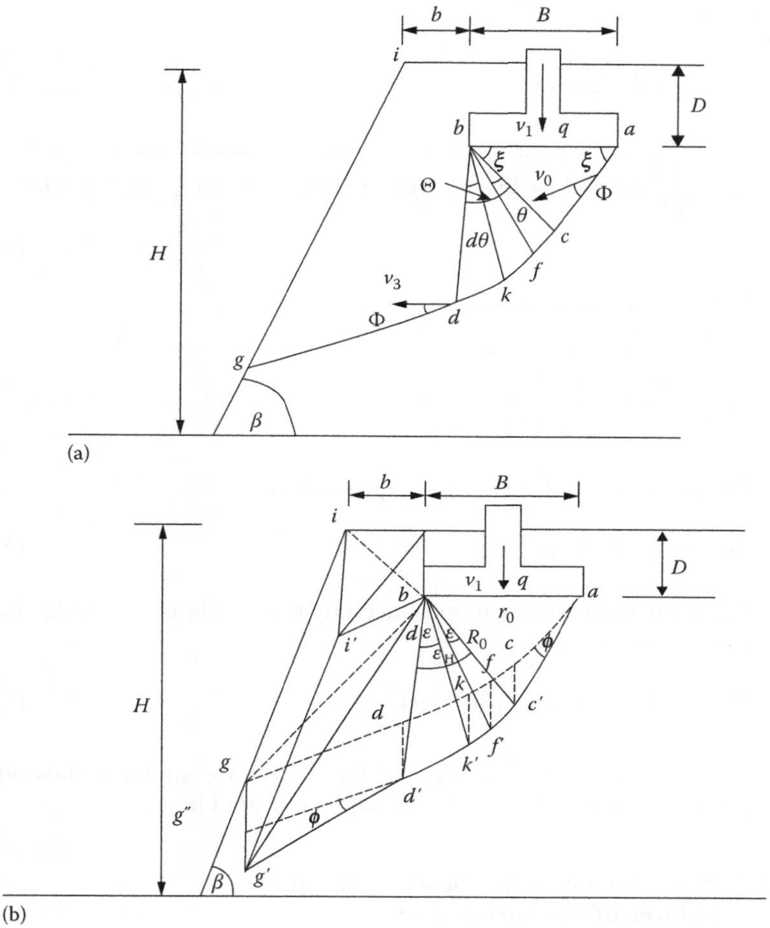

Figure 5.24 **Three-dimensional failure mechanism for buried load: (a) failure mechanism (excluding the two ends) and (b) failure mechanism for the two ends.**

And dg in Equation 5.69 is obtained as

$$dg = \sqrt{(x_g - x_d)^2 + (z_g - z_d)^2} \tag{5.72}$$

b. Resistance work done rate produced by the weight of the wedge *bdgih*. Area of *bdgih* is obtained as

$$S_{bdgih} = S_{bdg} + S_{big} + S_{bhi} \tag{5.73}$$

in which

$$S_{bdg} = \frac{1}{2} bd \cdot dg \cos\phi \tag{5.74}$$

$$S_{big} = \sqrt{\frac{1}{16}(bg + gi + bi)(bg + gi - bi)(bg - gi + bi)(-bg + gi + bi)} \tag{5.75}$$

where

$$bg = \sqrt{(x_b - x_g)^2 + (z_b - z_g)^2}$$
$$gi = \sqrt{(x_g - x_i)^2 + (z_g - z_i)^2}$$
$$bi = \sqrt{(x_b - x_i)^2 + (z_b - z_i)^2}$$

Weight of the wedge *bdgih* is expressed as

$$W_{bdgih} = \gamma \cdot L \cdot S_{bdgih} \tag{5.76}$$

Then, the work done rate produced by the weight of the wedge *bdgih* is given as

$$\dot{W}_{D4} = W_{bdgih} \cdot v_3 \cos(180° - \eta) \tag{5.77}$$

Other items such as $\dot{W}_{R1}$, $\dot{W}_{R2}$ and $\dot{W}_{D1} \sim \dot{W}_{D3}$ are similar to those given in the previous section and will not be repeated here.

5.3.1.7 Work done rate produced at two end failures of the buried load

As shown in Figure 5.24b, the coordinates of point d', g' and i' are similar to the case of $D = 0$. As point i' should be on the velocity discontinuity plane $bd'g'i'$, x, y, z are replaced by $x_{i'}$, $y_{i'}$, $z_{i'}$ in the Equation 5.57, then $y_{i'}$ is

$$y_{i'} = b\frac{z_{d'} \cdot y_{g'} - y_{d'} \cdot z_{g'}}{z_{d'} \cdot x_{g'} - x_{d'} \cdot z_{g'}} + D\frac{x_{d'} \cdot y_{g'} - y_{d'} \cdot x_{g'}}{z_{d'} \cdot x_{g'} - x_{d'} \cdot z_{g'}} \tag{5.78}$$

a. Velocity discontinuity plane $bd'g'i'$
   For the velocity discontinuity plane $bd'g'$, it is similar to the previous case except that

$$bi' = \sqrt{x_{i'}^2 + y_{i'}^2 + z_{i'}^2} \tag{5.79}$$

The resistance work done rate produced by c along the velocity discontinuity plane $bd'g'i'$ is also given by Equation 5.61.

b. Resistance work done rate produced by the tensile failure plane bhi'
Area of the tensile failure plane bhi' is expressed as

$$S_{bhi'} = \frac{1}{2} D \cdot hi' \tag{5.80}$$

in which $hi' = \sqrt{b^2 + y_{i'}^2}$. Usually, the tensile strength of soil mass can be taken as (1/4 ~ 1.0)c (Baker, 1980). The calculation shown in the later part of this chapter will demonstrate that the tensile strength of soil mass has only a small effect on the safety factor, so it is assumed to be equal to $c/3$ in the present study. As the tensile direction is along the direction of velocity v_3, the corresponding resistance work done rate produced by the tensile failure plane is written as

$$\dot{W}_{\text{RE7}} = \frac{c \cdot S_{bhi'} \cdot v_3}{3} \tag{5.81}$$

c. Driving work done rate produced by wedges $b - dd'g'g$, $b - ii'g'g$ and $b - hii'$
Weight of the wedge $b - dd'g'g$ is expressed as

$$W_{b-dd'g'g} = \frac{\gamma}{3} S_{dd'g'g} \cdot bd \cos\phi \tag{5.82}$$

Area of the slope surface $ii'g'g$ is expressed as

$$S_{ii'g'g} = \frac{1}{2}\left(ii' + gg'\right) \cdot gi \tag{5.83}$$

in which

$$ii' = y_{i'}, \quad gg' = dd' + dg \tan\phi, \quad \text{and} \quad gi = \sqrt{(x_g - x_i)^2 + (z_g - z_i)^2}$$

Weight of the wedge $b - ii'g'g$ is given as

$$W_{b-ii'g'g} = \frac{\gamma}{3} \cdot S_{ii'g'g} \cdot \left(b \cdot \sin\beta + D \cdot \cos\beta\right) \tag{5.84}$$

Area of triangle hii' is given as

$$S_{hii'} = \frac{1}{2} b \cdot y_{i'} \tag{5.85}$$

Weight of the wedge $b - hii'$ is written as

$$W_{b-hii'i} = \frac{\gamma}{3} \cdot S_{hii'} \cdot D \tag{5.86}$$

Then driving work done rate produced by the weight of these wedges is expressed as

$$\dot{W}_{DE3} = \left(W_{b-dd'g'gi} + W_{b-ii'g'g} + W_{b-hii'} \right) \cdot v_3 \cos\left(180° - \eta\right) \tag{5.87}$$

$\dot{W}_{RE1} \sim \dot{W}_{RE5}$ and $\dot{W}_{DE1} \sim \dot{W}_{DE2}$ are similar to the case for $D = 0$ m and will not be repeated here. Referring to Figure 5.24, the total resistance work done rate is expressed as

$$\begin{aligned} \dot{W}_R &= \dot{W}_{R1} + \dot{W}_{R2} + \dot{W}_{R3} \\ &\quad + 2\left(\dot{W}_{RE1} + \dot{W}_{RE2} + \dot{W}_{RE3} + \dot{W}_{RE4} + \dot{W}_{RE5} + \dot{W}_{RE6} + \dot{W}_{RE7} \right) \end{aligned} \tag{5.88}$$

The total driving work done rate of Figure 5.24 is obtained as

$$\dot{W}_D = \dot{W}_{D1} + \dot{W}_{D2} + \dot{W}_{D3} + \dot{W}_{D4} + 2\left(\dot{W}_{DE1} + \dot{W}_{DE2} + \dot{W}_{DE3} \right) \tag{5.89}$$

F will be obtained by setting Equation 5.88 equals Equation 5.89.

5.3.1.8 Comparison of Cheng's method with other analytical solutions

Referring to Figure 5.23, when $b = 0$, based on the slip-line fields by Sokolovskii (1965), it is possible to find the closed-form solution N_c to the ultimate load q_u for a weightless soil mass as

$$N_c = c \cdot \cot\phi \left\{ \tan^2\left(45° + \frac{\phi}{2}\right) \exp\left[\left(\pi - 2\beta\right)\tan\phi\right] - 1 \right\} \tag{5.90}$$

For a 2D plane problem with weightless soil mass, N_c values for different slope angles are calculated by using Equations 5.68 and 5.90 separately, and the results are shown in Figure 5.25. The general tends for the variations of the N_c and angle friction, which was predicted by both of the methods, are similar but N_c values by Equation 5.68 are slightly larger than the N_c values by Equation 5.90. This indicates that the 3D failure mechanism of Figure 5.23 is a reasonable upper-bound solution for 2D analysis (not exact, but good enough for engineering purposes).

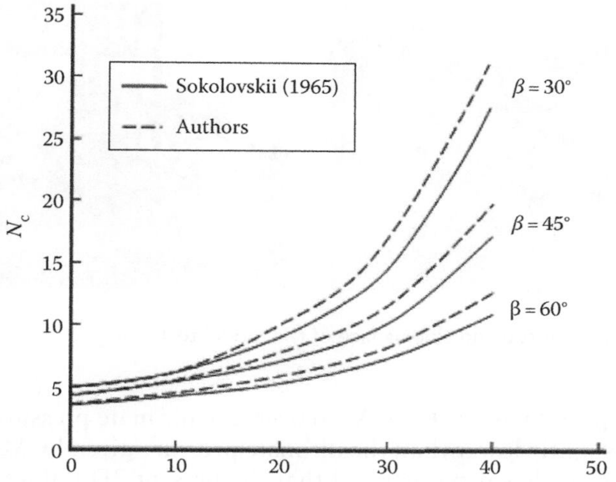

Figure 5.25 Comparison of N_c values between the Sokolovskii method and Cheng's method.

Further comparison has been carried out for a 3D slope stability analyses with the following soil properties ($c = 20$ kN/m^2, $\phi = 20°$ and $\gamma = 20$ kN/m^3) and slope geometry ($B = 2$ m, $b = 1$ m, $D = 0$ m and $H = 6$ m). The results of the dimensionless limit pressure q_u/c values with different L/B values given by Michalowski (1989) and Cheng are illustrated in Figure 5.26. The ultimate pressures decrease with the increase in L/B ratio. Until the L/B value is greater than 5, the normalized ultimate pressures are not sensitive to the variations of the L/B values. Such tends are shown in both the methods. This indicates that the effect of the patched pressure on the top surface of the slope increases rapidly as the L/B ratio is

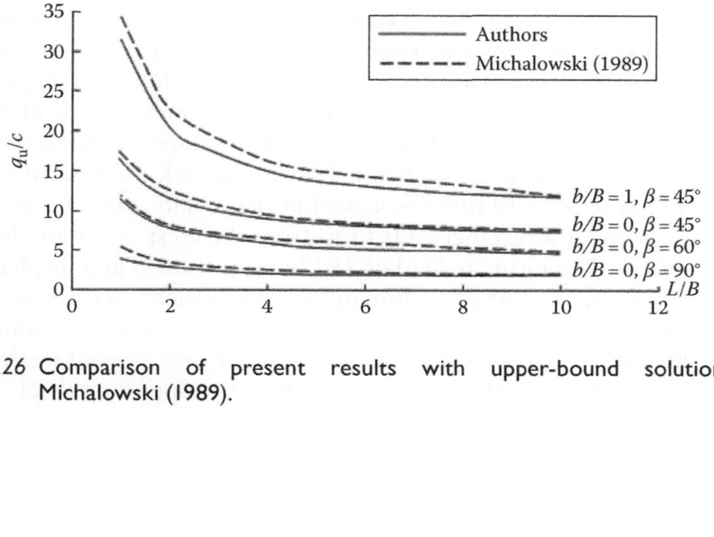

Figure 5.26 Comparison of present results with upper-bound solutions by Michalowski (1989).

Figure 5.27 LVDT at top and sloping face of the model test.

reduced, especially for $L/B < 5$. Moreover, the ultimate pressures given by Cheng are slightly lower than the ultimate pressure given by Michalowski (1989), and this has demonstrated that the present 3D failure modes are more reasonable and more critical than that by Michalowski (1989). The present results can give better predictions for the ultimate local pressure on the top surface of the slope.

A laboratory test complying exactly with the present problem as shown in Figure 5.27 has been conducted for the verification of the proposed method. A hydraulic jack applied a local pressure on top of a 0.8 m high 65° slope. The soil used for the model slope is classified as highly permeable poorly graded river sand. The unit weight and the relative density are $\gamma = 15.75$ kN/m^3 and 0.55, respectively. Shear strength parameters ($c' = 7$ kPa and $\phi' = 35°$) of the soil were determined by means of a consolidated drained triaxial test. The depth, breadth and height of the soil tank are 1.5 m, 1.85 m and 1.2 m, respectively. The soil is compacted by an electric compactor with a 0.2 m×0.2 m wooden end plate. Five linear variable differential transducers (LVDT) are set up to measure the displacement of soil at different locations, which included the upper right-hand side (RHS), upper left-hand side (LHS), lower right-hand side (RHS) and the lower left-hand side (LHS), as shown in Figure 5.27. The two pairs of transducers on the slope surface are placed symmetrically with a horizontal spacing of 300 mm. The first and second pairs of transducers are placed at vertical distances of 150 and 450 mm from the top of the slope, respectively.

The vertical displacement-controlled hydraulic jack exerts uniform distributed pressure on a 10 mm thick steel-bearing plate with size $B = 0.3$ m and $L = 0.644$ m at a distance of 0.13 m from the crest of the model slope ($b = 0.13$ m) until an ultimate load of 35 kN was attained at a displacement of about 6 mm. As a result, the ultimate bearing capacity of the slope under the current soil properties, geometrical conditions and boundary conditions is 181.2 kPa, which gives a safety factor of 1.021 by the present method and is very close to 1.0. For the slope surface, the corresponding displacement at the maximum pressure is about 2 and 1 mm at the top and the bottom of the slope, respectively. Beyond the peak load, the applied load decreases with increasing jack displacement. It is clear that the displacements of the slope are basically symmetrical. The failure surface of the present test is shown in Figures 5.28 and 5.29, and the sectional view at the middle of the failure mass is shown in Figure 5.30. After the maximum load is achieved, the load will decrease with increasing displacement. At this stage, the local triangular failure zone is fully developed while the failure zones at the two ends of the plate are not clearly formed. When the applied load decreases to about 25 kN, the load is maintained constant for a while and the failure zones at the two ends become visually clear. When the displacements are further increased, the applied load decreases further and the failure zone

Figure 5.28 Bearing capacity failure beneath the bearing plate.

Figure 5.29 Three-dimensional bearing failure beneath the bearing plate.

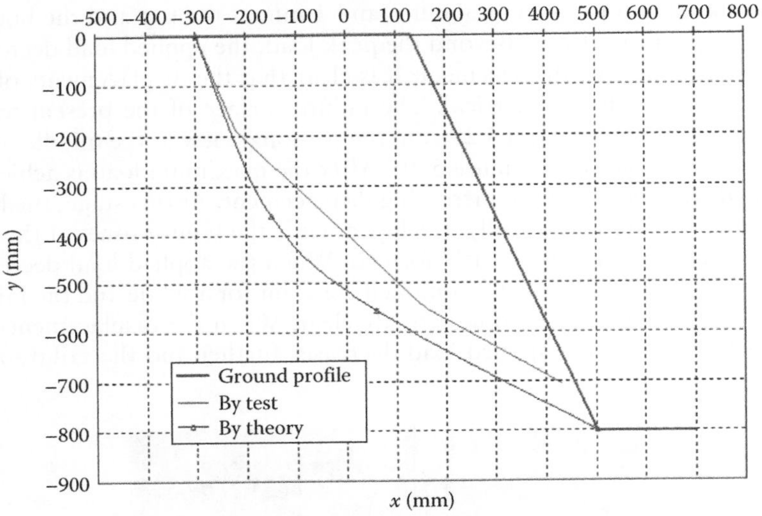

Figure 5.30 Comparisons between the measured and the predicted failure surface profile at midsection of failure.

propagate towards the slope surface until the failure surface as shown in Figure 5.30 is obtained. This 3D failure mechanism measured from the model test is basically similar to that given in the present chapter, and the prediction of the FOS from the present theory is also satisfactory.

For this test, there are several interesting phenomena worth discussing. The failure profile and cracks first initiated beneath the footing are shown in Figure 5.28, which is a typical bearing capacity failure with a triangular failure zone. This can also be observed from the upper part of the failure profile, as shown in Figures 5.28 and 5.29; ε and η are obtained as 67.5° and 83.8° from the critical FOS, as given by Equation 5.68. As the load increased, the failure zone extended and propagated towards the toe of the slope, and the final failure surface is shown in Figure 5.29. The failure mechanism of the physical model test is a local triangular failure beneath the bearing plate, and the failure surface propagates towards the slope surface until a failure mechanism is formed. The failure profile matched reasonably well with that predicted from the present formulation, as shown in Figure 5.30, and is also in compliance with that developed by Cheng and Au (2005) using the slip-line method. In addition, the prediction of the FOS is also close to the back analysis result of the laboratory test. In view of the difficulty in ensuring complete uniformity for the compaction of the model slope, the small discrepancy between the predicted and measured failure profiles as shown in Figure 5.30 can be considered acceptable.

5.4 LOCATION OF GENERAL CRITICAL NON-SPHERICAL 3D FAILURE SURFACE

Up to the present, there has been only limited research in determining the critical 3D slip surface due to the difficulties in performing large-scale global optimization analysis of an N–P type function. The 3D critical slip surface can be classified into two major groups: (1) the first assumes a slip surface to have a particular shape, for example, an extended circular arc (Baligh and Azzouz, 1975), a cylindrical surface (Ugai, 1985) or an ellipsoidal surface (Zhang, 1988); (2) the second is valid for an arbitrary slip surface. When the analysis of slope stability was carried out based on the first group of methods, the slip surface could be expressed analytically and the critical slip surface could be found easily through simple numerical computations.

Based on the methods that were valid for a slip surface of arbitrary shape, however, it was quite difficult to search for the critical surface because possible slip surfaces exist infinitely. Thomaz and Lovell (1988) recommended a procedure for 3D slope stability analysis using the method of random generation of surfaces. It has been indicated that the critical slip surface determined using the method was probably not the most critical one since the convergence criterion for solutions with required precision did not exist.

Leshchinsky et al. (1985) and Leshchinsky and Baker (1986) presented a mathematical approach based on limit equilibrium and variational calculus for 3D slope stability analysis. Solving the variational limit equilibrium equations made it possible to obtain the minimum FOS and the associated critical surface at the same time. However, the method based on variational analysis had not yet been applied to practical problems because of the required mathematical format.

Yamagami and Jiang (1997) proposed an approach based on dynamic programming and random number generation. The approach could determine the location and shape of the 3D critical slip surface as well as the associated FOS for a slope of arbitrary shape including layered soils and/or phreatic surface. The random number generation was employed to generate states and thus transformed the 3D dynamic programming problem into a 2D one, while 3D slope stability analysis could not directly be performed using a dynamic programming algorithm only. The dynamic programming approach by Yamagami and Jiang (1997) is generally applicable for simple problems, but there is no mechanism behind dynamic programming method to escape from the local minimum during the optimization. The problem of the local minimum for some complicated 2D problems has been studied by Cheng (2003), and this is a difficult problem for arbitrary slope problem. For complicated problems where the presence of a local minimum may be a critical factor in the search for a global minimum, the solution will be a difficult N–P type optimization problem for both 2D and 3D problems.

5.4.1 Three-dimensional NURBS surfaces

The success of 3D global optimization requires the description of a general 3D surface using limited control variables but one that is able to model arbitrary geometry. This is extremely difficult, and there is no simple way to ensure a very special shape can be generated for an arbitrary solution domain. However, for most of the normal cases, a relatively smooth function may be good enough to model the 3D surface. NURBS, or non-uniform rational B-splines (Les and Wayne, 1997; David, 2001), are now commonly adopted for describing and modelling curves and surfaces in solid modelling, computer-aided design and computer graphics. It is capable of representing a regular surface such as flat planes and quadric surfaces as well as complex fully sculptured surfaces with only few local and global controls. A NURBS surface is a special case of a general rational B-spline surface that uses a particular form of knot vector. For a NURBS surface, the knot vector has a multiplicity of duplicate knot values equal to the order of the basis function at the ends, that is, a NURBS surface uses an open knot vector. The knot vector may or may not have uniform internal knot points, and this can be controlled by the user easily. Non-uniform spaced knot points are important for slope stability analysis as the critical failure surface may have an arbitrary extent controlled by the topography and the use of non-uniform spaced knot point is necessary for general problems.

A Cartesian product of rational B-spline surface in four-dimensional (4D) homogeneous coordinate space is given by Les and Wayne (1997) and David (2001):

$$Q(u,w) = \sum_{i=1}^{n+1} \sum_{j=1}^{m+1} B_{i,j}^{h} N_{i,k}(u) M_{j,l}(w) \tag{5.91}$$

where

$B_{i,j}^{h}$s are the 4D homogeneous polygonal control vertices (3D coordinates and coordinate weight factor which are stored in matrix NURBSSurface, as discussed later)

$N_{i,k}(u)$ and $M_{i,l}(w)$ are the non-rational B-spline basis functions in the x and y directions, respectively, given in the following equations:

$$N_{i,1}(u) = \begin{cases} 1 & \text{if } x_i \le u < x_{i+1} \\ 0 & \text{otherwise} \end{cases} \tag{5.92a}$$

$$N_{i,k}(u) = \frac{(u - x_i) N_{i,k-1}(u)}{x_{i+k-1} - x_i} + \frac{(x_{i+k} - u) N_{i+1,k-1}(u)}{x_{i+k} - x_{i+1}}$$

$$u_{\min} \le u < u_{\max}, \quad 2 \le k \le n+1 \tag{5.92b}$$

$$M_{j,1}(w)=\begin{cases}1 & \text{if } y_j \le w < y_{j+1}\\ 0 & \text{otherwise}\end{cases} \tag{5.92c}$$

and

$$M_{j,l}(w)=\frac{(w-y_j)M_{j,l-1}(w)}{y_{j+l-1}-y_j}+\frac{(y_{j+l}-w)M_{j+1,l-1}(w)}{y_{j+l}-y_{j+1}}$$

$$w_{\min}\le w < w_{\max},\quad 2\le l \le m+1 \tag{5.92d}$$

Projecting back into 3D space by dividing with the homogeneous coordinate gives the rational B-spline surface as

$$Q(u,w)=\frac{\sum_{i=1}^{n+1}\sum_{j=1}^{m+1}h_{i,j}B_{i,j}N_{i,k}(u)M_{j,l}(w)}{\sum_{i=1}^{n+1}\sum_{j=1}^{m+1}h_{i,j}N_{i,k}(u)M_{j,l}(w)}=\sum_{i=1}^{n+1}\sum_{j=1}^{m+1}B_{i,j}S_{i,j}(u,w) \tag{5.93}$$

where

$B_{i,j}$s are the 3D control net vertices (3D coordinates which are a sub-matrix of matrix NURBSSurface)

$S_{i,j}$ (u, w) are the bivariate rational B-spline surface basis functions given by

$$S_{i,j}(u,w)=\frac{h_{i,j}N_{i,k}(u)M_{j,l}(w)}{\sum_{i1=1}^{n+1}\sum_{j1=1}^{m+1}h_{i1,j1}N_{i1,k}(u)M_{j1,l}(w)}=\frac{h_{i,j}N_{i,k}(u)M_{j,l}(w)}{\text{Sum}(u,w)} \tag{5.94}$$

where

$$\text{Sum}(u,w)=\sum_{i1=1}^{n+1}\sum_{j1=1}^{m+1}h_{i1,j1}N_{i1,k}(u)M_{j1,l}(w)$$

It is convenient, though not necessary, to assume $h_{i,j} \ge 0$ for all i, j. The smooth NURBS surface will be controlled by the coordinates of the control points but will not pass through the control points exactly. The greater the values of $h_{i,j}$, the closer will be the NURBS surface to the control points.

In the earlier-mentioned formulas, the symbols are as follows:

u and w are the NURBS surface's transverse and longitudinal directions, being similar to x axis and the y axis.

n and m are the number of control net vertices in the u and w directions.

k and l are the order in the u and w directions.

$b()$ is the array containing the control net vertex:

$b(,1)$ contains the x component of the vertex.

$b(,2)$ contains the y component of the vertex.

$b(,3)$ contains the z component of the vertex.

$b(,4)$ contains the homogeneous coordinate weighting factor, h.

$B_{i,j}$ are the 3D control net vertices. $B_{i,j}=b(n\times m,\ 1\sim 3)$.

$B^h_{i,j}$ are the 4D homogeneous polygonal control. $B^h_{i,j}=b(n\times m, 1\sim 4)$.

When Equations 5.91 through 5.94 are used, the number of control net vertices must be equal to $n\times m$. However, for slope stability analysis, it is not always practical to have $n\times m$ control net vertices. In the optimization analysis, the coordinates of the control nodes change and a $n\times m$ net vertices regular grid cannot be adopted. If the control net vertices are not arranged to form a regular grid, the NURBS surface may twist seriously (cusps). A cusp is highly unlikely to occur in reality and should be avoided in the generation of a non-spherical failure surface. In fact, restraining forces are provided by cusp (if present), and this situation can be eliminated in the generation of non-spherical failure surfaces. Excessive unacceptable failure surfaces generated from the NURBS points greatly reduce the efficiency of analysis, and this has been experienced by Cheng et al. (2005) in a preliminary study, and a simple method is proposed to generate a NURBS surface which can avoid this problem.

Firstly, four extreme fixed corners (net vertices) define a domain similar to a net of $n\times m$ points (see Figure 5.31). The z ordinates of these four extreme fixed corners can change during the optimization analysis, and

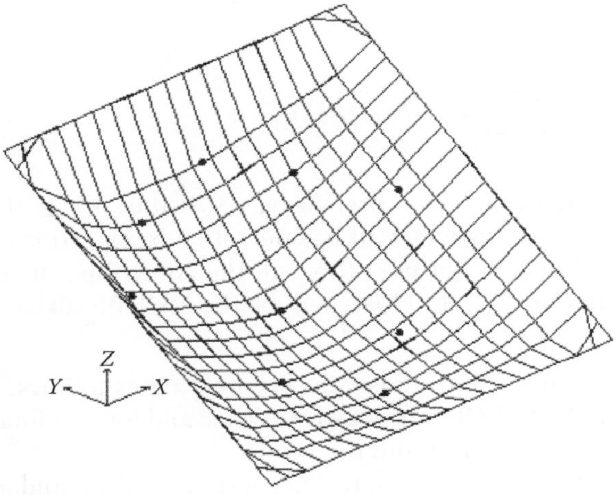

Figure 5.31 The NURBS surface with nine control nodes.

they are the control variables. In general, the user should define a solution domain large enough to cover all possible failure mechanisms, and this is usually not difficult. To eliminate the formation of a cusp, a NURBS surface can be viewed as a net stretched tightly with in-plane force. Secondly, each control node should affect the net in order in forming the 3D sliding surface. The coordinates of every point on the regular grid are controlled by the control node. This approach is equivalent to putting a stone on a net stretched tightly and every net point sinks with the stone. A NURBS surface with no kink is then generated even if the control points are not spaced regularly. Obviously, the more the control nodes used for modelling, the better will be the quality of the NURBS surface. Thirdly, the shape of the NURBS surface will change with the coordinates and weighting factors of control nodes. The greater the weight factor of a control node, the closer the NURBS net is to the control node. It is easy to modify the shape of a NURBS surface by change the weighting factors.

During the simulated annealing analysis, each control variable will vary sequentially, and Cheng (2003) has proposed a simple trick to avoid generating an unacceptable 2D failure surface by changing the requirement of *kinematically acceptable mechanism* to *dynamic boundaries* of control variables. Such a technique has been used by Cheng and Zhu (2005) for 3D analysis. Therefore, the shape of the trial failure surfaces is examined along the longitudinal as well as transverse directions so that the boundaries of the control variables are modified, which is effectively Cheng's approach (2003) applied along two directions. Using this technique, most of the failure surfaces generated are kinematically acceptable. By using the concept of stretched net and the requirement of a *kinematically acceptable mechanism*, most of the failure surfaces as generated from NURBS functions are suitable for optimization analysis.

5.4.2 Spherical and ellipsoidal surface

For simple problems, the use of spherical and ellipsoidal failure surfaces may be sufficiently good in application. For a spherical failure surface, it is simple to operate, and the number of control variables are four: (xc, yc, zc, r), where (xc, yc, zc) are the coordinates of the centre of the sphere and r is the radius of the sphere. For an ellipsoidal failure surface defined by global axes x, y, z, the equation will be

$$\frac{(x-xc)^2}{a^2}+\frac{(y-yc)^2}{b^2}+\frac{(z-zc)^2}{c^2}=1 \tag{5.95}$$

The ellipsoid can be defined in terms of the rotated axes x', y' and z' so that two more additional control variables for the rotation of the axes are required in the optimization process. The number of control variables are hence eight: (xc, yc, zc, a, b, c, θx, θy), where (xc, yc, zc) is the centre of the ellipsoid, a, b, c are the axis lengths of the ellipsoidal and θx and θy are the rotation of the xy and yz planes, respectively. The use of an ellipsoidal failure surface is attractive in that the number of control variables is not great and the solution time is acceptable for ordinary design. Every ellipsoid is convex in shape and includes a spherical shape as a special case so that it is suitable for ordinary problems. Example 5.3 will illustrate the advantage of using the ellipsoid in analysis.

5.4.3 Selection of sliding surfaces

For 3D limit equilibrium analysis, the potential failure mass of a slope is divided into a number of columns. The NURBS/spherical or ellipsoidal surface intersects with the ground profile and generates soil columns and sliding mass. For the 3D failure surfaces as generated, some surfaces are not unacceptable and should be removed from the analysis. The following failure surfaces should be eliminated in the optimization analysis:

1. The number of columns formed is too small. Cheng et al. (2005) have found that if the number of columns used for stability analysis is too small, the results on the safety factor will be greatly affected. This kind of problem may come up if the size of each column is too large (see Figure 5.32a). Generally, this problem can be avoided by a good pre-processing of the mesh for generating the soil columns and is only a minor problem.
2. The sliding surface is not a complete concave surface. Most sliding surfaces are completely concave surfaces. Any sliding surface which is composed of concave and convex portions can be eliminated in the analysis (see Figure 5.32b). This case is absent for spherical and ellipsoidal failure surfaces and is also not commonly found as the concave portion will induce additional restraining forces and is not critical in general. However, the user should be given the choice that this type of composite failure surface can be accepted in the optimization process. Based on the present proposal on the use of dynamic domains to the control variables in longitudinal and transverse directions, this situation is practically eliminated.
3. If the failure mass is divided into several unconnected parts as shown in Figure 5.32c, the failure surface will not be accepted. For the case shown in Figure 5.32c, only the larger failure mass is considered acceptable in the analysis, while the smaller failure zone is not considered.

An acceptable 3D failure mass for computation similar to that shown in Figure 5.33 will then be generated from the selection criteria as given earlier.

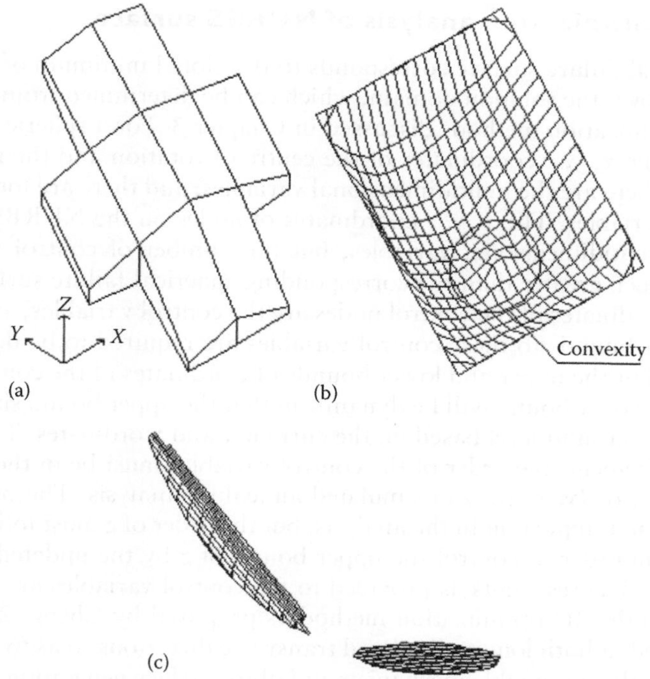

Figure 5.32 Three cases should be considered.

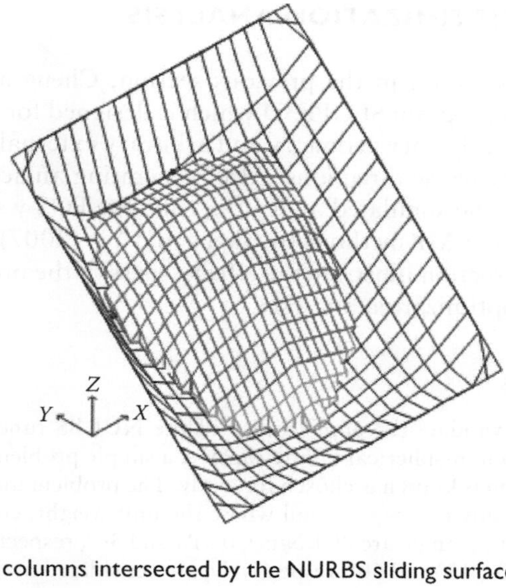

Figure 5.33 Sliding columns intersected by the NURBS sliding surface.

5.4.4 Optimization analysis of NURBS surface

The critical failure surface corresponds to the global minimum of the FOS function over the solution domain, which can be determined from the heuristic optimization methods discussed in Chapter 3. For a spherical failure surface, the x, y, z coordinates of the centre of rotation and the radius of sliding sphere are the multidimensional variables, and there are totally four control variables. The x, y, z coordinates of nodes on the NURBS surface are the multidimensional variables, but the number of control variables will be much greater than the corresponding spherical failure surface. The x, y, z coordinates of the control nodes are the control variables, and upper and lower bounds to these control variables are required to be defined by the user. For the upper and lower bounds of z ordinates of the control variables, the upper bound will be dynamic in that the upper bound should not exceed the ground level based on the current x and y ordinates. To achieve this requirement, the order of the control variables must be in the form of $(x, y, z,\ldots)$ or $(y, x, z,\ldots)$ in simulated annealing analysis. The order of x and y are not important in the analysis, but the order of z must follow (x, y) or (y, x) in order to control the upper bound of z by the updated x and y ordinates. The restraints as provided to the control variables are basically similar to the 2D optimization method as proposed by Cheng (2003) but are applied in both longitudinal and transverse directions so as to impose a kinematically acceptable mechanism in failure surface generation.

5.5 CASE STUDIES IN 3D LIMIT EQUILIBRIUM GLOBAL OPTIMIZATION ANALYSIS

Based on the discussion in the previous section, Cheng and Zhu (2005) have developed a program SLOPE3D which is designed for a general asymmetric slope with arbitrary geometry and arbitrary external load in the longitudinal and transverse directions. After generating an acceptable sliding surface based on the simulated annealing rule, the FOS was calculated by 3D Bishop, Janbu or MP methods by Cheng and Yip (2007). The numerical examples in this section illustrate the effectiveness of the proposed NURBS function in the optimization analysis.

Example 5.6

In order to validate the applicability of the NURBS function in the location of a non-spherical failure surface, a simple problem where the exact solution is known is chosen for study. The problem under consideration has only one type of soil where the unit weight, cohesion and internal friction angle are 20 kN/m^3, 0 kPa and 36°, respectively. Slope angles for 30°, 45° and 60°, respectively, are considered in the analysis.

Table 5.8 Minimum FOSs after optimization calculation

Factors of safety \ *Slope degree*	*30°*	*45°*	*60°*
Bishop[a]	1.25747	0.72625	0.41998
Janbu	1.25749	0.72627	0.42002
MP	1.25755	0.72575	0.41905
Theoretical value: $F = \frac{\tan\phi}{\tan\theta}$	$\frac{\tan 36°}{\tan 30°} = 1.25841$	$\frac{\tan 36°}{\tan 45°} = 0.72654$	$\frac{\tan 36°}{\tan 60°} = 0.41947$

[a] Spherical search is applied to the Bishop method, while the NURBS function simulated annealing search is used for Janbu's and MP's methods.

Theoretically, the critical failure surface is a very shallow symmetrical failure surface parallel to the slope surface and the FOS is equal to $\tan\phi/\tan\theta$, where ϕ is the friction angle of soil and θ is the slope angle of the slope. The minimum FOSs after optimization calculation are given in Table 5.8.

In the present example, 17 control nodes are used to generate the NURBS failure surface, and very good results are obtained from the analysis. All the critical failure surfaces are very shallow type surfaces parallel to the ground surface, which is in accordance with classical soil mechanics theory. The small differences between the optimized values and the theoretical values can be considered to be acceptable in view of the discretization required for computation. These results have demonstrated the effectiveness of the NURBS function under this simple condition. For the present problem, the results are practically independent of the number of control points (unless the number is very small) as the critical failure surface is a very shallow surface. As long as the control points are near the ground surface, a good shallow failure surface practically parallel to the ground surface is generated. For the following two examples where the critical failure surfaces are not parallel to the ground surface, the results are more sensitive to the number of control points.

Example 5.7

To illustrate the differences between the minimum FOSs from spherical and NURBS failure surfaces, a special problem is devised with 3D Janbu analysis. The geometry of the slope is shown in Figure 5.34. There are two kinds of soils with a groundwater table. The geological profile remains constant in the direction normal to the figure. The cohesion of the upper soil is much less than that of the lower soil so that the critical failure surface is controlled by the boundary between the two layers of soil. Firstly, the critical spherical sliding surface is evaluated by a simulated annealing algorithm. After 11,521 calculation steps, a minimum FOS, 0.6134, is obtained for 3D Janbu analysis. The x, y, z coordinates of the centre of rotation and the radius of the spherical sliding surface with the minimum FOS are (0.0000, −0.3462, 8.5384) and 6.1879 m, respectively. The spherical critical failure surface shown

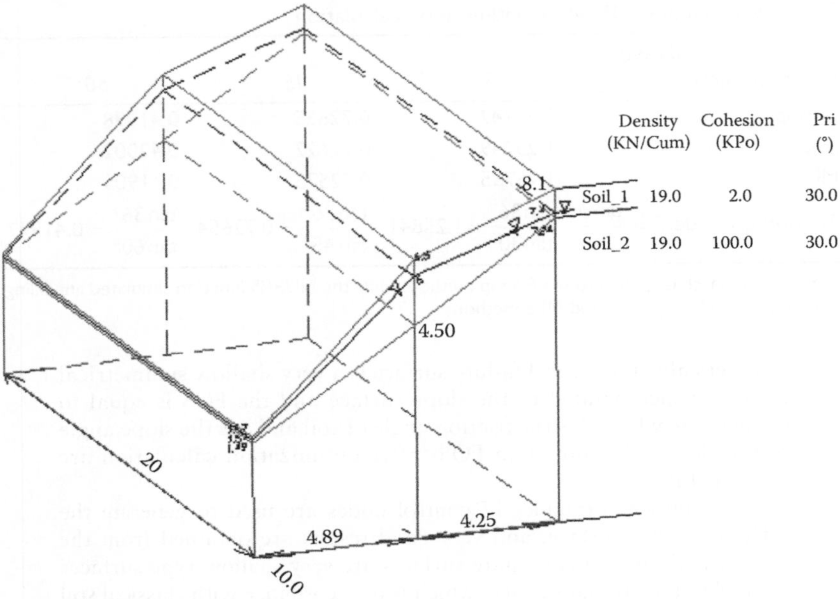

Figure 5.34 Slope geometry for Example 5.7.

in Figure 5.36b is tangential to the boundary between layer 1 and layer 2 at one point (geometry requirement).

Secondly, a NURBS sliding surface is constructed with 17 nodes. The dimension of the control net SS′T′T shown in Figure 5.35c is 8 m×6 m, and there are 361 columns used for analysis. After 78,943 calculation steps, a minimum FOS of 0.5937, which is smaller than that obtained by spherical failure surface search, is attained. The finer mesh shown in Figure 5.35c is the failure surface and the greater mesh is the NURBS grids as discussed before. The critical symmetrical NURBS failure surface formed by the control net SS"T"T and the NURBS surface is tangential to the boundary between layers 1 and 2 over a region instead of just a single point touch, which is the expected solution. In Figure 5.35d, the failure mass is the lower part of the complete NURBS surface formed by the control net. This example has illustrated the importance of a non-spherical search for a critical 3D failure surface under general condition. For five NURBS points, the minimum FOSs and number of trials are 0.751 and 46,082 from optimization search, while the corresponding figures for 10 NURBS points are 0.647 and 65,090. The minimum FOS is practically insensitive to the number of NURBS points if the number is 15 or above. Unlike the previous case which is practically a planar failure mode, the present critical failure surface requires 15 or more NURBS points for a good description of the 3D failure surfaces.

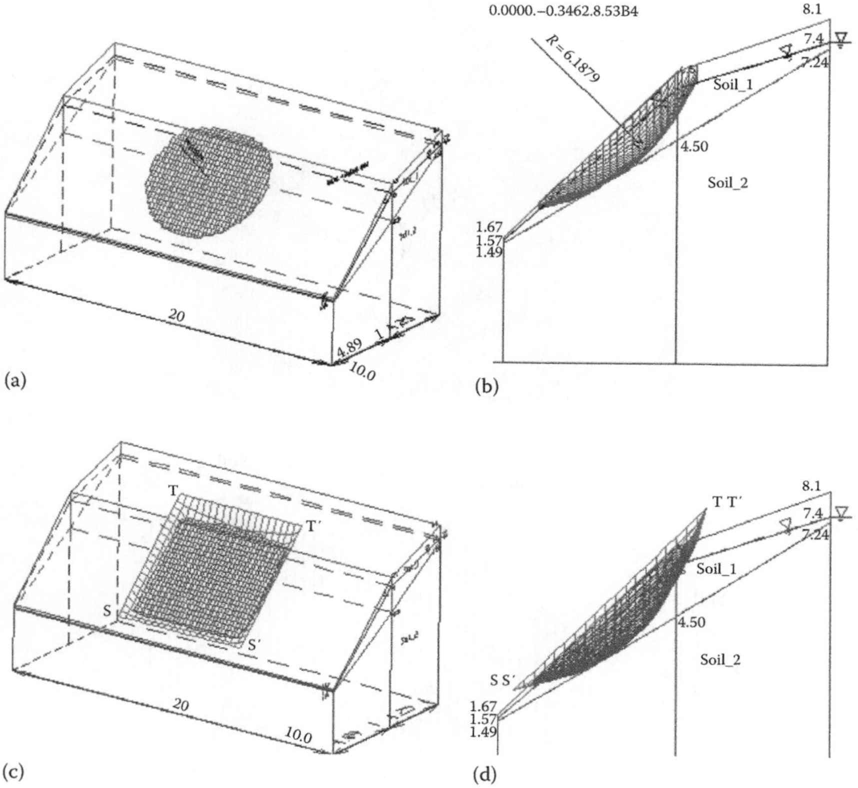

Figure 5.35 Sliding surface with the minimum FOS for Example 5.7: (a) spherical sliding surface, (b) section through middle of spherical failure mass, (c) NURBS sliding surface and (d) section through centre of the NURBS surface.

Example 5.8

In this example, a typical slope in Hong Kong with a soft band is considered. As shown in Figure 5.36, there are three kinds of soils and no groundwater table. The unit weight, cohesion and internal friction angle of the first layer of soil are 18 kN/m^3, 10 kPa and 30°, respectively. The soil parameters for the second layer which is thin in thickness are 18.5 kN/m^3, 2.0 kPa and 5°, respectively. The soil parameters for the third layer are 19 kN/m^3, 20 kPa and 25°, respectively. Actually this is a slope with a soft band soil, and each soil boundary surface is an irregular surface fluctuating in 3D space. Since the soil parameters for the second layer of soil are low, a major portion of the failure surface will lie within layer 2 which is thin in thickness and a spherical surface will not be adequate for the optimization analysis.

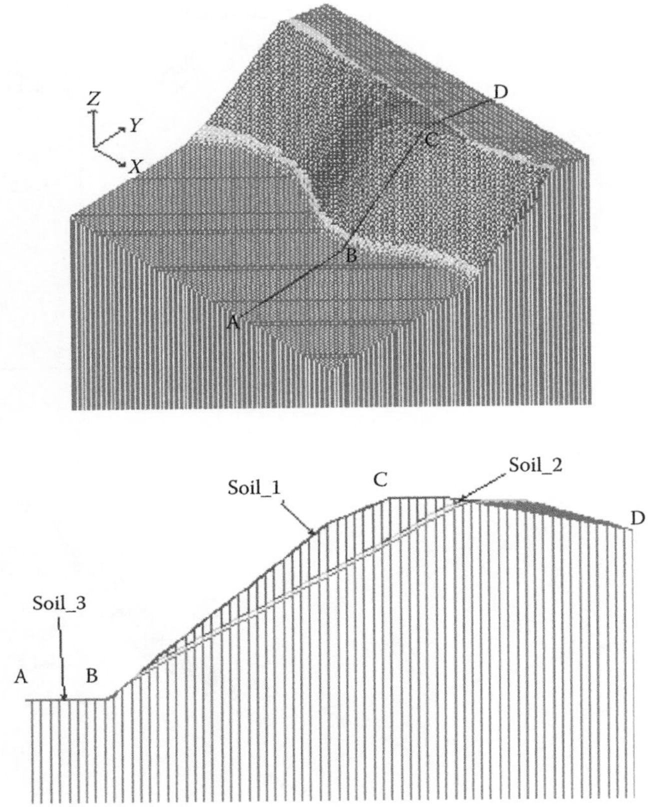

Figure 5.36 Slope geometry of Example 5.8.

For critical spherical asymmetric failure search which is obtained after 7488 calculation steps, the minimum FOS is 0.6177 from 3D Janbu analysis. The x, y, z coordinates of the centre of rotation and the radius of the spherical sliding surface with the minimum FOS are (59.8299, 52.8347, 42.2479) and 20.4227, respectively, and are shown in Figure 5.37a and b. For the critical spherical surface, the middle part of the failure surface lies within soil 3, while the outer part lies within soil 2.

For the NURBS sliding surface search, 15 nodes are used to form the NURBS surface. The dimension of the control net SS"T"T is 54.8 m × 34.4 m with 361 columns. In the present analysis, Cheng and Zhu (2005) have tried two options for the initial failure surface:

1. Fifteen nodes are used based on the most critical spherical failure surface and this initial solution is far from the critical solution (63,507 trials).

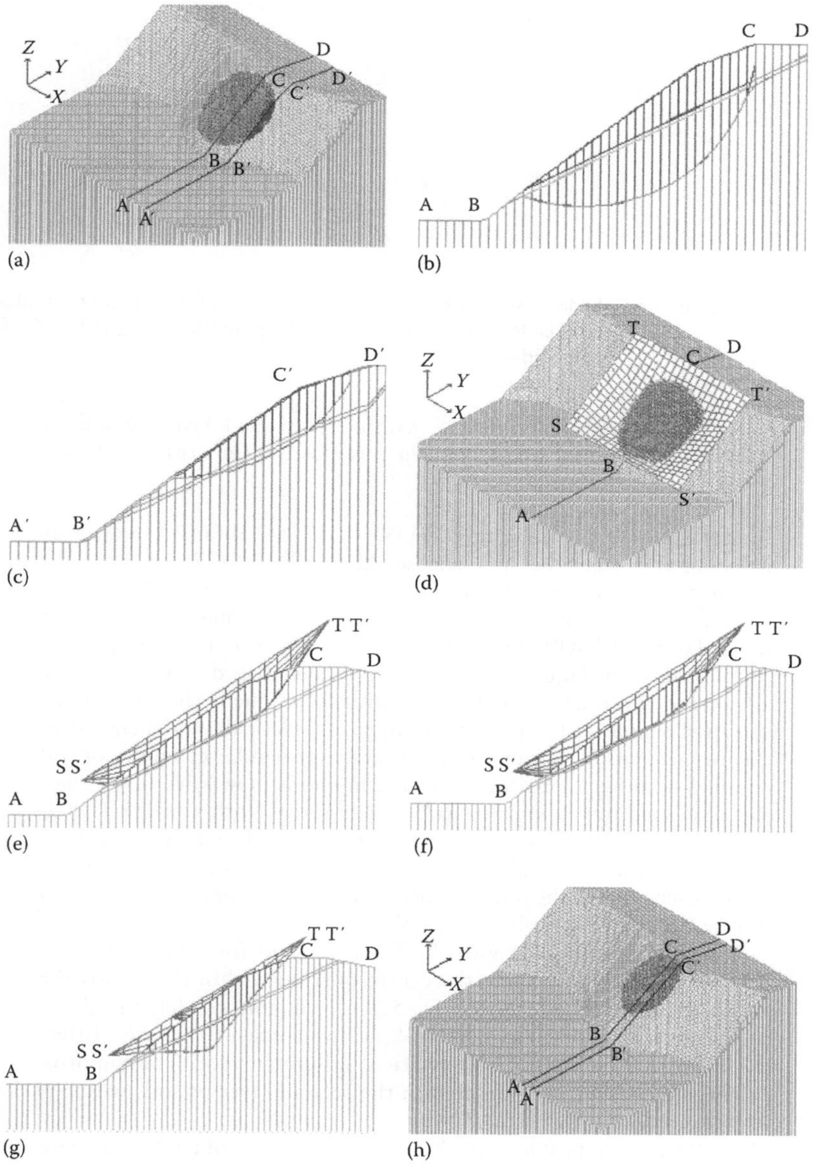

Figure 5.37 Sliding surfaces with the minimum FOS: (a) spherical sliding surface, (b) section along A–D for spherical search, (c) section along A′–D′ for spherical search, (d) NURBS sliding surface, (e) section along A–D for 15 points, (f) section along the middle for 10 points, (g) section along the middle for 5 points, (h) ellipsoid sliding surface

(continued)

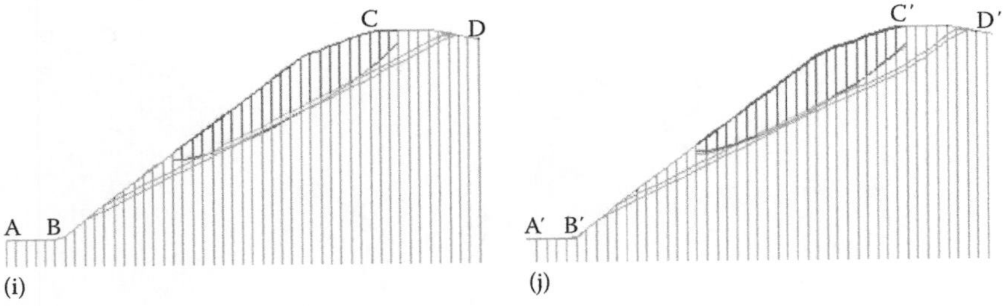

Figure 5.37 (continued) Sliding surfaces with the minimum FOS: (i) section along ABCD for ellipsoid search and (j) section along A′B′C′D′ for ellipsoidal search.

2. Fifteen nodes are chosen within the second layer of soil so that the initial NURBS surface is close to the critical solution (53,424 trials).

In both cases, the minimum FOS, 0.517, is obtained, which is much smaller than the spherical search result, 0.6177. As shown in Figure 5.37e, a major portion of the critical failure surface lies within soil 2, which is the expected result. The failure mass is the region formed by the NURBS surface from control net SS"T"T and the ground profile shown in Figure 5.37e. This study has also demonstrated the advantage of using a simulated annealing technique in the global optimization search as the global minimum is practically independent of the initial solution. It is true that for relatively regular geometry or soil conditions, other global optimization methods may work faster than the simulated annealing method, which agrees with Cheng's (2003) findings. For a problem where a good initial solution is difficult to be established, the simulated annealing method has the advantage of being insensitive to the initial solution and escape from local minimum during the search in its basic formulation.

The mesh shown in Figure 5.37d is the grid for computation of the NURBS surface, while the centre portion within the grid is the actual failure mass. From Figure 5.37e, it is noticed that the shape of the critical asymmetric NURBS failure surface is greatly different from that of the critical spherical failure surface. The majority of the sliding surface is located in the second layer of soil, which is as expected.

For the present problem, the FOS and the number of trials to achieve the critical solution are 0.9345 and 8425 for 5 NURBS points and 0.5611 and 63611 for 10 NURBS points. As shown in Figure 5.37f, the critical failure surface by 10 points is basically acceptable except that the extent of failure surface within soil 2 is not sufficient, which is the limitation in using insufficient NURBS point to form the critical failure surface. For the critical failure surface shown in Figure 5.37g, the number of NURBS points is too small so that only the outer edge of the critical failure surface lies within soil 2. The critical solution is practically insensitive to the number of NURBS points if the number exceeds 15, and this is similar to the situation for the second case.

For the critical ellipsoidal failure surface, the results are shown in Figure 5.37h–j. The minimum FOS is 0.521 and the number of trials is 10605. *Xc*, *Yc* and *Zc* in Equation 5.95 for the critical ellipsoidal failure surface are (57.72, 63.96, 36.21), while a, b, c are 2.41, 15.1 and 5.27 m, respectively. This critical ellipsoidal failure surface is greatly different from the critical spherical failure surface, and this is obvious. The FOS from ellipsoidal search is very close to that from NURBS search, but the number of trials in the optimization analysis can be greatly reduced. It appears that for a normal problem, ellipsoidal search is sufficiently good for design purposes. It should be noted that there are noticeable differences in the critical NURBS surface and the ellipsoidal surface even though the FOSs from these surfaces are close to each other.

5.6 EFFECT OF CURVATURE ON FOS

Many highway slopes have curvatures which can affect the stability of slopes, but this problem was seldom considered in the past. Xing (1988) has considered the case of a concave slope and has demonstrated that curvature can play an important part in the stability of the slope. In the present chapter, the effect of curvature is investigated in more detail. Consider a simple slope with a typical section in Figure 5.38.

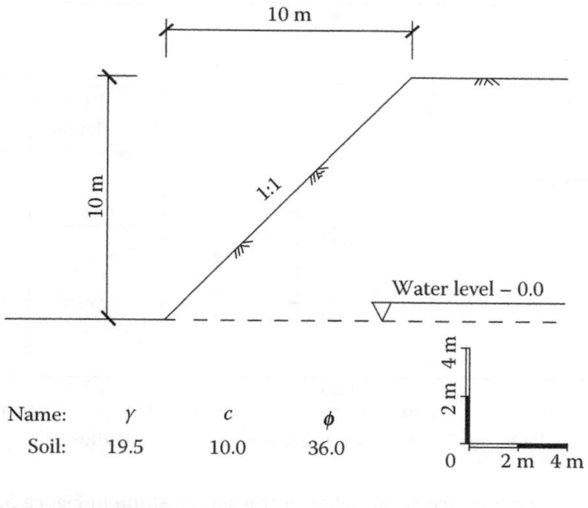

Figure 5.38 A simple slope with curvature.

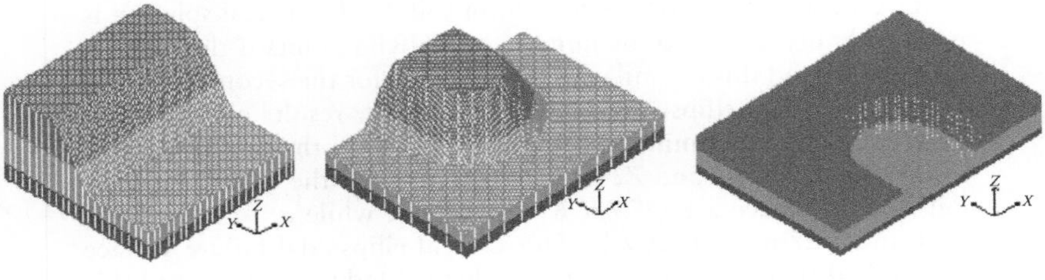

Figure 5.39 Layout of concave and convex slopes.

The typical section of the slope is shown in Figure 5.38, while the 3D view of the slope is shown in Figure 5.39 for clarity. For the present problem, the curvature at the bottom of slope and the corresponding FOSs for Bishop and Janbu analyses using spherical failure surfaces are shown in Figure 5.40. Conceptually, a concave slope should possess the greatest FOS, while a convex slope should possess the smallest FOS, and such results are given in Figure 5.40. Curvature plays a major role in determining the FOS. This is expected as a concave slope provides more confinement from the two ends of failure mass, while the convex slope has no confinement from the two ends. Xing (1988) has obtained the results for a concave slope with similar behaviour, which is further extended for a convex slope. Interestingly, the relation between FOS and the radius of curvature of slope appears to be approximately linear for the present problem.

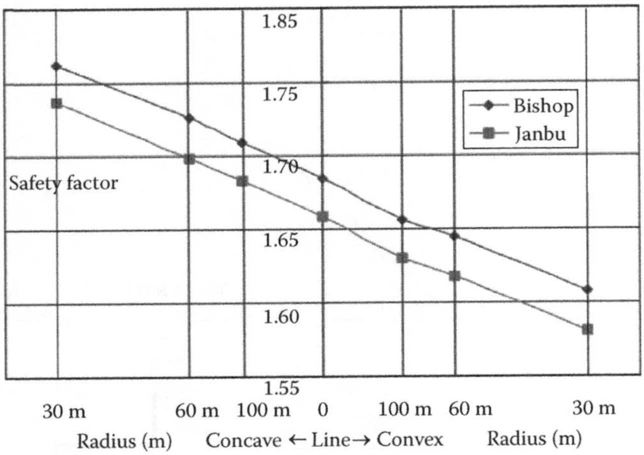

Figure 5.40 Effect of curvature on stability of the simple slope in Figure 5.39.

5.7 THREE-DIMENSIONAL SRM ANALYSIS

Extension of the SRM to 3D analysis appears to be simple in principle. Currently, there are only a limited number of commercial programs which can provide the 3D SRM option (fewer than the 2D SRM analysis). Cheng et al. (2009) have found many practical limitations in using the 3D SRM in the present study, and these may be some of the possible reasons for the limited range of computer programs available for 3D SRM. To date, both 3D LEM and the 3D SRM have various limitations. The lack of a suitable method for locating the critical general 3D failure surface for a complicated problem is one of the major limitations of 3D LEM, and in this respect, 3D SRM appears to be more effective for engineering use than 3D LEM. For 3D SRM, Wei and Cheng (2008) have pointed out the importance of the mesh design for the presence of a soft band material.

Consider a vertically cut slope with a 0.5 m thick planar weak layer (Figure 5.41). The FOS obtained by the 3D SRM model shown in Figure 5.42 is 0.58, and the slip surface is along the weak layer (Figure 5.42c). Since the slope fails along the weak layer ($c=0$, $\phi=10°$) and the weak layer is inclined at a 35° angle, the FOS can be obtained by a simple computation as tan 10°/ tan 35° = 0.2518. The FOS obtained by 3D Janbu is 0.2539, which agrees well with the expected result from basic soil mechanics. The FOS obtained by 3D SRM is 0.58, which is much larger than the expected value of 0.25, and Cheng et al. (2009) have carefully checked the model in Figure 5.43a. When the vertically cut slope is about to fail, a large shear strain is mobilized

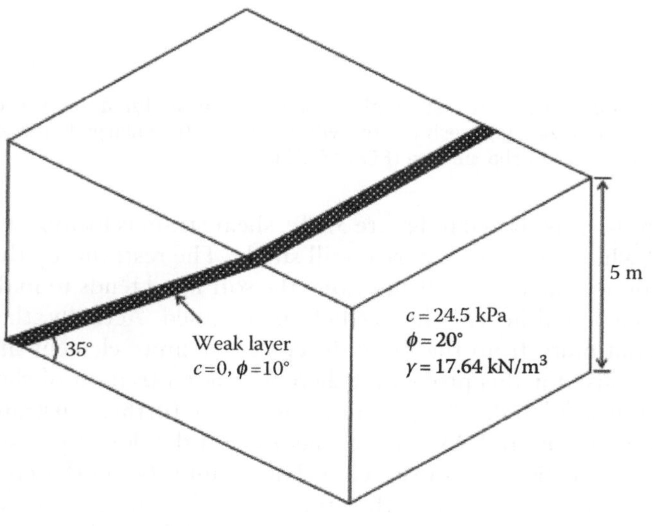

Figure 5.41 Geometry of a vertical cut with an inclined weak layer.

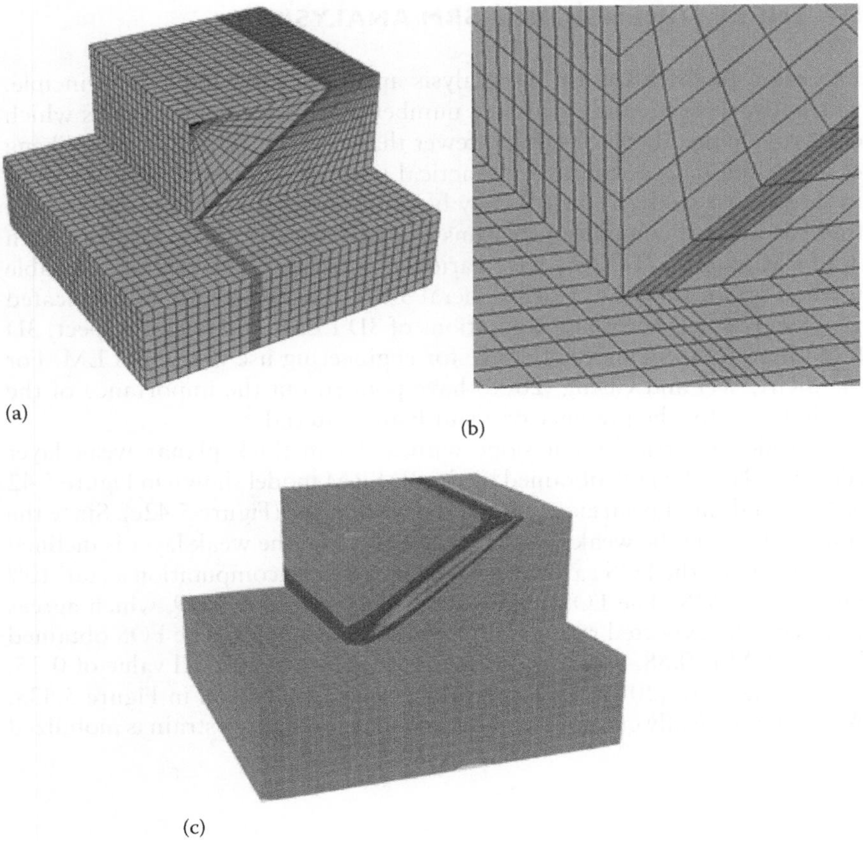

Figure 5.42 Model one of the strength reduction analysis for a vertical cut with a weak layer: (a) mesh for the whole model, (b) enlarged view of mesh at toe and (c) slip surface (FOS = 0.58).

at the slope toe. As shown in Figure 5.42c, shear strain is localized along the soft band while the bottom part is still stable. The restraint by the bottom elements on the inclined elements along the soft band tends to make failure more difficult, and hence a higher FOS is obtained. Actually, the requirement of continuity from the finite difference or finite element methods is the major cause for this problem if there is major variation of shear strain across a thin soft band. This phenomenon can be further illustrated by the model shown in Figure 5.43. If the element size at the slope toe is very small, a FOS of 0.40 is obtained for this case. This result is better than for the first model, but it is still much larger than the expected value. Cheng et al. (2009) have developed a third model as shown in Figure 5.44, and the weak layer is raised slightly (by 0.12 m) while the thickness is still 0.5 m. The soft band

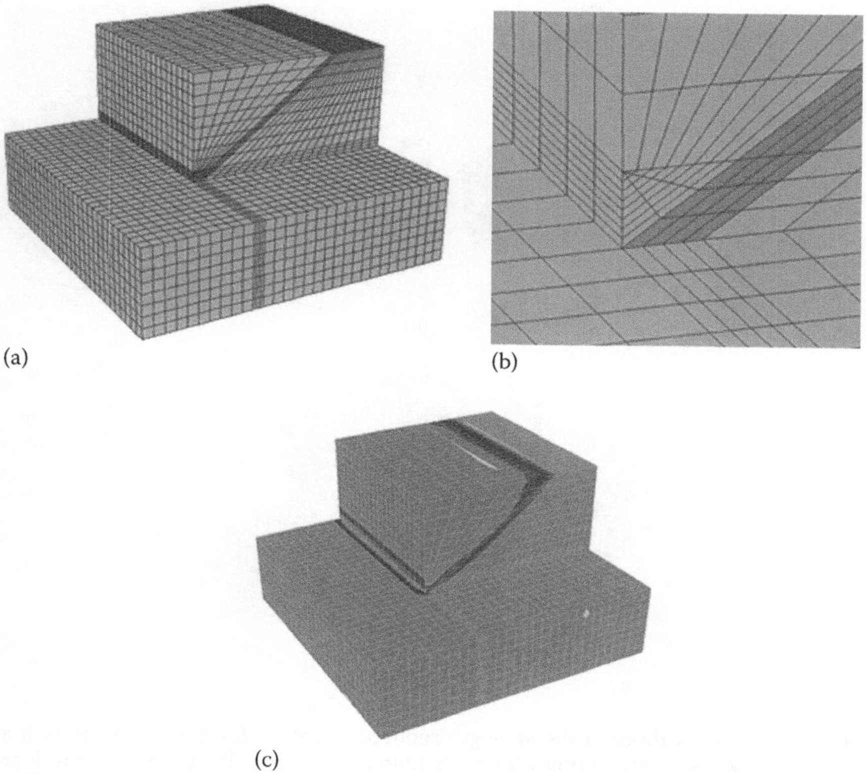

Figure 5.43 Model two of the strength reduction analysis for a vertical cut with a weak layer: (a) mesh for the whole model, (b) enlarged view of mesh at toe and (c) slip surface (FOS = 0.40).

is modelled by four layers of inclined elements, and only the top layer of elements is not connected to the bottom, which may generate the resistance to failure. Even though this model is practically the same as the previous model, a continuous shear band can now be formed easily in the top layer of the soft band elements, and a FOS 0.25 is obtained, which is the expected value. The generation of a highly localized shear failure zone similar to that of a soft band needs great care in 3D SRM mesh design. For highly complicated 3D problems, it is not easy to generate a good mesh automatically without any trial and error. Cheng et al. (2009) suggest that 3D LEM can be carried out to determine the critical failure surface, then a graded mesh based on the result from 3D LEM can be designed for further analysis in highly complicated problems.

In this soft band example, the cohesion of the soft band is zero. In order to investigate the results when the friction angle of the soft band is very small,

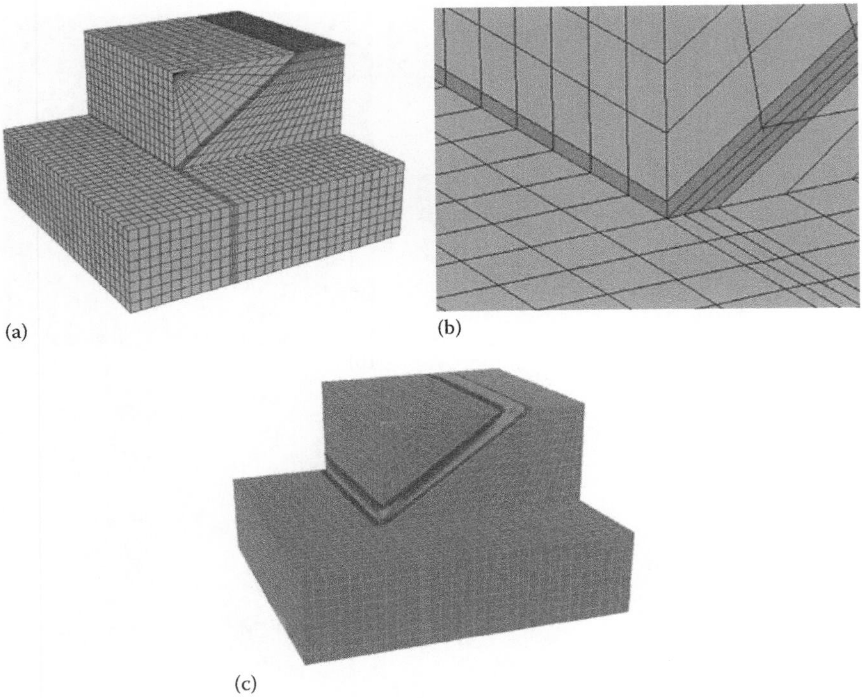

Figure 5.44 Model three of the strength reduction analysis for a vertical cut with a weak layer: (a) mesh for the whole model, (b) enlarged view of mesh at toe and (c) slip surface (FOS = 0.25).

another model is developed in which cohesion is equal to 10 kPa and the friction angle is zero. The FOSs obtained by the first model (Figure 5.42a), the second model (Figure 5.43) and the third model (Figure 5.44a) are 0.75, 0.56 and 0.47, respectively. The FOS for this problem is obtained as 0.4826. The third numerical model gives a FOS of 0.47 which is very close to the LEM result of 0.4826. The first model is poor as the FOS obtained is much larger than the expected value. The FOS obtained by the second model is larger than the expected value, but the situation is better than in the previous example where the cohesion strength is 0, so when the friction is zero for the soft band, the influence of the mesh design is slightly smaller than in the previous case.

Cheng and Yip (2007) have proposed a new 3D asymmetrical LEM formulation. Cheng and Yip (2007) considered an earthquake coefficient of up to 0.5, but a higher earthquake coefficient which may not be realistic will also be studied in this section. The slope model under consideration is shown in Figure 5.45. The slope height is 6 m, the slope angle is 45°,

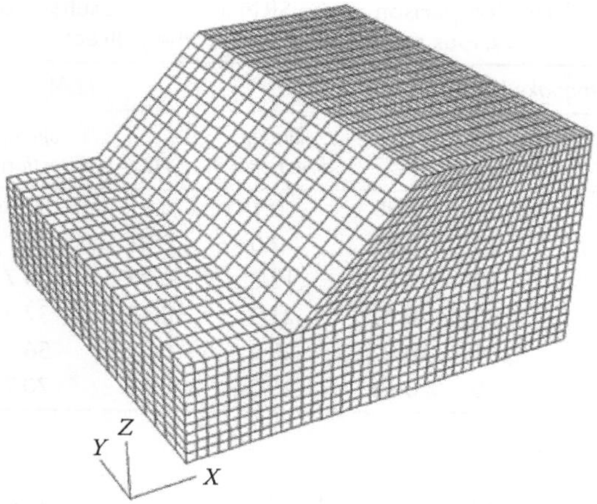

Figure 5.45 Mesh for a slope with transverse earthquake load.

the soil unit weight is 20.0 kN/m^3, the cohesion is 5 kPa and the friction angle is 35°.

The results of the transverse earthquake load analyses are given in Tables 5.9 and 5.10. In Table 5.9, the earthquake coefficient in the y direction (transverse) is 0.5, while the earthquake coefficient load in the x direction varies from 0.1 to 1.0. Although the FOS obtained by the SRM are larger than those obtained by the LEM, the differences are not great. With an increase in the earthquake load in the x direction, the FOS becomes smaller, and this is also a very reasonable result. For the SRM analysis, the average of the sliding direction is used in Tables 5.9 and 5.10. Although with increasing earthquake load in the x direction the sliding angle obtained by the SRM also gets smaller, the angle is smaller than that obtained by the LEM. This difference may be caused by the different mechanisms in the

Table 5.9 Comparison of SRM and LEM results with various earthquake loads in the x direction

Earthquake load		*SRM*		*LEM*	
Q_x	Q_y	*FOS*	*Sliding direction (°)*	*FOS*	*Sliding direction (°)*
10	50	1.0	27.6	0.90	42.5
30	50	0.8	21.3	0.69	37.9
50	50	0.64	15.8	0.54	32.4
80	50	0.47	9.4	0.37	27.1
100	50	0.38	6.0	0.28	24.4

Note: Q_x and Q_y are the earthquake loads (in % of soil weight) in the x and y directions, respectively.

Table 5.10 Comparison of the SRM and LEM results with various earthquake loads in the y direction

Earthquake load		*SRM*		*LEM*	
Q_x	Q_y	*FOS*	*Sliding direction (°)*	*FOS*	*Sliding direction (°)*
30	1	0.90	0.1	0.81	0.8
30	10	0.89	2.0	0.81	8.1
30	30	0.86	12.8	0.77	24.7
30	50	0.80	21.3	0.69	37.9
30	100	0.62	29.2	0.52	56
30	200	0.42	29.0	0.33	70.7

LEM and SRM. In LEM analysis, the sliding mass is assumed to be a rigid block, and it moves in the same direction at initiation of slope failure. In the SRM analysis, the stress–strain relation is used, and the velocity and displacement for each element of the sliding mass are different.

In Table 5.10, the earthquake load in the x direction is 30% of the soil weight, while the earthquake load in the y direction varies from 1% to 200%. When the earthquake load in the y direction varies from 1% to 200%, the two results obtained by the SRM and LEM agree well even though the earthquake load is very large and the FOS gets smaller with the increase in earthquake load. It is also reasonable that the FOS gets smaller slowly but the sliding direction gets larger drastically as the earthquake load increases in the y direction.

Cheng et al. (2009) have tried several hundred cases for both 3D LEM and SRM, and the results from the two analyses agree reasonably well for normal situations where the slope angles and external loads are not high and where there is no special geological feature with very high or very low shear strength parameters. In general, the FOSs obtained by the SRM are slightly higher than those from the LEM, which is a similar result to the corresponding 2D analysis by Cheng et al. (2007a). It is also found that while there is no major difficulty in adopting either 2D SRM or LEM, there is still room for improvement in 3D SRM and LEM, in particular for 3D SRM.

For the simple, homogeneous infinite slope, it appears that the formation of a distinct 3D failure mode is more difficult than that of a 2D failure mode. With careful thought and assessment, such results can be explained as the soil parameters under the 2D and 3D cases are considered to be the same. Furthermore, slopes are seldom homogeneous with respect to soil properties or geometry. The normal understanding that 3D failures are more commonly found while 2D failures are seldom observed in practice (contrary to those cases in the present study) can be attributed to the heterogeneity in the soil properties under 2D and 3D conditions. For heterogeneous ground conditions, 3D failures are actually formed easily by the SRM or LEM analysis. The case with a soft band has also demonstrated that the design of the mesh can have a major impact on the analysis, and the general guideline in finite element mesh design for non-linear problems is also applicable to the SRM analysis.

Chapter 6

Implementation

6.1 INTRODUCTION

When the British first occupied Hong Kong in the 1840s, they immediately put into place a network of military roads in Hong Kong Island. Roads would require the formation of a corridor by cut and fill. Both the resulting cut and fill slopes were designed using the rules of thumb, namely, 10 on 6 for cut soil slopes and 1 on 1.5 for fill slopes (Hong Kong Government, 1972). The fill slopes of course were formed by end tipping and not compacted. This practice would allow the roads to be built quickly and it was accepted that some of the slopes might fail from time to time. Corridors and platforms for other developments were also created in a similar fashion. As Hong Kong developed and grew, such slopes would become too hazardous for civilian use. One of the first engineered cut slopes in Hong Kong was the aviation checkerboard at Kai Tak airport and was analysed by using Bishop's method of slices in the 1960s.

Two disastrous landslides that took place on 18 June 1972 following 653 mm of rainfall from 16 to 18 June 1972 were to fundamentally change the control of the design and implementation of engineered slopes in Hong Kong. The first landslide was in Sau Mau Ping in eastern Kowloon that killed 71 people and the second one at Kotewall Road in Mid-levels, Hong Kong Island that killed 67 people. On the afternoon of 18 June 1972, an earth embankment failed, leaving tons of landslide debris in the resettlement area in Sau Mau Ping. By nightfall, on Hong Kong Island, landslide debris originating from Po Shan Road completely destroyed a 12-storey Kotewall Court and a 6-storey house, partially damaging an unoccupied block on its way. The Hong Kong government then decided to start controlling all man-made cut slopes. This was not the case, however, for loose fill slopes. The relatively gentle slope angle gave the false impression that the slopes were safe. It is however not necessarily true, as fill is meta-stable and may collapse easily under high water pressure, which has been happening in the past in Hong Kong.

If the soil is fully saturated, the slope may generate sufficient pore water pressure to liquefy. On 25 August 1976, a loose fill slope 40 m away from the previous landslide failed. The rainstorm associated with tropical storm Ellen triggered many landslides across the territory, but the worst failure was in Sau Mau Ping. At around 9 a.m. on 25 August 1976, the fill slope behind Block 9 of Sau Mau Ping Estate in Kwun Tong collapsed, killing 18 people and injuring 24. A comprehensive report on these landslides can be found in the Civil Engineering Development Department publication (CEDD, 2005). An extensive programme of rehabilitating and upgrading the man-made slopes to meet the modern safety standard was put in place, after major landslides for Po Shan Road in 1972 for cut slopes and Sau Mau Ping in 1976 for fill slopes in Hong Kong.

It is of note that since soil nailing first became popular in the late 1980s, there have been no significant failures of permanently nailed cut slopes. Apart from a few exceptional cases, for example, Ching Cheung Road, most cut slopes were upgraded by soil nailing. It can be concluded that soil nailing is a very cost-effective and robust means to stabilize over-steepened cut slopes, and in so doing also maintains the stability of the more deep-seated overall failure mode of a slope.

There are, however, a number of aspects for which a critical review of existing practice is in order. First, it is the uncontrolled grouting pressure on the soil nails and second, the soil nail head design. The first problem arises from the configuration and arrangement of the soil nails, that is, a nail and grout tube within a hole. In the soil nailing design, it has been assumed that the bond strength between the grout and the ground is a function of the overburden pressure only (Watkins and Powell, 1992). Recent researches have suggested that it is actually dependent on the grout pressure and not the overburden pressure (Yeung et al., 2007). If we step back from the problem for a while and ask ourselves what is the best method to pressure grout the ground, tube-a-manchette would immediately come to mind. And the response would be, why not. A tube-a-manchette would allow pressure grouting to be carried out at a particular location at a particular time. This flexibility would be most welcomed by practitioners. If we go back to the nail proper, there is no reason why the nail cannot be in the form of a pipe or for that matter, constructed in either steel or other materials like fibre-reinforced plastic (FRP). If the nail is in the form of a pipe, it can be doubled up as a tube-a-manchette pipe and so grouting can be done to a designed grouting pressure. Currently, the authors are considering the use of FRP pipes as a new possible soil nail material. FRP pipes are manufactured by a pultrusion process as shown in Figures 6.2 through 6.4. As an example, an FRP nail can be grouted up to eight bars (Yeung et al., 2005; Cheng, 2007a).

The other aspect that may be further improved is the nail head design. At the moment, it can either be through a bearing capacity-type

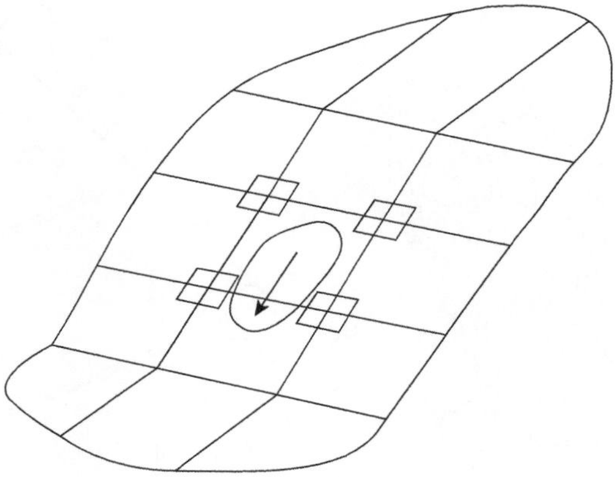

Figure 6.1 Failure of soil mass in-between soil nail heads as observed in Hong Kong.

calculation or prescriptive based on past experience (Pun and Shiu, 2007). Recent soil nailed slope failure has suggested that the weak points of a nailed slope would be in the vicinity of the nail heads (Figure 6.1). This begs the second question: why local failure between soil nail heads is not normally considered in a design at present. With the advent of the high-tensile alloyed steel wire (3 mm diameter, with tensile strength 1770 N/mm^2) and net weaving technology, there is no reason why local slip involving soil body between soil nail heads should not be designed properly against failure. One of the solutions is to do away with the soil nail heads and replace them with a high-tensile alloyed steel wire mesh (Ruegger and Flum, 2001). One example is the TECCO system developed by Geobrugg. The system consists of a TECCO wire mesh, TECCO spike plates (to facilitate force transmission from mesh to nails), TECCO compression claws (to connect mesh sheets and for fixing along the outer edges) and soil nails (grouted anchor bars) (Figure 6.2). What we are proposing here is that the practitioner should deal with local failure, shallow failure and global failure simultaneously.

6.2 FRP NAIL

Corrosion protection is of paramount importance for the durability of steel soil nails installed in slopes. Provision of 2 mm sacrificial steel thickness is the most widely used method for corrosion protection of soil nails in Hong Kong. Corrosion protection of steel bars by hot-dip galvanization or epoxy coating is also commonly adopted in Hong Kong under some corrosive conditions (Shiu and Cheung, 2003). Anyhow, there is a reduction of 4 mm in diameter for tensile capacity of the soil nails. When the required

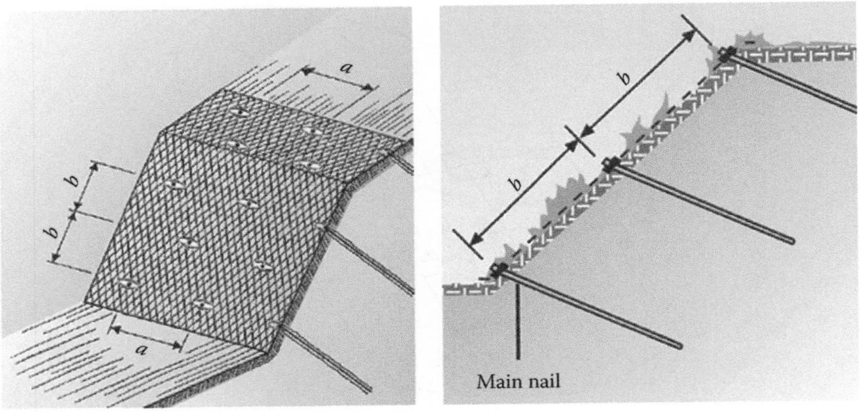

Figure 6.2 The TECCO system developed by Geobrugg.

stabilizing force is large, soil nails with 40 mm diameter steel reinforcement are often installed at very close spacing (1–1.5 m in Hong Kong). Such steel bars are heavy and thus difficult to manoeuvre on site. As a result, the zinc coating can be damaged easily. The length of each 40 mm diameter steel bar that can be handled on site is limited by its weight and individual site conditions, and the typical length of each segment is approximately 3–5 m (due to lack of adequate working space in Hong Kong). Couplers are often used to connect bars to the required total length. When soil nails are required in aggressive ground, a double corrosion system similar to that for pre-stressed ground anchors is required, resulting in a significant increase in construction costs and time. The total cost of soil nail construction, including the steel nails, couplers, handling and transportation costs, drilling and corrosion protection system, in Hong Kong is hence much higher than that in many other countries.

Usually, no pressure is applied during grouting of conventional soil nails (gravity flow of grout) as it is difficult to apply pressure with the current soil nail system. There are many cases reported in Hong Kong where shrinkage of grout has resulted in significant reduction in bond stress between cement grout and soil, and remedial works are required. Many engineers also have reservation on the bond stress transfer of soil nails within loose fill slopes. In particular, it is found that even though good compaction has been carried out to loose fill, the compacted dry density of the fill will decrease with time, possibly due to washout of fines by groundwater. In view of this concern, expensive and visually unpleasing concrete grillage is commonly used in Hong Kong for loose fill slopes.

In view of the various problems associated with the use of reinforcement bars as soil nails, there are various researches being carried out in

Hong Kong, China and many other countries. The features that are required for soil nails in Hong Kong include

1. Lightweight and high strength
2. Application of pressure to control the grouting zone, quality of grouting and bond strength
3. Free from corrosion problems
4. Acceptable cost
5. Ease of construction – handling, joining and cutting

Recently, there have been rapid developments in the use of FRP for various structural purposes (Dolan, 1993; Dowling, 1999). The authors have carried out research works on the use of glass fibre-reinforced polymer (GFRP) and carbon fiber reinforced polymer (CFRP) bars as soil nails for the project at Sanatarium Hospital in Hong Kong as bar-type FRP can be found easily in the market. From the pilot studies by the authors, the limitations of GFRP bars have been found to be as follows: (1) pressure grouting system is complicated, (2) joining of bars is not easy and (3) low shear strength. The limitations of CFRP bars are as follows: (1) pressure grouting is complicated, (2) joining of bars is not easy and (3) cost is high. An innovative system of GFRP pipes has been devised by Dae Won Soil Company Limited (Korea) and is used for the present study. The system can fulfil the five criteria for new soil nail systems as mentioned above and may be suitable for use in Hong Kong and other countries to improve the economy and constructability of soil nails.

GFRP is a material of lightweight, high corrosion resistance and high strength. For the present system, GFRP pipes of 37 mm internal diameter and 5 mm thick are utilized. They are fabricated by a pultrusion process during which glass fibres are drawn through die and bundled together through a resin matrix (Figure 6.3). The fibres are coated with sheeting and

Figure 6.3 Glass fibre drawn through die and coated with epoxy.

Figure 6.4 Fibre drawn and coated with sheeting to form a pipe bonded with epoxy.

are pulled through a shaping die (Figure 6.4) to form the pipes. Pultrusion is a continuous moulding process that uses fibre reinforcement in polyester or other thermosetting resin matrices. Pre-selected reinforcements such as fibre glass, mat or cloth are drawn through a resin bath in which all materials are thoroughly impregnated with a liquid thermosetting resin. The wetted fibre is formed to the desired geometric shape and pulled through a heated steel die. The resin is cured inside the die by controlling the precise temperature of curing. The laminates solidify to the shape of the die and they are slowly but continuously pulled by the pultrusion machine. Typical FRP lamination formed by pultrusion process is shown in Figure 6.5. This process of manufacturing has the advantages of forming various shapes suitable for different engineering uses. The mechanical properties and strengths of the laminates can be controlled easily by using different types of resin. For example, suitable filler, catalysts, ultraviolet inhibitors and

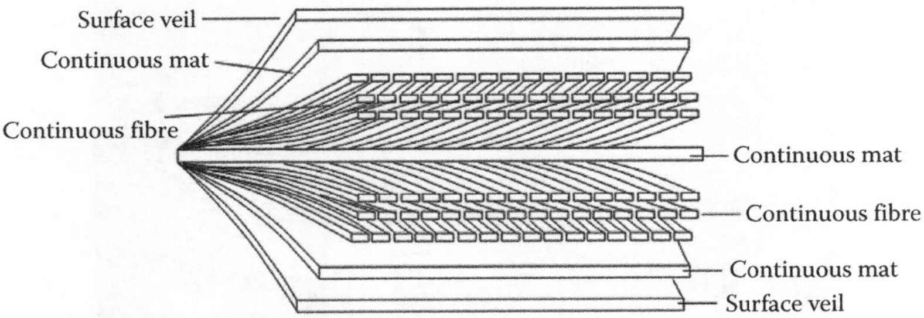

Figure 6.5 Lamination of FRP as produced by pultrusion process.

pigments can be used to form the resin matrix, binding the fibres together and providing structural corrosion resistance as well as strength.

Field tests of the effect of the tube-a-manchette grouting technique have demonstrated major beneficial improvement in soil properties. The effective cohesive strength of soil can be increased greatly, which is highly beneficial to the stability of slopes. Furthermore, the deformation and elastic modulus of soil are also greatly improved by the soil improvement process, while the FRP pipe acts as the grouting tube as well as the reinforcement to the grouted soil mass. Since the quality of grouting is good due to the use of pressure grouting, this new material and grouting technique would be useful in loose fill where bond stress is always a problem. Use of expensive and visually unpleasing concrete grillage in loose fill slopes can also be avoided by the use of tube-a-manchette grouting and additional cost-saving can be possible.

The installation of the system in Hong Kong indicates that there are no insurmountable installation difficulties encountered on site. The nails can be installed and grouted easily and high strengths have been obtained from pull out tests. The maximum test loads for four pull out tests are all about 21.1 tons. While the allowable tensile strength of FRP pipes was set at 11.5 tons in design, it can be seen that the design value for the FRP pipes is quite safe and conservative.

Since there are doubts about the transfer of bond stress from soil nails in poor soil, grillage system is also adopted commonly in conjunction with the use of soil nails for loose fill slopes or slopes with very poor quality soil in Hong Kong, Taiwan and Japan. There are various possible methods to stabilize slopes in different kinds of soil, and only limited systems commonly adopted in Hong Kong are covered in this book.

6.3 DRAINAGE

Inadequate surface drainage design and detailing are quite common. This is mainly due to the local concentration of flow. It is commonly observed that, while most of the storm drainage provisions for slopes in Hong Kong are adequate, some are under-designed by a wide margin. For example, drainage lines are something we have to take note of and ensure that, if they are present, they must be properly accounted for in the drainage system design. Such concentrated flow may also have an impact on the slope groundwater table level if not diverted from the recharge zones where open discontinuities are present. Horizontal drains and sub-soil drains are useful in drawing down the groundwater table and relieving artesian water pressure.

In Hong Kong, slopes normally tend to have lower hydraulic conductivity as they get deeper. When there are zones with large difference in hydraulic conductivity like colluvium on top of Grade-V granite, perched water may develop. Rise in transient perched and regional groundwater levels should be taken into account in design. Persistent clay layers and kaolin-infilled discontinuities may also have an adverse impact on the design groundwater table assumptions. Ideally, an accurate hydrogeological model should be set up, failing which, sensitivity analysis of the groundwater regime assumptions should be carried out and where necessary, more pessimistic assumptions should be used in the design.

6.4 CONSTRUCTION DIFFICULTIES

Difficulties during installation of soil nails should also be addressed. Certain geological features may also cause construction difficulties during installation of the soil nails. For example, volcanic tuff would be very hard to drill through, resulting in excessive wear and tear of the drill bits. The presence of core stones may catch the drill bits and stop them from being withdrawn easily. This may slow down the drilling process and cause drill hole collapse. The presence of a network of soil pipes may result in excessive loss of grout during the grouting stage. All such problems can be overcome if these had been identified early on. Examples are as follows: (1) to use a drill bit with harder cutting beads; (2) to use the Odex (or under-reaming) drilling system where a temporary casing can be introduced to avoid drill hole collapse and (3) drilling and grouting can be performed in two stages. After the first-stage drilling, quick-set cement grout should be injected, followed by a second-stage drilling before the second- and final-stage grouting. Following these steps can avoid excessive grout loss for most cases.

Chapter 7

Routine assessment of feature and design of landslip preventive measures

7.1 INTRODUCTION

In Chapter 6, we note that there is a legacy of substandard man-made cut slopes, fill slopes and retaining walls, known as features in this book, which the engineers in Hong Kong have to deal with. Under the Hong Kong Government Systematic Inspection of Features in the Territory (SIFT), a feature would be registered with a unique reference number if it meets the Geotechnical Engineering Office's feature registration criteria. A SIFT class would then be identified and a report of the feature would be stored in the Slope Information System (SIS). The report would indicate facilities like roads, buildings, etc., in the vicinity of the feature. According to Works Bureau (1999) and the Geotechnical Engineering Office (2004a), a Consequence-to-Life Category from 1 to 3 and Economic Consequence Categories from A to C would also be assigned to the feature. The SIS also records any landslip incidents in the vicinity of the feature.

The total number of registered features in Hong Kong is about 55,000. In this chapter, we attempt to briefly introduce the Hong Kong practice in routine assessment and where necessary, upgrading of these pre-existing man-made features. We believe the practice would be of some reference value and interest to the readers.

The process would start from a Stage-2 study, which essentially is a screening process to identify the status of each feature so that appropriate line of actions can be taken. For substandard features for which the government is responsible, the features would be incorporated into Stage 3 of the Government Landslip Preventive Measures (LPM) programme, and for substandard features for which private owners are responsible, the Buildings Department would issue the Dangerous Hillside Order (DHO) stipulating the owners to investigate and, where necessary, implement required upgrading works.

7.2 GEOTECHNICAL ASSESSMENT

In order to be able to evaluate the stability of the feature and recommend appropriate upgrading works, if necessary, the following steps known as a Stage-3 study are usually taken:

1. Desk study
2. Aerial photograph interpretation
3. Site reconnaissance
4. Ground investigation (GI)
5. Laboratory testing
6. Stability analysis
7. In case the stability analysis indicates that the feature cannot meet the current geotechnical safety standard, upgrading works would be needed, which may include:
   a. Trimming of over-steepened cut slopes.
   b. Installation of soil nails.
   c. Installation of prescriptive raking drains.
   d. Application of sprayed concrete over the surface of the soil nailed slope if the slope angle is steeper than 50° and application of erosion control mat and hydroseeding if the slope is shallower than 50°.
   e. If loose fill exists, this should usually be replaced by compacted fill or compacted cement-enhanced fill.
   f. Installation of a surface drainage system.
8. Implementation of upgrading design on site
9. Long-term maintenance

7.3 DESK STUDY

The key characteristics of a feature are its location, face direction and construction: cut/fill/wall, length, dip angle and maximum height. Other points of interest would be facilities in the vicinity, surface protection and drainage provisions. A 1:1000 site plan complete with some site photographs would be essential. Apart from the physical characteristics, it is also necessary to systematically identify the party who has the maintenance responsibility. The process is known as Systematic Identification of Maintenance Responsibility (SIMAR).

According to the 1:20,000 geological maps (Hong Kong Geological Survey, 1987), the feature geology can be assessed. Part print of the geological map centred at the feature should be studied carefully. Based on the geological map and the previous GI, the soil strata of the feature and its surroundings can be established. This information together with the current GI information should form the basis to establish the soil strata of the feature and its surroundings. Past instability like the presence of any major instability at or near the feature should be noted. Even a minor landslip scar should be taken into account.

All previous GI reports relevant to the study would also be retrieved from the Geotechnical Information Unit (GIU). Apart from the GI reports, previous Stage 2 or 3 study reports, if available, would also be kept at the GIU. Based on the information, the geology of the feature as well as the soil parameters can be inferred. After the critical sections are analysed, depending on the results of the stability assessment, say, if the minimum factor of safety (FOS) is higher or lower than 1.2, the need for any upgrading works would be decided upon.

The file kept by the Geotechnical Engineering Office (GEO) on the feature may reveal relevant correspondence like memos concerning the feature. For example, it might be revealed that non-inclusion of the feature in the earlier LPM programme might have been a result of the possible development of a facility closely related to the feature. The file may also contain information on the maintenance responsibilities of the feature.

A list of utilities undertakers who would need to be consulted, if the site is on Hong Kong Island, would be as follows:

a. Water Supplies Department
b. Drainage Services Department
c. The Hong Kong and China Gas Company Limited
d. The Hong Kong Electric Corporation Limited
e. Hong Kong Cable Television Limited
f. Other telecommunication operators

The enquiry would confirm if there is any existing service or utility in the vicinity of the feature. All such recorded services and utilities, including unrecorded services and utilities revealed during site reconnaissance, must be recorded clearly and shown on a plan so that these could be properly accounted for in the design of upgrading works for the feature.

7.4 AERIAL PHOTOGRAPH INTERPRETATION AND GROUND-TRUTHING

The Hong Kong Government regularly takes aerial photographs of the whole of Hong Kong. For example, on Hong Kong Island, the earliest photographs were taken in 1924 (flight height, 11,000 ft). Since then, aerial photographs were taken in 1945 (20,000 ft), 1949 (8,600 ft) and more recently, in 2002 (4,000 ft), 2004 (2,500 ft), etc. From the aerial photographs, one can find out the detailed history of the site such as when was the feature under study developed. It can also be clarified if the feature was formed in association with any road work or site formation for building development. Detailed field inspection for the feature is of paramount importance. The inspection should confirm or otherwise the recorded information plus any other information like unrecorded exposed water pipes, signs of seepage or distress noted on the feature. Where appropriate, field mapping on any existing landslide should also be carried out. Rock outcrops observed may indicate a shallow rock head in the vicinity.

7.5 GI AND FIELD TESTING

Confirmatory GI works should be carried out to verify the assumed geological and hydrogeological setting of the feature. Even though some existing laboratory testing results are available for use in the current study. Laboratory tests on the samples retrieved from the confirmatory GI should be ordered accordingly to further confirm the validity of the adopted soil parameters used in the analysis. For ease of future maintenance, the results of the laboratory tests from the confirmatory GI should also be recorded in the maintenance manual.

Previous GI reports on the site containing logs of boreholes, core holes, trial pits and surface stripping, if available, should be studied. The information might reveal if the geology of the area is mantled with fill and colluvium.

Confirmatory GI for the current study should comprise boreholes, core holes, trial pits and surface stripping. In addition, the following are also required:

1. Sampling of Mazier and standard penetration test (SPT) liner samples in the boreholes
2. Sampling of block samples, undisturbed samples and large disturbed samples in the trial pits
3. In situ density tests in fill materials encountered in the trial pits
4. Installation of standpipe piezometers in the boreholes and trial pits

Inspection pits should be opened up to verify the foundation conditions, if present, during the confirmatory GI. The existence of any other buried utilities is to be verified by the confirmatory GI before the commencement of the proposed landslip preventive works. As part and parcel of the GI field works, a topographical survey should be conducted. Based on the height and gradient of the section obtained from the survey, a representative critical section should be prepared.

7.6 LABORATORY TESTING

Laboratory tests should be carried out on the samples of soil recovered from the GI for the feature and its vicinity.

1. Single-stage and multi-stage isotropically consolidated undrained triaxial compression tests with pore water pressure measurement
2. Moisture content determination
3. Atterberg limit determination
4. Particle size distribution
5. Specific gravity determination
6. Chemical tests

Common chemical tests are soil resistivity and redox potential determination. Other tests would be those for pH value as well as sulphate, chloride, carbonate and organic matter content.

7.7 MAN-MADE FEATURES

A cut slope is normally very steep and shallow failure is usually critical. In order to balance the earthwork, it is likely that there would be fill deposited during the construction of the road corridor or building platform. If in doubt, GI in the form of trial pitting should be planned to determine the extent and thickness of the fill within the feature. In situ density tests should be carried out for the top 3 m of fill. The results should indicate if there is any fill in a loose and *meta-stable* state with an *average* relative compaction of significantly less than 95% when compared with the *average* maximum dry density obtained from Proctor test using the standard hammer at 2.5 kg.

7.8 RAINFALL RECORDS

Rainfall data at 5-min intervals can be obtained from the nearest GEO automatic rain gauge. Throughout the monitoring period, one should check if any red rainstorm warning signals and thunderstorm warning signals are hoisted in accordance with the Hong Kong Observatory Warning and Signals Database. Rainstorms that induce notable groundwater responses within the monitoring period should then be analysed. The maximum rolling rainfall for various durations should be identified and the corresponding return periods assessed based on the statistical parameters derived and given in Geotechnical Engineering Office (2001). The most critical date would normally fall on the day when a red rainstorm signal is hoisted.

7.9 GROUNDWATER REGIME

Standpipe piezometers with Halcrow buckets should be installed in boreholes.

Daily readings from these over a period of time of 14 days would reveal the groundwater table level during that period. In addition to groundwater monitoring, visual inspection of the feature may reveal signs of seepage and inspection records may also reveal signs of seepage. This information should be taken into account when deciding the design of groundwater table level.

According to Endicott (1982), the groundwater level rises linearly proportional to the rainfall depth up to a threshold value. No further rise in groundwater level would occur thereafter due to attainment of the saturation rate of infiltration. The threshold value depends on the geology, surface condition, degree of saturation, etc., and is therefore, site-dependent. The graphical approach suggested by Endicott is therefore usually used to estimate the rise in groundwater level due to rainfall infiltration at a specific site. From the monitoring of the installed standpipe piezometers, the rise in groundwater level in response to the rolling rainfall as recorded in the nearest rain gauge can be plotted. A linear best fit curve can then be generated for a, say 1 in 10-year return period, rainfall at a particular strata interface. From the plot, one can estimate the rise in groundwater table at the interface for a 1 in 10-year return period rainfall.

If site-specific piezometric data are not available, which is usually the case, the wetting band approach should be used instead. According to the Geotechnical Control Office (1984), the thickness of the perched water table or the rise in the main groundwater table can be assumed to be approximately equal to the thickness of the descending wetting band. The relationship between rainfall intensity and the depth of the wetting band for bared surface slopes is given by Lumb (1975) as

$$\text{Depth of wetting band}, h = \frac{kt}{n}\left(s_{\mathrm{f}} - s_0\right),$$

where

k is coefficient of permeability
n is porosity $= e/(1+e)$, where e is the voids ratio
s_0 is the initial degree of saturation
s_{f} is the final degree of saturation
t is the duration of rainfall

The intensity adopted for the design rainstorm should be the one causing the soil to become fully saturated (Lumb, 1962) multiplied by 2 to take account of the surface runoff. The duration of the storm of the calculated intensity for a 1 in 10-year return period can be calculated based on Peterson and Kwong (1981).

Response test results in the existing boreholes can be used to estimate the permeability of the soil mass and to calculate the wetting band thickness. If the runoff, soil porosity and degree of saturation are known, the wetting band can be estimated. If the feature is located in the vicinity of probable relict landslide, a veneer of colluvium could be present and should be verified by GI. In addition, if there is any localized reduction in the permeability of the colluvium compared with the underlying residual soils, a perched water table at 1 m above the interface between colluvium and residual soils should also be allowed.

As a result, two groundwater conditions for a 1 in 10-year return period rainfall should be adopted for the stability analysis.

1. Design groundwater table after taking into account the wetting band and
2. A perched water table at the geological strata interface

A review of the design groundwater levels should be conducted and presented in the Maintenance Manual when more monitoring data are collected from the confirmatory GI.

7.10 STABILITY ASSESSMENT OF THE EXISTING FEATURE

The stability of the existing cut slope can be checked using the computer programme SLOPE. The method used may be Janbu's method with variable inclined inter-slice force. Only circular failure mode is considered. The stability analysis is normally carried out on a section with inferred geological information, groundwater conditions and design soil shear strength parameters. A surcharge of loading of 30 kPa may need to be applied to take into account a three-storey building (i.e., 10 kPa per storey) at the slope crest. Meanwhile, the additional surcharge and lateral loading induced by the wind load on the building should also be assessed in accordance with the code of practice on Wind Effects in Hong Kong (2004) and incorporated in the SLOPE model. All conceivable failure modes need to be considered. A minimum FOS of 1.0–1.2 for a 10-year return period rainfall is required for the feature depending on the Consequence-to-Life Category (1–3) if rigorous groundwater monitoring has been carried out for the feature. Results of the stability analysis may indicate whether or not the critical slips under the two groundwater conditions considered are below the current geotechnical standard. If yes, then landslip preventive measures are required.

According to Works Bureau (1999), the required relative compaction of a fill slope should be at least 95% of the maximum dry density derived from a Proctor test. In addition, Geotechnical Control Office (GCO) probing tests should also be carried out. If most of the GCO probe values (number of blows per 100 mm penetration) in the fill are less than 10 within a depth of 2.5 m, it would imply that the fill materials are not well compacted. Based on the results, the feature is considered below the current geotechnical standards and has liquefaction potential. Slope-upgrading works would be required.

For masonry retaining walls, they should be designed and checked according to the Geotechnical Control Office (1993). The matrix suction of the retained materials must be ignored. The frictional angle at the rear face of the masonry retaining walls should be $\delta' = \frac{2}{3}\varphi'$. The base friction angle should be $\delta_b = 0.9\varphi'$. Coefficients of active and passive earth pressure can be evaluated based on the guidelines provided by the Geotechnical Control Office (1993). The design groundwater table should be assumed at one third of the wall height. The thickness of the masonry retaining walls should be based on the GI results. The required FOS should be 1.5 for sliding and 2.0 for overturning.

7.11 DESIGN OF LANDSLIP PREVENTIVE WORKS

7.11.1 Design options for masonry retaining walls

Typical assessment results of the masonry retaining walls are as summarized below. The masonry retaining walls have to be upgraded to the modern standard.

Sections	*Minimum FOS of existing masonry retaining walls*
1	0.8
2	0.66
3	1.02
Required FOS	1.4

Walls	*Sliding*	*Overturning*
Masonry retaining walls at toe	0.39	0.22
Masonry retaining walls at crest	0.92	0.58
Masonry retaining walls at stream course	0.91	1.10
Required FOS	1.5	2.0

Some explanations are warranted here for the extremely low FOS as intuitively we would have expected to see a FOS close to 1.0, considering that the retaining walls have been standing up for a long time. The explanation lies in the suction in the soil matrix. The standard required that the unreliable matrix suction be totally ignored in any design and all soil must be assumed to be fully saturated.

Option 1: Soil nailing
Option 2: Thickening of walls at the front
Option 3: Thickening of walls at the back

7.11.2 Design options for fill slopes

Eyewitness report of the Sau Mau Ping loose fill landslide on 18 June 1972 suggested the slip was in the form of a 2 m thick slab of soil. It is, therefore, now a standard practice to remove the top 3 m of the loose fill from the loose fill slope and replace it with compacted soil or cement-enhanced soil (see Figure 7.1a). The drawback of this method is that all mature trees established on the slope face have to be felled during the upgrading works. To avoid felling of matured trees, engineers have developed two alternative methods: backfill with lightweight concrete (see Figure 7.1b) and soil nailing in loose fill.

Option 1 (Figure 7.1a)

- Excavate loose fill pit by pit and backfill resulting voids with compacted soil or cement-enhanced soil
- Erosion control mats complete with wire mesh and hydroseeding

Option 2 (Figure 7.1b)

- Excavate loose fill pit by pit and backfill resulting voids with lightweight concrete
- Existing fill to remain in place locally around mature trees
- Erosion control mats complete with wire mesh and hydroseeding
- 300-mm-thick top soil for hydroseeding on the backfilled slope surface

Option 3

- Soil nails plus reinforced concrete grillage
- Erosion control mats complete with wire mesh and hydroseeding

Common items

- Construction of maintenance access staircases and drainage channels is recommended for all options.

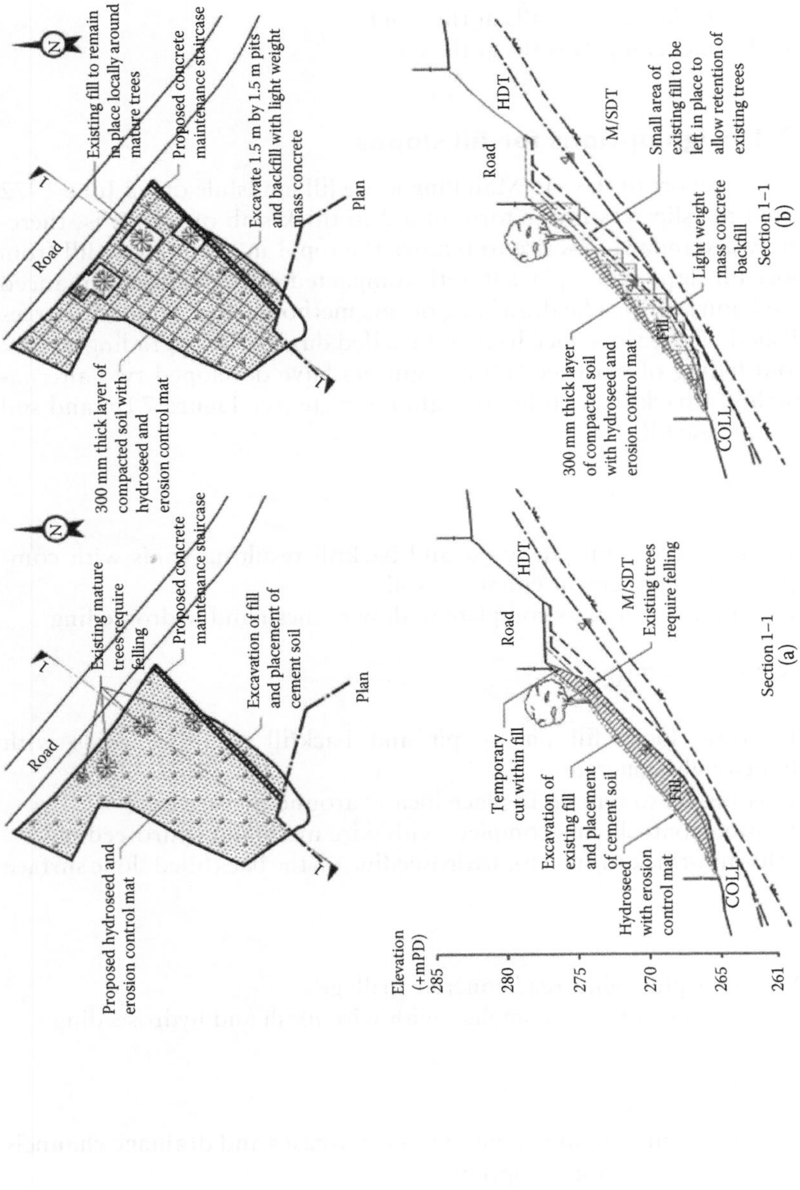

Figure 7.1 Proposed landslip preventive (upgrading) works.

An option would be chosen based on consideration of construction costs, duration and various impacts. Supplementary GI involving surface stripping and trial pitting is usually required in order to ascertain the extent of the loose fill materials.

7.11.3 Design options for cut slopes

Stability analysis of the upgraded feature should meet the current geotechnical standards, namely minimum FOS higher than 1.4 or 1.2 at the most critical section depending on whether the groundwater regime is well understood. There are two main upgrading options for pre-existing substandard cut slopes in soil:

Option 1: Trimming of slope profile to less than 55°

Protect with erosion control mats, wire mesh, hydroseed grass cover
Construct drainage system

Option 2: Provision of soil nails

Protect with erosion control mats, wire mesh, hydroseed grass cover
Construct drainage system

There should also be provision of maintenance access staircase and prescriptive raking drains. The final option would be chosen based on construction costs, duration and environmental impact resulting from the works.

7.12 SOIL NAILING

Soil nailing design generally follows Watkins and Powell (1992). A typical section is shown in Figure 7.2. Prior to the installation of the working soil nails, pull out tests (the minimum number is stipulated by the Geotechnical Engineering Office [2003]) are required to be conducted on site to check

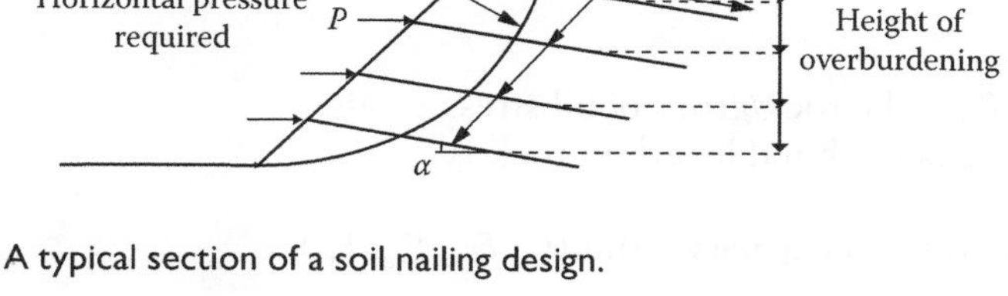

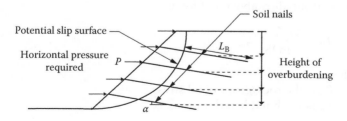

Figure 7.2 A typical section of a soil nailing design.

the pull-out capacity, which in turn checks the workmanship and design assumptions (mainly soil/grout bond strength) of the test loads.

1. Allowable tensile force of the rebar

$$F_s = \frac{\pi D_{eff}^2}{4} \times f_{st},$$

where
D_s is the diameter of rebar in millimetres
Corr is corrosion allowance in millimetres
D_{eff} is the effective diameter of rebar (D_s – 2 × Corr)
f_{st} is the allowable tensile stress of rebar

2. Grout/steel allowable bond force

$$F_{gs} = \frac{\beta \pi D_{eff} L_B \mathrm{sqr}(f_{cu})}{\mathrm{FOS}},$$

where
β is the grout/steel factor
L_B is the bond length
f_{cu} is the grade of grout
FOS is factor of safety

3. Soil/grout allowable bond force

$$F_{sg} = \frac{\pi D L_B c' + 2 D L_B \gamma' h \tan\varphi'}{\mathrm{FOS}},$$

where
D is the diameter of the drill hole
c' is the cohesion of bonded soil
h is the overburden height
L_B is the bond length
φ' is the friction angle of bonded soil

4. Rock/grout allowable bond force

$$F_{rg} = \frac{\pi D L_R R_b}{\mathrm{FOS}},$$

where
R_b is the rock/grout bond stress
L_R is the bond length in rock

Allowable soil nail capacity = min(F_s, F_{gs}, F_{sg}, F_{rg})

Allowable soil nail capacity per metre run = allowable soil nail capacity/horizontal spacing of soil nail

$$\text{Required horizontal pressure per metre run} = \frac{PH}{\cos\alpha}$$

where

P is the horizontal pressure required (say from SLOPE) in kPa
H is the horizontal span of loaded area or spacing of soil nails in metres
α is the inclination of soil nail

Required horizontal pressure per metre run must be smaller than the allowable soil nail capacity per meter run.

Typical parameters are

$H = 1.5$ m
$c' = 3.0$ kPa
$\varphi' = 33°$
$\gamma_{soil} = 19$ kN/m^2
$D_{eff} = 25$ mm
$D = 0.12$ m
$\alpha = 20°$
Corr = 2 mm
$F_{st} = 0.55$ fy or 230 MPa
$F_{cu} = 30$ MPa
$\beta = 0.5$ for type-II deformed bar
FOS = 2

In essence, soil nailing is the insertion of a rebar into a slope or a retaining wall. The rebars are usually inserted into downwardly inclined pre-drilled holes and then fully grouted. The grout used should be of non-shrink type. Although in a limit equilibrium analysis the spacing of the soil nails is not important, we should always bear in mind that a soil nailed feature is essentially a type of reinforced earth structure. It is therefore essential that soil nails always be regularly and closely spaced. The maximum spacing of soil nails should not normally be larger than 2 m c/c both vertically and horizontally. The soil nail head would typically be 400 mm by 400 mm by 250 mm thick and sunken into the slope surface.

Modification of FOS as a result of the presence of a perched groundwater table would be necessary by taking into account the existence of a perched groundwater table as shown in Figure 7.3.

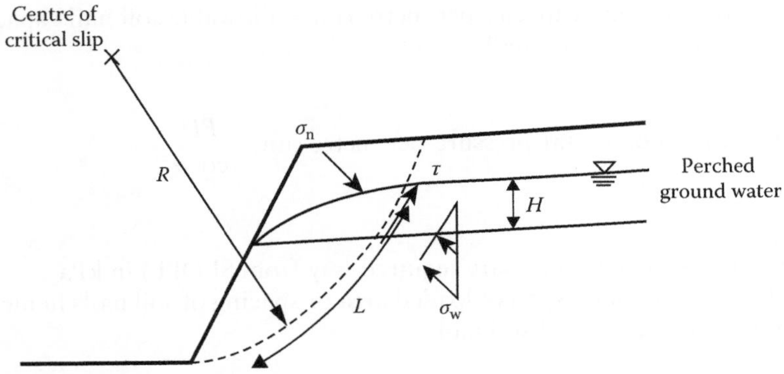

Figure 7.3 Perched groundwater table.

$$\text{Modified FOS} = \frac{M_R - \Delta M_R}{M_D},$$

where

M_D is the disturbing moment
M_R is the restoring moment
ΔM_R is the reduction on restoring moment due to perched groundwater table

$$= \Delta T \times R$$
$$= \Delta\sigma_n \times \tan\varphi' \times R$$
$$= 0.5 \times \sigma_w \times L^2 \times \tan\varphi' \times R$$

for a critical slip surface.

7.13 SOIL NAILING IN LOOSE FILL

For analysis of soil nails in loose fill, steady-state parameters for loose fill are adopted if liquefaction may occur. According to the GEO and Hong Kong Institution of Engineers (2011), a value of steady-state undrained shear strength equal to 0.2 times the mean effective stress $(C_{ss} = 0.2p'_{peak})$ at the onset of undrained behaviour is to be adopted. For slope with no positive pore water pressure, the ratio of undrained shear strength to vertical stress would be about 0.13.

Soil nails should be designed following the recommendations given by the GEO and Hong Kong Institution of Engineers (2011). The design has to satisfy two criteria simultaneously:

Criterion 1: The proposed soil nailing should be designed using the *soil nailing in loose fill* approach using the undrained condition
Criterion 2: The proposed soil nailing should also satisfy the conventional design under the drained condition.

In order to satisfy both design criteria, the soil nails and grillage system should first be designed assuming the loose fill material to behave as an undrained material. A conservative value of $C_{ss}/p'_{peak} = 0.2$ is adopted to take account of the gradual increase in undrained shear strength with depth of the fill material.

The design outputs would be

1. Confirmation of the overall stability of the soil nails and grillage system
2. Reinforced concrete design of the grillage system
3. Design for support for the grillage system
4. Design of soil nail heads

As a final step, the adequacy of the soil nails must be checked again assuming the fill has a drained shear strength of $c' = 0$ and $\varphi' = 33°$.

Other assumptions are

1. FOS is on the shear strength
2. Static analysis
3. Janbu's method with variably inclined inter-slice force
4. Hydraulic pore water pressure
5. Circular slip surface
6. Minimum slip weight for each failure mechanism of the slip surface is 50 kN (this is to avoid the minor surface slip, which should be dealt with by erosion control)
7. The reduction in restoring moment due to the perched water table should be duly taken into account
8. FOS for Criterion 1 is 1.1 (overall stability)
9. FOS for Criterion 2 is 1.4 (with perched water table)

7.14 SURFACE AND SUB-SOIL DRAINAGE

A detailed site reconnaissance on the feature should be carried out. Feature characteristics like length, facing angle, height and construction should be noted. The presence of any boulder field when present and the existing stream course runs in the neighbourhood of the feature should be recorded. The stream course seasonal flow should also be noted. The thickness of the masonry wall if not known, should be ascertained by trial pitting or sinking horizontal/inclined core holes.

The capacity of the existing surface drainage system has to be assessed. If any part of the system is found to be inadequate to accommodate the surface runoff during a 1 in 200-year rainfall, a new surface channel system complete with catch pits should be proposed. In accordance with the Geotechnical Engineering Office (2002), an appropriate type (type 1, 2 or 3) of prescriptive raking drains may be proposed as a contingency measure to account for any unforeseen adverse groundwater condition. A row of raking drains 5 m long may be proposed to enhance the reliability and robustness of the upgraded slope.

7.15 SURFACE EROSION CONTROL AND LANDSCAPING

Erosion in the form of a gully is quite common. Such erosion may intensify as a result of concentrated runoff from any man-made construction in the vicinity. Hydroseeding is a planting process whereby slurry consisting of seed and mulch is sprayed directly onto a slope surface. The slurry may also include ingredients like fertilizer, tackifying agents, green dye, etc. The mulch would facilitate grass growth by maintaining moisture around the seeds. The sprayed concrete can be of wet-mix or dry-mix. In construction, shotcrete usually means wet-mix, whereas gunite, a trade name, means dry-mix. The wet-mix method would produce less rebound (i.e., waste), whereas the dry-mix method would allow the nozzleman to adjust the water content instantaneously. Operation can also be stopped frequently. During placement, the slurry is propelled from the nozzle. The impacting force would compact the placed concrete.

The sprayed concrete serves two purposes:

1. Prevents infiltration
2. Prevents surface erosion

Where necessary, the sprayed concrete surface may be coated with two water-base paint coats to improve the aesthetics of the feature. Planting of creepers like *Ficus pumila* in a row of 150 mm diameter planter holes at 1 m centre to centre may also be proposed.

A 300 mm layer of compacted soil with an erosion control mat should be provided on the surface of the backfilled pits in order to support a green slope surface. All existing mature trees should be retained as far as practicable. Small shrubs should be planted. It would be a good practice to provide a planter box along the feature toe.

The entire feature is usually covered with 75 mm thick sprayed concrete with weep holes at a very steep angle of about 70°. For a cut slope steeper than 50°, it would be difficult to sustain growth of vegetation like grass. One would have to check if the slope surface would be obscured by other features/building at its toe. This would soften the adverse visual impact to the surrounding environment. A comprehensive treatment of the subject can be found in the Geotechnical Engineering Office (2011).

7.16 SITE SUPERVISION DURING IMPLEMENTATION

Under the Hong Kong Buildings Ordinance, construction works in the private sector need to be supervised according to a minimum standard as stipulated by the Buildings Department (2009a,b). Under the system, the project leader (called authorized person) has to prepare a site supervision plan. The plan would have four streams: authorized person, registered structural engineer, registered geotechnical engineer and registered specialist contractor. The supervision would cover site safety, workmanship and material quality, and last but not the least, design verification by a qualified geotechnical specialist. There are two levels of supervision: (1) periodic site supervision by a registered professional engineer in the geotechnical discipline (T5) at weekly intervals. The T5 person should be familiar with the geotechnical aspect of the design. (2) Full-time site supervision by a graduate engineer with a minimum of 2 years geotechnical experience.

7.17 CORROSIVENESS ASSESSMENT

Corrosiveness assessment of the soils should be based on chemical test results. The Geotechnical Engineering Office (2008) provides the following Classification of Soil Aggressivity:

Classification of soil aggressivity	*Total marks as per the soil aggressivity assessment scheme*
Non-aggressive	≥0
Mildly aggressive	−1 to −4
Aggressive	−5 to −10
Highly aggressive	≤−11

As an example, if the assessment indicates that the soils are *mildly aggressive* or *unlikely to be aggressive*, only hot-dip galvanization of the proposed soil nails is required (Eyre and Lewis, 1987).

If the feature is influenced by buried foul water drains and/or salt water mains, the ground should be regarded as potentially aggressive, and corrugated plastic sheath in addition to galvanization and sacrificial steel thickness should be provided. The corrosiveness of the soil should be verified by the laboratory tests from the confirmatory.

7.18 PRECAUTIONARY MEASURES AND OTHER CONSIDERATIONS

During the upgrading works, usually no ground/building settlement monitoring is required except for building at the crest of the feature. In order to avoid causing excessive ground movement when a building is founded on a shallow foundation near the feature to be soil nailed, no adjacent soil nail should be constructed within a 3 m horizontal distance until the grout of the soil nail has achieved the initial set.

Where undue excessive grout intake/loss defined as a situation where one or more nails cannot be fully grouted after injecting a volume of grout equal to five times the calculated gross volume of the drill hole in soil nailing works is observed, all site operations should be stopped and a technical review should be conducted. Practical recommendations on how to deal with the situation can be found in the Geotechnical Engineering Office (2004b).

7.19 LONG-TERM MAINTENANCE

On completion of the upgrading works, the feature should continue to be regularly inspected and maintained. There are two types of maintenance: routine maintenance inspection and engineering inspection for maintenance. Both should be scheduled on a regular basis in accordance with the Geotechnical Engineering Office (1995). The proposed raking drains if installed, which are prescriptive in nature, do not require any special monitoring for performance.

Chapter 8

Numerical implementation of slope stability analysis methods

In the previous chapters, the author introduced the concepts of limit equilibrium methods, limit analysis methods, SRMs and DEMs. At present, there are various commercial softwares targeting these methods of analysis, and some numerical procedures for the stability analysis are introduced in this chapter so that students can appreciate the actual procedures required in the numerical analysis.

8.1 NUMERICAL PROCEDURES FOR SIMPLIFIED LIMIT EQUILIBRIUM METHODS

The limit equilibrium method is currently the most commonly used method adopted by engineers because of its simplicity and the possibility of assessing the results by manual calculation. To start with the basic implementation of the limit equilibrium method, a simple slope with a water table as shown in Figure 8.1 is considered.

In this problem, the first step is the division of the solution domain into a series of slices, and 10 slices have been chosen as illustrations for this example. In general, for any point with a major change in the geometry of the ground profile, a further division in the slice should be added. Other than these turning points, the slices can be divided evenly for simplicity. There is a minor problem in this method of division of slices for cases where very thin slices may be generated. In the author's SLOPE 2000 manual, another option where uniform division is enforced is also available to avoid the formation of very thin slices. It should however be noted that thin slices usually do not affect the computation of the factor of safety. Using 10 slices for the present circular slip surface problem, the basic slice details are generated in Tables 8.1 and 8.2.

For the Bishop method (1955), the denominator $\Sigma w_i \sin \alpha_i$ is a constant and is evaluated as 200.366. For the numerator $\Sigma[c'b_i + (w_i - u_i b_i)\tan\varphi]m_\alpha$, where $m_\alpha = \sec \alpha/(1 + \tan \alpha \tan \varphi/F)$, using an initial factor of safety 1.0, the results of iteration analysis are given in Table 8.3.

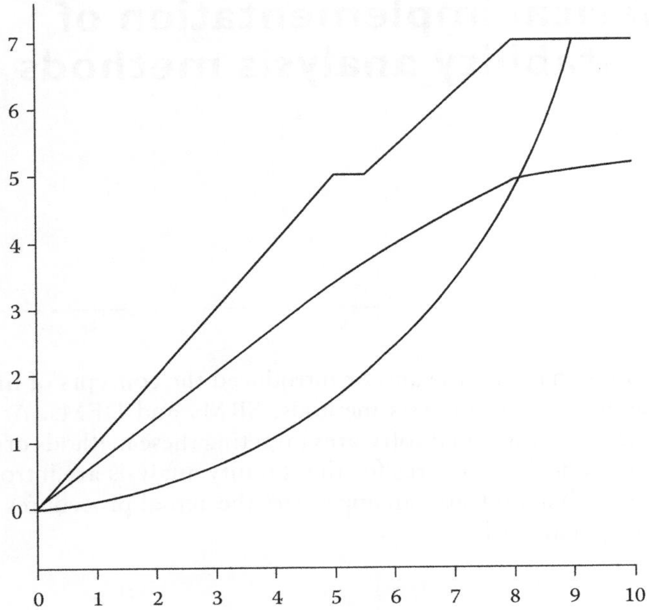

Figure 8.1 A simple slope for illustration.

Table 8.1 Coordinates of the circular slip surface and corresponding ground surface level

No.	*X(m)*	*Y(m)*	*Ground level*
1	0.000	0.000	0.000
2	0.625	0.054	0.625
3	1.250	0.147	1.250
4	1.875	0.282	1.875
5	2.500	0.460	2.500
6	3.750	0.954	3.750
7	5.000	1.663	5.000
8	5.500	2.018	5.000
9	6.750	3.144	6.000
10	8.000	4.789	7.000
11	9.000	7.000	7.000

Table 8.2 Basic slice details

No.	*Weight (kN)*	*V (kN)*	*H (kN)*	*α (°)*	*l (m)*	*u (kPa)*	*w sin α*	*Xc*	*Yc*
1	3.48	0.00	0.00	4.92	0.627	1.87	0.298	0.416	0.223
2	10.19	0.00	0.00	8.52	0.632	5.37	1.510	0.907	0.511
3	16.39	0.00	0.00	12.17	0.639	8.45	3.454	1.581	0.856
4	22.07	0.00	0.00	15.86	0.650	11.11	6.030	2.2	1.226
5	58.61	0.00	0.00	21.58	1.344	14.11	21.555	3.157	1.849
6	74.09	0.00	0.00	29.55	1.437	16.60	36.538	4.393	2.734
7	30.54	0.00	0.00	35.43	0.614	17.14	17.701	5.245	3.272
8	70.23	0.00	0.00	42.01	1.682	14.34	47.005	6.119	3.879
9	59.83	0.00	0.00	52.77	2.066	6.53	47.634	7.346	5.097
10	20.46	0.00	0.00	65.66	2.427	0.68	18.640	8.333	6.261

V, vertical surcharge; *H*, horizontal load, inward as positive; *α*, base inclination of slip surface; *l*, base length of slice; *u*, pore water pressure; *Xc*/*Yc*, coordinates of the centroid of slice.

Table 8.3 Numerators in iteration analysis for Figure 8.1 using the Bishop method

Slice	*F = 1.0*	*F = 0.897*	*F = 0.863*	*F = 0.852*	*F = 0.847*	*F = 0.846*
1	2.720	2.702	2.695	2.693	2.692	2.692
2	5.522	5.462	5.440	5.432	5.429	5.428
3	8.025	7.906	7.862	7.846	7.840	7.838
4	10.265	10.072	10.002	9.976	9.967	9.964
5	26.271	25.631	25.400	25.316	25.285	25.273
6	32.792	31.753	31.382	31.246	31.197	31.179
7	13.418	12.923	12.748	12.684	12.660	12.652
8	32.295	30.918	30.433	30.258	30.193	30.170
9	33.270	31.529	30.925	30.707	30.627	30.600
10	15.095	14.107	13.771	13.651	13.607	13.591
	Σ = 179.672	Σ = 173.003	Σ = 170.658	Σ = 169.809	Σ = 169.498	Σ = 169.384
	F = 0.897	F = 0.863	F = 0.852	F = 0.847	F = 0.846	F = 0.845 converged

Usually, the factor of safety during iteration analysis changes in a monotonic way. For the present problem, when the seventh iteration is evaluated, the factor of safety remains stationary at 0.845. A large fluctuation in the temporary factor of safety is usually associated with the non-converging case, which is, however, highly uncommon for circular slip surface with the simplified methods. To accelerate the iteration analysis, the author usually adopts an over-shooting method. For example, from $F = 1.0$ to $F = 0.897$ in the first iteration, the second trial factor of safety adopted by the author can be taken as $0.897 + (0.897 - 1.0)/2$ or 0.846. Using this new trial factor of safety, convergence is achieved in the second iteration!

For the same problem, if Janbu method (1957) is used, the denominator $\Sigma w_i \tan\alpha_i$ is a constant and is evaluated as 285.749. For the numerator $\Sigma[c'b_i+(w_i - u_i b_i)\tan\varphi]\eta_\alpha$, where $\eta_\alpha = \sec^2\alpha/(1+\tan\alpha\tan\varphi/F)$, using an initial factor of safety 1.0, the results of iteration analysis are shown in Table 8.4.

If over-shooting method is used in the Janbu method, only four iterations will be required to achieve convergence (1957). It is interesting to note that the Janbu method (1957) usually requires more iterations to achieve convergence than the Bishop method, and the factor of safety from the Janbu method (excluding f_0) appears to be always on the low side as compared with those from other methods (except under the action of nails). If the correction factor which is 1.061 is adopted for the present problem, the Janbu method (1957) will give an overall factor of safety of 0.82, which is comparable to 0.845, 0.859, 0.854 and 0.854 for the Bishop method (1955), load factor method, Sarma method (1973) and Spencer method (1967), respectively.

To determine the internal forces, the base normal force P as given by Equation 8.1 can be used for both the Janbu and Bishop methods, using the converged factor of safety.
or

$$P = \left\{ W - \frac{(c'l - ul\tan\phi')\sin\alpha}{F} \right\} \Big/ \cos\alpha\left(1 + \frac{\tan\alpha\tan\phi'}{F}\right) \tag{8.1}$$

The base shear force T will then be given by Equation 8.2 by Mohr–Coulomb relation as:

$$T = \frac{\left[(P - ul)\tan\phi' + c'l\right]}{F} \tag{8.2}$$

Slice	*P*	*T*	*For Bishop method*
1	3.224	3.185	
2	9.341	6.422	
3	14.767	9.274	
4	19.589	11.788	
5	51.207	29.900	
6	64.255	36.883	
7	26.827	14.966	
8	62.381	35.685	
9	51.263	36.188	
10	14.108	16.072	

Table 8.4 Numerators in iteration analysis for Figure 8.1 using the Janbu method

Slice	*F* = *1.0*	*F* = *0.856*	*F* = *0.806*	*F* = *0.787*	*F* = *0.779*	*F* = *0.775*	*F* = *0.774*
1	2.730	2.704	2.693	2.688	2.686	2.686	2.685
2	5.583	5.496	5.459	5.443	5.437	5.434	5.433
3	8.210	8.033	7.959	7.928	7.915	7.909	7.907
4	10.671	10.382	10.262	10.211	10.190	10.181	10.178
5	28.251	27.259	26.852	26.682	26.610	26.580	26.567
6	37.695	35.979	35.286	34.998	34.878	34.827	34.806
7	16.467	15.597	15.249	15.105	15.045	15.020	15.009
8	43.465	40.815	39.769	39.339	39.159	39.084	39.052
9	54.984	50.889	49.303	48.657	48.388	48.275	48.227
10	36.628	33.238	31.962	31.447	31.233	31.144	31.106
	$\Sigma = 244.684$	$\Sigma = 230.392$	$\Sigma = 224.794$	$\Sigma = 222.500$	$\Sigma = 221.542$	$\Sigma = 221.140$	$\Sigma = 220.970$
	$F = 0.856$	$F = 0.806$	$F = 0.787$	$F = 0.779$	$F = 0.775$	$F = 0.774$	$F = 0.773$

Slice	P	T	For Janbu method
1	3.200	3.461	
2	9.262	6.950	
3	14.611	9.998	
4	19.341	12.663	
5	50.395	31.953	
6	62.966	39.156	
7	26.223	15.815	
8	60.729	37.519	
9	49.236	37.728	
10	13.004	16.571	

It should also be noted that the base normal and shear forces are not sensitive to the use of different methods of analysis, which is also supported by the results as shown earlier. On the other hand, there are greater differences in the interslice shear and normal forces among the different methods of analysis.

The interslice shear force is simply assumed as zero for the Janbu and Bishop methods. For the interslice normal force under the Bishop formulation (1955), some commercial programmes produce the output for the users. Since the Bishop method (1955) cannot satisfy horizontal force equilibrium, the use of horizontal force equilibrium from left to right or from right to left will produce different internal normal forces. Hence, to avoid users making mistakes in the interpretation, the author does not give the interslice normal force for the Bishop method (1955) in SLOPE 2000. The interslice normal force under the Janbu formulation (1957) can be determined using Equation 8.3 as:

$$\left(E_L - E_R\right) = W \tan\alpha - T \sec\alpha \tag{8.3}$$

Once E_R for slice 1 is determined using the fact that E_L is zero for this slice, E_R from slice 1 will become E_L for slice 2 and Equation 8.3 can be used to determine E_R for slice 2. Furthermore, because the number of internal forces is 1 less than the number of slices, the interslice normal force is given by:

Section I	E	For Janbu method
1	3.17	
2	8.67	
3	15.37	
4	22.26	
5	33.45	
6	36.45	
7	34.14	
8	21.37	
9	5.0	

In the presence of earthquake and soil nail force, the problem is more complicated. Consider Figure 8.1 with the application of a horizontal earthquake coefficient 0.1, a horizontal inward pressure of 10 kPa applied on the ground surface from $x=3$ to $x=4$ and a soil nail with a length of 5.8 m and at an angle of inclination of 10° applied at $x=4$.

Consider the soil nail with a nail head coordinate (4,4), at an angle of inclination of 10° and intersecting with the slip surface at (6.998, 3.471). Behind the slip surface, the effective nail length is 2.755 m. The effective bond load using the Hong Kong Practice is 6.04 kPa with a bond load factor of safety 2.0. The effective nail load is hence given by 6.04 × 2.755 = 16.65 kN. Using the coordinate (6.998, 3.471), the nail load should be applied at slice 9, and the nail load is decomposed to a vertical load of 16.4 kN and an inward horizontal load of 2.89 kN. Given the coordinates of the centre of rotation and radius of rotation as (−0.544, 9.985) and 10 m, respectively, the restoring moment arising from the horizontal load and the soil nail and the overturning moment arising from the earthquake load can be determined, as shown in Table 8.5.

The overall moment is summed up to be 102.741, and this term is added to the denominator of the Bishop equation (2.25) in the form of $102.741/R$, where $R=10$ m for the present circular slip surface. The final value will be 200.366 + 102.741/10 = 210.64. On the other hand, the vertical component of the nail load (16.4 kN) will be added to W under the vertical force equilibrium. The horizontal force due to the earthquake or nail load are not added in Equation 2.25 directly, as the horizontal force equilibrium is not considered in the Bishop method. These two terms enter indirectly into the Bishop equation through the overall moment term in the denominator of Equation 2.25. Based on these results, the factor of

Table 8.5 Overturning and restoring moment from earthquake and soil nail

Slice	*Overturning moment*	*Restoring moment*
1	3.401	0
2	9.654	0
3	14.963	0
4	19.328	0
5	47.687	49.576
6	57.323	15.275
7	20.499	0
8	42.883	0
9	29.248	85.013
10	7.619	0

Table 8.6 Iteration analysis for the Bishop method with earthquake, external load and soil nail using over-shooting method

Slice	*F = 1.0*	*F = 0.794*
1	2.720	2.680
2	5.522	5.389
3	8.025	7.761
4	10.265	9.841
5	26.271	24.873
6	32.792	30.542
7	13.418	12.353
8	32.295	29.351
9	35.011[a]	31.139
10	15.095	13.040
	$\Sigma = 181.414$	$\Sigma = 166.97$
	$F = 0.863$	$F = 0.794$ converged

[a] Is changed by soil nail compared with Table 8.3.

Table 8.7 Horizontal load from earthquake and soil nail and external load

Slice	*Earthquake*	*Soil nail/external load*
1	0.348	0
2	1.019	0
3	1.639	0
4	2.207	0
5	5.861	7.5
6	7.409	2.5
7	3.054	0
8	7.023	0
9	5.983	16.4
10	2.046	0

safety of this slope will be 0.794 for the Bishop method, which is shown in Table 8.6.

For the Janbu method, the denominator becomes $\Sigma(\tan\alpha + H)$, and the results for the horizontal load on each slice are given in Table 8.7.

From Table 8.7, the net horizontal load is 10.189 kN. The vertical component of the soil nail load is added to W for vertical force equilibrium, and the denominator is 299.743 (consider also the vertical component of the soil nail in slice 9). The iteration analysis using the Janbu method is given in Table 8.8.

Table 8.8 Iteration analysis for Figure 8.1 using the Janbu method

Slice	*F = 1.0*	*F = 0.739*	*F = 0.727*	*F = 0.725*
1	2.730	2.676	2.672	2.672
2	5.583	5.402	5.391	5.390
3	8.210	7.846	7.825	7.821
4	10.671	10.080	10.045	10.040
5	28.251	26.240	26.124	26.107
6	37.695	34.255	34.061	34.032
7	16.467	14.735	14.639	14.625
8	43.465	38.238	37.953	37.910
9	57.862	49.477	49.033	48.967
10	36.628	30.153	29.824	29.775
	Σ = 247.561	Σ = 219.102	Σ = 217.568	Σ = 217.339
	F = 0.826	F = 0.731	F = 0.726	F = 0.725 converged

There are some simplified stability methods which adopt a given interslice force function but can fulfill only force but not moment equilibrium in the basic formulation. For example, in the Corps of Engineers method, $f(x)$ is assumed to be constant and is equal to the slope angle defined by the two extreme ends of the failure surface. In the Lowe–Karafiath method (1960), $\lambda f(x)$ is assumed to be the average of the slope angle of the ground profile and the failure surface at the section under consideration. For the problem in Figure 8.1, $f(x)$ is given by Table 8.9.

Based on this $\lambda f(x)$, the factor of safety is determined to be 0.867 using the method outlined in Section 2.2.4. It should be noted that λ is not an unknown in this method, and if necessary, the maximum value from Table 8.9 can be taken as λ, and $f(x)$ can be determined accordingly. For this problem, if the China-type Load factor method is used, the factor of

Table 8.9 $f(x)$ for the Lowe–Karafiath method

Section	$\lambda f(x)$
1	0.0856
2	0.148
3	0.209
4	0.270
5	0.360
6	0.476
7	0.554
8	0.633
9	0.744

safety will be 0.859 while the Corps of Engineers method gives a value of 0.881. In general, the author finds that the results from the Lowe–Karafiath method (1960) are comparable to those from the other rigorous methods and appear to be more reliable in general. On the other hand, the Swedish method gives a value of 0.756, which is much smaller than the other methods. The Swedish method is well known to be conservative and is generally not recommended.

8.2 NUMERICAL PROCEDURES FOR "RIGOROUS" LIMIT EQUILIBRIUM METHODS

Whereas the simplified method for a single slip surface is simple enough for hand or spreadsheet calculation, the calculation required for the rigorous method is much more tedious in computation. In general, a computer program is required for the rigorous analysis, but the detailed procedures required in the rigorous analysis can help the readers to fully understand the individual rigorous method.

8.2.1 Spencer and Morgenstern–Price method

In this section, the Spencer method will be considered, as this method is the most popular version used by engineers, for normal routine design work. The extension of the Spencer method to the Morgenstern–Price method is also discussed in this section. Consider the example in Figure 8.1 again. For the Spencer method, resolving vertically gives

$$P\cos\alpha + T\sin\alpha = W - (X_R - X_L) \tag{8.4}$$

Rearranging, and substituting for T gives

$$P = \frac{\left[W - (X_R - X_L) - 1/F\,(c'l - ul\tan\phi')\sin\alpha\right]}{m_\alpha} \tag{8.5}$$

where

$$m_\alpha = \cos\alpha\left(1 + \tan\alpha\frac{\tan\phi'}{F}\right)$$

Resolving horizontally gives

$$T\cos\alpha - P\sin\alpha + E_R - E_L = 0 \tag{8.6}$$

Table 8.10 Base normal and shear forces in the first step, which is actually the Janbu analysis (using initial factor of safety 1.0 and $\lambda=0$)

Slice	*Base normal*	*Base shear*
1	3.200	3.460
2	9.262	6.949
3	14.611	10.000
4	19.341	12.661
5	50.397	31.950
6	62.968	39.153
7	26.224	15.814
8	60.732	37.516
9	49.239	37.726
10	13.006	16.570

Rearranging and substituting for T gives

$$E_R - E_L = P\sin\alpha - \frac{1}{F}\left[c'l + \left(P - ul\right)\tan\phi'\right]\cos\alpha \tag{8.7}$$

Because $\Delta X=\lambda\Delta E$, combining Equation 8.7 and Equation 8.5 can yield an expression for P, which depends on λ and F only. ΔE is then determined by Equation 8.7, and hence, E is formed successively. X is then formed by using $X=\lambda E$. In the first step of the solution, the Janbu simplified analysis is used to determine the first trial factor of safety (the Bishop factor of safety can be used for circular surface), with λ being zero in the first step. The base normal and shear forces can then be determined accordingly. In the present example, the factor of safety is 0.773 as given previously. The base normal and shear forces as shown in Table 8.10 are stored for use when a new λ value is tried.

Based on Equations 8.5 and 8.7, using an initial $\lambda=0$, base normal and shear forces P and T can be determined accordingly as given in Table 8.11. With a moment point at (4.91, 8.585), P and T and W can take moments about the moment point to evaluate the factor of safety with respect to moment.

The factor of safety 0.853 has not converged with respect to moment consideration, and in the next step, this value is used for the next iteration analysis. The results of the analysis for the second step of iteration are shown in Table 8.12.

The factor of safety has changed from 0.853 to 0.862 in the second step, and the process continues until the value becomes stable at step 4. Based on the converged base normal force, the factor of safety with respect to force as given by Equation 8.8 is used, which gives $F_f=0.773$. This factor of safety is different from the moment factor of safety F_m, which is 0.862. $F_m - F_f$ is, therefore, a positive value that must be assigned for λ until this

Table 8.11 Base normal, shear and ΔE in the first iteration (FOS with respect to moment = 0.853, $\lambda = 0$)

Slice	*P*	*T*	ΔE
1	2.029	2.675	3.174
2	5.867	5.372	5.499
3	9.210	7.728	6.693
4	12.123	9.788	6.893
5	31.435	24.699	11.179
6	39.122	30.267	3.006
7	15.706	12.225	−2.316
8	36.613	29.002	−12.771
9	35.750	29.164	−16.375
10	11.363	12.810	−5.021

Table 8.12 Base normal, shear and ΔE in the second iteration (FOS with respect to moment = 0.862, $\lambda = 0$)

Slice	*P*	*T*	ΔE
1	2.055	2.693	2.871
2	5.954	5.433	4.914
3	9.382	7.848	5.880
4	12.396	9.979	5.895
5	32.327	25.324	8.752
6	40.539	31.260	0.132
7	16.371	12.690	−3.462
8	38.431	30.274	−15.488
9	37.983	30.728	−19.180
10	12.581	13.662	−6.358

value becomes zero. The user can assign a value of 0.1 (or any other value) and increase it gradually until the term $F_m - F_f$ has a change in sign, and a more precise value of λ can then be interpolated. The author has chosen λ to be 0.635 (converged value) for illustration purposes. The results of iteration analysis are shown in Tables 8.13 and 8.14.

$$F_f = \frac{\sum\left(c'l + (P - ul)\tan\phi'\right)\cos\alpha}{\sum P\sin\alpha} \tag{8.8}$$

Based on the results in Table 8.14, the factor of safety with respect to force is 0.857, and hence the factor of safety of this system is given as 0.857 with $\lambda = 0.635$.

Table 8.13 Base normal, shear and ΔE in the first step (FOS with respect to moment=0.848, λ=0.635)

Slice	*P*	*T*	ΔE
1	5.048	4.789	5.640
2	10.356	8.515	8.855
3	13.961	11.054	9.900
4	16.395	12.780	9.448
5	36.925	28.543	13.784
6	39.136	30.280	3.010
7	13.970	11.009	−2.591
8	29.500	24.021	−12.800
9	27.762	23.571	−14.393
10	8.874	11.067	−3.682

Table 8.14 Base normal, shear and ΔE in the fourth step (FOS with respect to moment=0.857, λ=0.635)

Slice	*P*	*T*	ΔE
1	5.097	4.823	5.074
2	10.500	8.616	7.887
3	14.217	11.233	8.684
4	16.767	13.040	8.088
5	38.001	29.297	10.857
6	40.616	31.313	0.001
7	14.595	11.447	−3.669
8	31.072	25.122	−15.148
9	29.648	24.892	−16.762
10	9.974	11.837	−4.890

If the double QR method is used, a λ value is assumed and the factor of safety with respect to force determined *directly* without any iteration analysis. The net moment (in Table 8.15) based on the corresponding λ and F_f is then used to evaluate the net moment of the system, and the factor of safety will be 0.854 with λ=0.617 for the present problem. These results differ slightly from the results obtained by using iteration analysis because of the truncation involved in the tedious calculations required in both analyses. Such a small difference is however acceptable from a practical point of view.

The procedures mentioned earlier can easily be extended to the Morgenstern–Price method by using the fact that $X_i = \lambda f(x_i)E$. Other than this small change, there is virtually no difference between the Spencer method and the Morgenstern–Price method.

Table 8.15 Temporary results during double QR analysis

Trial	λ	*Net moment*
1	0	10.167
2	0.05	9.471
3	0.1	8.732
4	0.15	7.957
5	0.2	7.150
6	0.25	6.319
7	0.3	5.470
8	0.35	4.608
9	0.4	3.740
10	0.45	2.871
11	0.5	2.004
12	0.55	1.144
13	0.6	0.293
14	0.65	−0.545
15	0.617	−0.001

8.2.2 Janbu rigorous method

The Janbu rigorous method differs from other rigorous methods in that the thrust line is assumed to be known, the resultant interslice forces E_i are assumed to be acting on a 'line of thrust', and β is the inclination of the line of thrust to the horizontal at the interface.

By taking moments about the base centre of each slice and assuming width b is small, the following equations can be obtained:

$$E_{i+1} b \tan\beta + \frac{(X_i + X_{i+1})b}{2} + (E_{i+1} - E_i)\left(h_i - \frac{b\tan\alpha}{2}\right) = 0$$

$$\frac{X_i + X_{i+1}}{2} = -E_{i+1}\tan\beta - \frac{E_{i+1}h_i}{b} + \frac{E_{i+1}\tan\alpha}{2} + \frac{E_i h_i}{b} - \frac{E_i \tan\alpha}{2} \tag{8.9}$$

where
- b is l cosα
- h is the height of the line of thrust
- β is the slope of the line of thrust

If the width of a slice is small, $X_i \approx X_{i+1}$ and $E_i \approx E_{i+1}$ X_i can be approximated as

$$X_{i+1} = -E_{i+1}\tan\beta - h\frac{dE}{dX} \tag{8.10}$$

For the requirement of overall horizontal equilibrium

$$\sum(E_i - E_{i+1}) = \sum \Delta E = 0 \tag{8.11}$$

Summation of ΔE is given by Equation 8.11 as

$$\sum(\Delta E) = \sum\{[W - \Delta X]\tan\alpha\} - \frac{1}{F}\sum\left[c'l + (P - ul)\tan\phi'\right]\sec\alpha = 0 \tag{8.12}$$

$$F_f = \frac{\sum\left[c'l + (P - ul)\tan\phi'\right]\sec\alpha}{\sum(W - \Delta X)\tan\alpha}$$

$$= \frac{\sum\left\{c'b + \left[(W - \Delta X) - ub\right]\tan\phi'\right\}\sec\alpha / m_\alpha}{\sum(W - \Delta X)\tan\alpha} \tag{8.13}$$

In Equation 8.11 of the Janbu method, the width of the slice should be small enough for it to be valid. The author has relaxed this requirement and has found that this will slightly improve the convergence and the results obtained by the Janbu rigorous formulation. In the first step of the Janbu rigorous method, the Janbu simplified method is performed. The results in Section 8.1 for the Janbu simplified method is, hence, adopted as the initial values in the Janbu rigorous analysis. Based on the initial internal forces and a thrust ratio of 0.333 from Equation 8.14, P is determined and is used in Equation 8.15a to give ΔE. Using the fact that E_1 is zero for the first slice, E can then be determined successively, and the results are given in Table 8.16. X is then determined directly from Equation 8.10 (Table 8.17).

Table 8.16 E from successive iterations in the Janbu rigorous analysis

Slice	*Iteration 1*	*Iteration 2*	*Iteration 3*
1	3.174	3.839	3.957
2	8.674	11.625	12.656
3	15.367	20.791	22.592
4	22.261	29.454	31.329
5	33.443	41.364	42.814
6	36.454	42.880	43.491
7	34.141	38.132	38.458
8	21.376	23.490	23.308
9	5.013	7.473	6.772

Table 8.17 X from successive iterations in the Janbu rigorous analysis

Slice	*Iteration 1*	*Iteration 2*	*Iteration 3*
1	1.231	1.489	1.534
2	5.572	7.560	8.258
3	11.246	15.054	16.154
4	16.651	21.094	21.971
5	23.719	28.225	28.748
6	26.244	28.734	28.575
7	6.784	5.973	5.629
8	7.080	8.312	7.749
9	3.211	−1.277	−2.005

$$P = \frac{\left[W - (X_R - X_L) - 1/F\left(c'l \sin\alpha - ul \tan\phi' \sin\alpha\right)\right]}{m_\alpha} \tag{8.14}$$

$$E_R - E_L = \left(W - (X_R - X_L)\right)\tan\alpha - \frac{1}{F}\left(c'l + (P - ul)\tan\phi'\right)\sec\alpha \tag{8.15a}$$

Starting from an initial factor of 1.0, the Janbu simplified method gives the first trial factor of safety, 0.773. In the Janbu rigorous analysis using the original Janbu formulation, the factor of safety starts from 0.826 and converged at 0.849. Using the author's modification where the width of each slice is taken as finite with a minor modification of Equation 8.9, the factor of safety varies from 0.845 to 0.848 and converges at 0.849, and the results of the analysis are shown in Tables 8.18 and 8.19.

Table 8.18 E from successive iterations in the Janbu rigorous analysis using Cheng's formulation

Slice	*Iteration 1*	*Iteration 2*	*Iteration 3*	*Iteration 4*
1	3.174	4.418	4.993	5.252
2	8.674	11.685	12.819	13.122
3	15.367	19.964	21.384	21.663
4	22.261	28.002	29.578	29.885
5	33.443	39.713	41.070	41.142
6	36.454	39.774	40.148	39.961
7	34.141	36.284	36.902	36.931
8	21.376	21.407	22.122	22.267
9	5.013	4.044	3.878	3.443

Table 8.19 X from successive iterations in the Janbu rigorous analysis using Cheng's formulation

Slice	*Iteration 1*	*Iteration 2*	*Iteration 3*	*Iteration 4*
1	2.189	3.046	3.442	3.621
2	5.990	7.829	8.340	8.297
3	10.451	12.920	13.441	13.568
4	15.094	18.027	18.692	18.793
5	23.441	26.040	26.211	25.912
6	21.490	19.130	18.207	17.987
7	19.627	21.243	23.830	24.786
8	11.135	8.291	7.051	6.428
9	4.485	4.929	5.861	5.428

8.2.3 Sarma method

In 1973, Sarma (1979) proposed a completely different approach to compute the factor of safety. His suggestion is based on the critical acceleration that is required to bring a soil mass to a state of limiting equilibrium. He also assumed that the slope of internal tangential force distribution $X(x)$ is known and is given by

$$X(x) = \lambda f_3(x)\left[c_{avg}H(x) + \left(K_{avg} - Ru\right)\frac{\gamma_{avg}H^2}{2}\tan\phi_{avg}\right] \tag{8.15b}$$

where

$X(x)$ is the interslice force parallel to slice interface
λ is the scaling factor determined from Equation 8.27
$f_3(x)$ is the scaling function determined by the user, usually chosen as 1.0
$\gamma_{avg}, c_{avg}, K_{avg}$, and $\tan\phi_{avg}$ are the average soil strength parameters along the interface of the slice (see Equation 8.35)
H is the height of the slice
Ru_i is the pore pressure parameter equal to the ratio of pore water pressure to the total overburden pressure

Considering the vertical and horizontal equilibrium of the slice i, with vertical and horizontal surcharge V and Q, the vertical and horizontal equilibrium equations are obtained as

$$P_i\cos\alpha + T_i\sin\alpha_i = W_i + V_i - \Delta X_i \tag{8.16}$$

$$T_i\cos\alpha - P_i\sin\alpha_i = KW_i + \Delta E_i + \Delta Q_i \tag{8.17}$$

where T is the shear force at slice base. It is assumed that under the action of KW, the full shear strength of the soil is mobilized. Hence

$$S = c' + (\sigma - u)\tan\phi' \Rightarrow T_i = (P_i - U_i)\tan\phi' + c'b_i \sec\alpha \tag{8.18}$$

From Equations 8.16 and 8.18

$$P_i \cos\alpha_i + (c'b_i \sec\alpha_i + P_i \tan\phi' - U_i \tan\phi_i')\sin\alpha_i = W_i + \Delta V_i - \Delta X_i$$

$$\Rightarrow P_i = \frac{[W_i + \Delta V_i - \Delta X_i - c_i'b_i \tan\alpha_i + U_i \tan\phi_i' \sin\alpha_i]\cos\phi}{[(\cos(\phi_i' - \alpha_i))]}$$

$$\Rightarrow P_i = \frac{[W_i + \Delta V_i - \Delta X_i - c_i'b_i \tan\alpha_i + Ru_i W_i \tan\phi_i' \sin\alpha_i]\cos\phi}{[(\cos(\phi_i' - \alpha_i))]}$$

and

$$T_i = \frac{(W_i + \Delta V_i - \Delta X_i - P_i \cos\alpha_i)}{\sin\alpha_i} \tag{8.19}$$

Substituting P_i in Equation 8.19 gives

$$T_i = \frac{[(W_i + \Delta V_i - \Delta X_i)\sin\phi_i' + c_i'b_i \cos\phi_i' - Ru_i W_i \sin\phi_i']}{[\cos(\phi_i' - \alpha_i)]} \tag{8.20a}$$

Put Equation 8.20a in Equation 8.17,

$$\Delta X_i \tan(\phi_i' - \alpha_i) + \Delta E_i + \Delta Q_i = D_i - KW_i \tag{8.20b}$$

$$D_i - \Delta Q_i = (W_i + \Delta V_i)\tan(\phi_i' - \alpha_i) + \frac{c_i'b_i \cos\phi_i' \sec\alpha_i - U_i \sin\phi_i'}{\cos(\phi_i' - \alpha_i)}$$

$$D_i - \Delta Q_i = (W_i + \Delta V_i)\tan(\phi_i' - \alpha_i) + \frac{(c_i'b_i \cos\phi_i' - Ru_i W_i \sin\phi_i')\sec\alpha_i}{\cos(\phi_i' - \alpha_i)} \tag{8.21}$$

Consider the horizontal equilibrium of the whole mass where $\Sigma\Delta E_i$ cancel out:

$$\Sigma\Delta Q_i + \Sigma\Delta X_i \tan(\phi_i' - \alpha_i) + \Sigma KW_i = \Sigma D_i \tag{8.22a}$$

For the moment equilibrium, we can take the moment about the centre of gravity of the whole sliding soil mass (including vertical surcharge) to eliminate the weight terms W_i and the vertical surcharge terms ΔV_i.

$$\Sigma\Delta Q_i\left(y_Q - y_g\right) + \Sigma\left(T_i\cos\alpha_i - P_i\sin\alpha_i\right)\left(y_g - y_i\right)$$
$$-\Sigma\left(P_i\cos\alpha_i + T_i\sin\alpha_i\right)\left(x_g - x_i\right) = 0 \quad (8.22b)$$

where

(x_g, y_g) are the coordinates of the centre of gravity of the whole soil mass
(x_i, y_i) are the coordinates of the midpoint of the slice base

Equations 8.16, 8.17, 8.20b and 8.22b give

$$\Sigma\Delta Q_i\left(y_Q - y_g\right) + \Sigma\left(KW_i + \Delta E_i + \Delta Q_i\right)\left(y_g - y_i\right)$$
$$-\Sigma\left(W_i + \Delta V_i - \Delta X_i\right)\left(x_g - x_i\right) = 0$$

$$\Sigma\Delta Q_i\left(y_Q - y_g\right) + \Sigma\left[D_i - \Delta X_i\tan\left(\phi_i' - \alpha_i\right)\right]\left(y_g - y_i\right)$$
$$-\Sigma\left(W_i + \Delta V_i - \Delta X_i\right)\left(x_g - x_i\right) = 0$$

$$\Sigma\Delta X_i\left[\left(y_i - y_g\right)\tan\left(\phi_i' - \alpha_i\right) + \left(x_g - x_i\right)\right]$$
$$= \Sigma W_i\left(x_g - x_i\right) + \Sigma D_i\left(y_i - y_g\right) - \Sigma\Delta Q_i\left(y_Q - y_g\right) \quad (8.23)$$

Sarma (1979) then assumed that the resultant shear force can be expressed as Equation 8.15b and f_3 is assumed to be known. This is equivalent to assuming the shape of the distribution of the interslice shear force. However, the magnitude of the interslice shear force is not assumed to be known. Therefore, substituting Equation 8.15b into 8.22 and 8.23 gives

$$\Sigma\Delta Q_i + \lambda\Sigma F_3\tan(\phi_i' - \alpha_i) + K\Sigma W_i = \Sigma D_i \quad (8.24)$$

$$\lambda\Sigma f_3\left[\left(y_i - y_g\right)\tan\left(\phi_i' - \alpha_i\right) + \left(x_g - x_i\right)\right]$$
$$= \Sigma W_i\left(x_g - x_i\right) + \Sigma D_i\left(y_i - y_g\right) - \Sigma\Delta Q_i\left(y_Q - y_g\right) \quad (8.25)$$

With an estimated value of f_3, Equations 8.24 and 8.25 can be solved to obtain λ and K which can be expressed as

$$\lambda = \frac{S_2}{S_3} \quad (8.26a)$$

$$K = \frac{S_1 - \lambda S_4}{\Sigma W_i} \quad (8.26b)$$

where

$$S_1 = \Sigma D_i - \Sigma \Delta Q_i \quad (8.27a)$$

$$S_2 = \Sigma W_i (x_g - x_i) + \Sigma D_i (y_i - y_g) - \Sigma \Delta Q_i (y_Q - y_g) \quad (8.27b)$$

$$S_3 = \Sigma f_3 \left[(y_i - y_g) \tan(\phi_i' - \alpha_i) + (x_g - x_i) \right] \quad (8.27c)$$

$$S_4 = \Sigma f_3 \tan(\phi_i' - \alpha_i) \quad (8.27d)$$

The lever arm of the interslice normal force is given by

$$h_{i+1} = \frac{\begin{bmatrix} E_i h_i - 0.5_i \tan\alpha_i (E_i + E_{i+1}) - 0.5_i (X_i + X_{i+1}) \\ - \Delta Q_i (y_Q - y_i) + \Delta V_i (X_V - X_i) \end{bmatrix}}{E_{i+1}} \quad (8.28)$$

This K value gives the critical acceleration for the soil mass under analysis while the corresponding λ gives the change in the interslice shear force of Equation 8.15b. To determine the factor of safety for the soil mass, an adjusted factor of safety is applied to the properties of the soil in the calculation of the mobilized shear force at the slice base by the equation of (8.27) until the value of K is zero (i.e. acceleration = 0).

For the choice of the assumed function $f_3(x)$, Sarma (1979) has carried out a detailed study and found that a value of 1 is applicable for most cases. If there are some local sections where X and E violate the failure criterion, f_3 can be slightly reduced.

For non-homogeneous soil, the normal force acting on the interslice surface i in soil layer j, which can be used as the initial value for estimating E in the iteration analysis, is given by

$$E_{i,j} = a_{i,j} W_{i,j} (h_{i,j} - h_{i,j+1}) + \frac{a_{i,j} \gamma_{i,j} (h_{i,j} - h_{i,j+1})^2}{2} + b_{i,j} (h_{i,j} - h_{i,j+1}) + d_{i,j} P_{wij} \quad (8.29)$$

where

$$a_{i,j} = \frac{1 - \sin(\beta_{i,j}) \sin\phi_{i,j}'}{1 + \sin(\beta_{i,j}) \sin\phi_{i,j}'} \quad (8.30a)$$

$$b_{i,j} = \frac{-2c_{i,j}' \cos\phi_{i,j}' \sin(\beta_{i,j})}{1 + \sin(\beta_{i,j}) \sin\phi_{i,j}'} \quad (8.30b)$$

$$d_{i,j} = \frac{2 \sin\phi_{i,j}' \sin(\beta_{i,j})}{1 + \sin(\beta_{i,j}) \sin\phi_{i,j}'} \quad (8.30c)$$

$$\beta_{i,j} = 2\alpha_{i,j} - \phi_{i,j}' \quad (8.30d)$$

$W_{i,j}$ is the weight per unit area at the level j along the ith interslice surface

$$= \sum_{k=1}^{j-1} \gamma_{i,k} \left(h_{i,k} - h_{i,k+1} \right) \tag{8.31}$$

$V_{i,j}$ is the piezometric height at level j along interslice surface i.

$$P_{wij} = \frac{1}{2} \gamma_w \left(V_{i,j} + V_{i,j+1} \right) \left(h_{i,j+1} - h_{i,j} \right) \tag{8.32}$$

where
$h_{i,j}$ is the y-coordinate of the jth soil boundary at the ith slice
$\gamma_{i,j}$ is the density of the soil layer j at slice i
γ_w is the density of water

The shear force acting on the interslice surface i for the non-homogeneous case is given by:

$$X_i = \lambda f_3(x) \left[\left(K_{i,\text{ave}} - Ru_{i,\text{ave}} \right) \left(\frac{\gamma_{i,\text{ave}} H_i^2}{2} \right) \tan \phi'_{i,\text{ave}} + c_{i,\text{ave}} H_i \right] \tag{8.33}$$

where

$$\gamma_{i,\text{ave}} = \frac{\sum \gamma_j \left(h_{i,j} - h_{i,j+1} \right)}{H_i} \tag{8.34a}$$

$$Ru_{i,\text{ave}} = \frac{\sum P_{wij}}{\gamma_{i,\text{ave}} H_i^2 / 2} \tag{8.34b}$$

$$K_{i,\text{ave}} = \frac{\sum E_{i,j}}{\gamma_{i,\text{ave}} H_i^2 / 2} \tag{8.34c}$$

$$\tan \phi'_{i,\text{ave}} = \frac{\sum \left(E_{i,j} - P_{wij} \right) \tan \phi'_j}{\sum \left(E_{i,j} - P_{wji} \right)} \tag{8.34d}$$

$$c_{i,\text{ave}} = \frac{\sum c'_j \left(h_{i,j} - h_{i,j+1} \right)}{H_i} \tag{8.34e}$$

H_i is the height of interslice i

For the actual solution required in the Sarma (1979) method, the author has adopted a bracket method, where an upper bound and lower bound of the factor of safety is defined. K will then be determined from Equation 8.26b. If there is a change in the sign of K based on the upper and lower bound of the factor of safety, then the chosen bound will be adequate, and the bound will be narrowed. If there is no change in the sign of K based on the upper and lower bound of the factor of safety, another range will be chosen. For the example in Figure 8.1, the bounds during the trial analysis are given in Table 8.20.

The critical situation (when factor of safety = 1) is obtained as −0.0777 while the corresponding λ is given by 0.000611.

During the initial analysis, Ru and K_i from Equations 8.34b and c are given in Table 8.21 while E is determined from Equation 8.29.

When the factor of safety is determined, X can be determined from Equation 8.20b. Base normal and shear force will then be given by Equation 8.19.

Table 8.20 Bounds for the Sarma analysis

Trial	*Lower bound*	*Upper bound*	λ
1	0.8	1	0.000718
2	0.8	0.9	0.000784
3	0.85	0.9	0.000750
4	0.85	0.875	0.000766
5	0.85	0.863	0.000775
6	0.85	0.856	0.000780
7	0.853	0.856	0.000777
8	0.853	0.854	0.000779
9	0.853	0.855	0.000778
10	0.854	0.854	0.000778

Table 8.21 Ru, Ki and E during iteration analysis for Sarma's method

Interface	*Ru*	*Ki*	E
1	0.335	1.560	4.710
2	0.329	1.251	14.068
3	0.322	1.067	25.031
4	0.314	0.857	33.002
5	0.301	0.646	46.712
6	0.280	0.525	54.097
7	0.281	0.441	36.255
8	0.255	0.287	21.658
9	0.136	0.174	7.863

Although the Sarma method (1979) can converge well for many cases, the thrust line back calculated from this method may sometimes be greatly outside the soil mass, which is not acceptable. The implementation of the Sarma method (1979) by the author does not require iteration analysis, and using an initial factor of safety from the Janbu or Bishop method, the results can be bracketed within a narrow range easily.

8.3 THREE-DIMENSIONAL ANALYSIS

For three-dimensional analysis, it is necessary to divide the soil mass into a series of columns. For simplicity, only spherical soil mass is considered in this section. For three-dimensional problems, there are two ways to divide the soil mass. The first method is the use of the grid in the longitudinal and transverse directions, which is useful for the determination of the internal forces along the transverse direction. Under this approach, the number of columns for each section will be different. This method of division requires attention to some columns along the periphery of the failure zone as incomplete columns will be formed. The second method is the division along the longitudinal direction simply according to the number of divisions; every section will have a similar number of divisions, which makes it simple for geometric calculation. The internal forces along the transverse direction, however, cannot be determined under this approach. The author's SLOPE 2000 manual adopts the second approach whereas his SLOPE3D manual adopts the first approach.

Consider the problem in Figure 8.1 again. Because of symmetry, the sliding direction is zero. The geometry of the failure zone at different sections is given using Equation 8.35. α_x and α_y (slopes along x and y direction) can be determined from the derivative of the equation of sphere when the centre coordinates of each column is known. γ_z, which is the angle between the z-axis and the normal at each soil column, is then determined from Equation 8.35 whereas the base area of each column is determined from Equation 8.36, where Δx and Δy are the widths of the columns.

$$\cos\gamma_z = \left(\frac{1}{\tan^2\alpha_x + \tan^2\alpha_x + 1}\right)^{1/2} \tag{8.35}$$

$$A = \Delta x \Delta y \frac{\left(1 - \sin^2\alpha_x \sin^2\alpha_y\right)^{1/2}}{\cos\alpha_x \cos\alpha_y} \tag{8.36}$$

			Slip surface coordinates	
No.	*X(m)*	*Y(m)*	*Thrustline (m)*	*Ground surface level*
Coordinates at section z = −6.008				
1	2.663	2.663		
2	2.955	2.798		
3	3.247	2.948		
4	3.539	3.113		
5	3.831	3.295		
6	4.124	3.496		
7	4.416	3.716		
8	4.708	3.959		
9	5.000	4.226		
10	5.500	4.754		
11	6.009	5.407		
Coordinates at section z = −4.673				
1	1.349	1.349		
2	2.262	1.601		
3	2.718	1.768		
4	3.175	1.964		
5	4.087	2.454		
6	5.000	3.099		
7	5.500	3.533		
8	6.045	4.091		
9	6.591	4.765		
10	7.136	5.607		
11	7.682	6.746		
Coordinates at section z = −3.338				
1	0.632	0.632		
2	1.724	0.836		
3	2.816	1.178		
4	3.908	1.676		
5	5.000	2.361		
6	5.500	2.751		
7	6.125	3.323		
8	6.750	4.014		
9	7.375	4.872		
10	8.000	6.003		
11	8.397	7.000		

No.	X(m)	Y(m)	Thrustline (m)	Ground surface level
Coordinates at section z = −2.003				
1	0.217	0.217		
2	1.413	0.385		
3	2.609	0.709		
4	3.804	1.206		
5	5.000	1.907		
6	5.500	2.274		
7	6.125	2.808		
8	6.750	3.444		
9	7.375	4.217		
10	8.000	5.191		
11	8.787	7.000		
Coordinates at section z = −0.668				
1	0.024	0.024		
2	1.268	0.173		
3	2.512	0.487		
4	3.756	0.982		
5	5.000	1.690		
6	5.500	2.046		
7	6.125	2.564		
8	6.750	3.177		
9	7.375	3.915		
10	8.000	4.832		
11	8.977	7.000		

The columns details are given

Columns details

No.	Weight (kN)	Surcharge load (kN)	Horiz. load (kN)	Gamma Z (°)	Base area (m²)	Base pore p. (kPa)	Alpha y (°)
Columns information at section z = –6.008							
1	0.57	0.00	0.00	43.48	0.538	0.00	24.800
2	1.65	0.00	0.00	44.65	0.548	0.00	27.129
3	2.62	0.00	0.00	45.92	0.561	0.00	29.508
4	3.47	0.00	0.00	47.29	0.575	0.00	31.944
5	4.20	0.00	0.00	48.76	0.592	0.00	34.447
6	4.79	0.00	0.00	50.34	0.611	0.00	37.026
7	5.23	0.00	0.00	52.04	0.634	0.00	39.697

No.	*Weight (kN)*	*Surcharge load (kN)*	*Horiz. load (kN)*	*Gamma Z (°)*	*Base area (m²)*	*Base pore p. (kPa)*	*Alpha y (°)*
8	5.49	0.00	0.00	53.87	0.662	0.00	42.475
9	6.30	0.00	0.00	56.58	1.212	0.00	46.454
10	1.55	0.00	0.00	60.51	1.381	0.00	51.993
Columns information at section z=–4.673							
1	7.45	0.00	0.00	31.54	1.430	0.00	15.414
2	9.11	0.00	0.00	33.86	0.734	0.33	20.072
3	12.32	0.00	0.00	35.68	0.750	1.54	23.255
4	32.56	0.00	0.00	38.80	1.564	2.76	28.180
5	40.46	0.00	0.00	43.69	1.685	3.33	35.133
6	21.07	0.00	0.00	48.11	1.000	2.67	40.947
7	19.03	0.00	0.00	51.79	1.177	1.05	45.602
8	16.53	0.00	0.00	56.12	1.307	0.00	50.913
9	12.20	0.00	0.00	61.15	1.509	0.00	56.918
10	4.73	0.00	0.00	67.29	1.886	0.00	64.105
Columns information at section z=–3.338							
1	12.23	0.00	0.00	22.06	1.573	1.11	10.528
2	35.28	0.00	0.00	25.89	1.620	5.43	17.371
3	54.11	0.00	0.00	30.92	1.699	8.63	24.481
4	67.94	0.00	0.00	36.94	1.824	10.16	32.020
5	31.21	0.00	0.00	41.96	0.898	10.12	37.927
6	35.23	0.00	0.00	45.89	1.199	8.42	42.402
7	32.79	0.00	0.00	50.70	1.318	5.12	47.785
8	28.01	0.00	0.00	56.17	1.499	0.91	53.798
9	20.26	0.00	0.00	62.66	1.817	0.00	60.837
10	4.89	0.00	0.00	69.36	1.505	0.00	68.042
Columns information at section z=–2.003							
1	15.81	0.00	0.00	14.01	1.645	2.61	7.975
2	45.23	0.00	0.00	18.94	1.688	8.20	15.116
3	69.41	0.00	0.00	25.16	1.764	12.20	22.508
4	87.54	0.00	0.00	32.25	1.888	14.34	30.321
5	37.43	0.00	0.00	37.81	0.845	14.69	36.255
6	43.52	0.00	0.00	41.79	1.119	13.29	40.451
7	41.84	0.00	0.00	46.58	1.214	10.44	45.446
8	38.17	0.00	0.00	51.87	1.351	6.42	50.931
9	31.75	0.00	0.00	57.90	1.571	1.25	57.159
10	17.60	0.00	0.00	66.34	2.620	0.00	65.820

No.	Weight (kN)	Surcharge load (kN)	Horiz. load (kN)	Gamma Z (°)	Base area (m²)	Base pore p. (kPa)	Alpha y (°)
Columns information at section z = −0.668							
1	17.69	0.00	0.00	7.84	1.677	3.43	6.848
2	50.34	0.00	0.00	14.62	1.717	9.51	14.118
3	77.27	0.00	0.00	21.95	1.791	13.90	21.630
4	97.65	0.00	0.00	29.78	1.914	16.35	29.558
5	40.40	0.00	0.00	35.68	0.822	16.87	35.500
6	47.45	0.00	0.00	39.73	1.085	15.60	39.574
7	46.11	0.00	0.00	44.53	1.171	12.95	44.404
8	42.92	0.00	0.00	49.78	1.292	9.21	49.672
9	37.19	0.00	0.00	55.68	1.480	3.98	55.588
10	26.15	0.00	0.00	64.92	3.076	0.47	64.858

Based on the geometry described earlier, the factors of safety are obtained as 0.960, 0.971, 0.915 and 0.906 for 3D Bishop (1955), Spencer (1967), Corps of Engineer and Lowe–Karafiath methods (1960) of analysis. Again, the factors of safety arising from different methods of analysis vary within a narrow range between the methods of analysis.

Appendix

GENERAL INTRODUCTION TO SLOPE 2000

The 2D and 3D formulations as well as the optimization search outlined in this book have been coded into two general-purpose programs: SLOPE 2000 and SLOPE3D. SLOPE3D is under development and a relatively nice 3D interface has been completed recently. It can be obtained from Cheng for testing and evaluation. SLOPE 2000 is a mature 2D program that supports some simple 3D cases. It has received government approval in Hong Kong and China. Most of the examples in this book are based on version 2.1 and 2.2 of this program. It has also been used for many projects in different countries, and the latest English and Chinese versions can be downloaded from Cheng's website at http://www.cse.polyu.edu.hk/~ceymcheng/download.htm. This program (version 2.5) is also incorporated into the Geo-Suite 1.0/2.0 delivered by Vianova Finland System Oy. SLOPE 2000 has many important and useful features including the following:

1. Location of critical failure surface with evaluation of a *global* minimum factor of safety for both circular and non-circular failure surfaces under general conditions. Very difficult problems with multiple soft band problems have also been tested with satisfying results. The verification examples in the user guide have demonstrated the power of the modern optimization methods in SLOPE 2000 as compared with other slope stability programs.
2. Generation of graphics files in the form of Autocad DXF, bitmap BMP, postscript, HP plotter format or vector format (CGM and Lotus PIC) for incorporation into other programs. For the Windows version, a clipboard and a Windows print manager are also supported.
3. Bishop simplified, Fellenius, Swedish, Janbu simplified and Janbu rigorous, China Load Factor (including the simplified version required in some Chinese codes), Sarma, Morgenstern–Price, Corps

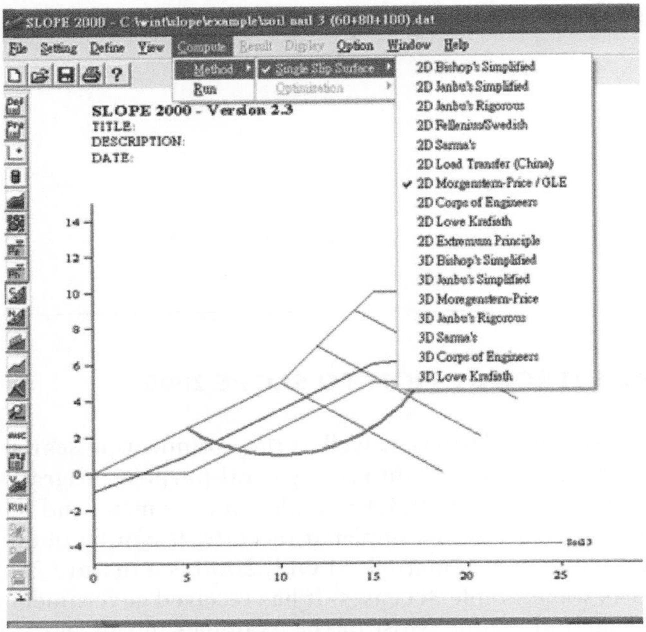

Figure A.1 Various types of stability methods available for analysis in SLOPE 2000.

of Engineers, Lowe Karafiath, GLE methods and extremum principles are implemented under the 2D analysis (12 methods) (see Figure A.1). The load factor method, extremum and GLE methods are not implemented while all the other corresponding 2D analyses are extended to 3D analyses. A true 3D slope stability analysis for spherical and non-spherical failure surfaces is covered by a separate program SLOPE3D by Cheng.

4. f_0 for the Janbu simplified method is incorporated into the program (optional) so that the user need not determine it from the design graph by Janbu.
5. The China load factor method is available.
6. Janbu rigorous method and Sarma method (2D and 3D) are available, which are not present in many commercial programs.
7. Windows and Linux versions are available. For the older version of SLOPE 2000, Dos and other platforms are available as well.
8. Water table, pore pressure coefficient, perch water table and excess pore pressure contour can all be defined.
9. Earthquake loading in the form of horizontal/vertical acceleration is accepted in this program.

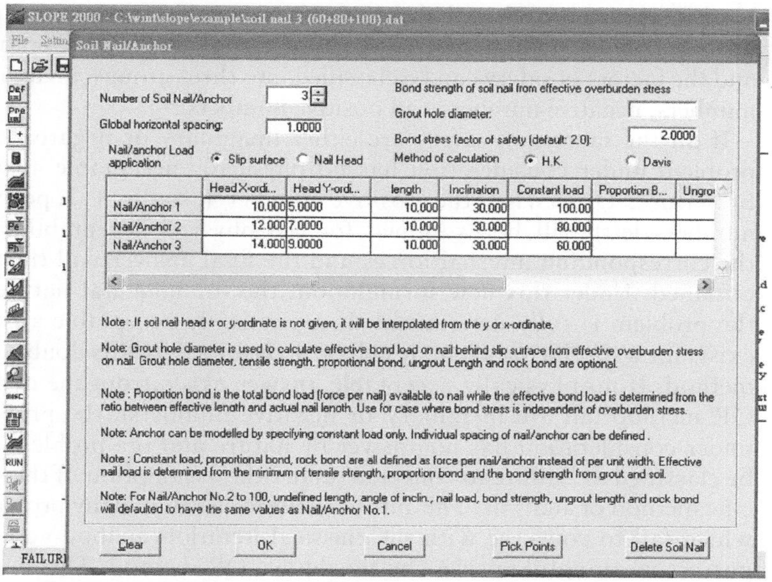

Figure A.2 Extensive options for modelling soil nails.

10. The program is able to accept vertical surcharge and horizontal loads with the presence of a rock boundary. Loading can be applied on or below ground level.
11. Many options are available for soil nail modelling, which is shown in Figure A.2. Nail load can be controlled by tensile strength of the nail, bond length proportional to the effective zone or bond stress from vertical overburden stress (unique). The bond stress from overburden stress can be determined from Hong Kong practice or the US Davis method. Furthermore, the horizontal spacing of the nail can be controlled for each individual row, which is also unique among other similar programs. The nail load can be applied to the failure surface by default or to the nail head if necessary. If part of the nail is not grouted or the end of the nail is socketed into rock, these options can also be modelled by SLOPE 2000.
12. Cheng (2003) has formulated the slope stability problem in a matrix approach and the factor of safety can be determined directly from a complex double QR matrix method. The special advantage of this double QR method is that the factor of safety and internal forces for *rigorous* methods can be determined directly from the matrix equation and no initial factor of safety is required. Cheng (2003) has proved that there are N factors of safety for a problem with N slices.

In this new approach, all the N factors of safety are determined directly from the tedious matrix equation *without* using any iteration, and the factors of safety can be classified into three groups: imaginary numbers, negative numbers and positive numbers.

If all the factors of safety are either imaginary or negative, the problem under consideration has no physically acceptable answer by nature. Otherwise, the positive number (usually 1–2 positive numbers left) will be examined for the physical acceptability of the corresponding internal forces and the final answer will then be obtained. Under this new formulation, the fundamental nature of the problem is *fully* determined. If a physically acceptable answer exists for a specific problem, it will be determined by this double QR method. If no physically acceptable answer exists from the double QR method (all are imaginary or negative numbers), the problem under consideration has no answer by nature and the problem can be classified as 'failure to converge' under the assumption of the specific method of analysis. The authors have found that many problems which fail to converge with the classical iteration method actually possess meaningful answers by the double QR method. This means that the phenomenon of *failure to converge* that emerges from the use of the iteration method of analysis may be a false phenomenon in some cases. The authors have also found that many failure surfaces that fail to converge are normal in shape and should not be neglected in ordinary analysis and design.

13. The China earthquake code for dam design is available. The coefficient varies with height according to the formula $a_h \xi a_i$, which is different between different slices.
14. Pond water (water table above ground) can be modelled automatically.
15. Interslice force function $f(x)$ can be determined from the lower bound/extremum principle. This is unique among all existing slope stability programs.
16. Soil parameters can vary with depth from ground surface or from contour lines.

ILLUSTRATION

For the slope as shown in Figure A.3, the bond load on the soil nail is defined by the overburden stress acting on the soil nail according to Hong Kong practice. To perform the analysis, choose the extremum principle from the method of analysis and select the parameters as shown in Figure A.4. Choose type 4 extremum formulation and select option 1 for type 4 formulation (maximum extremum). Click the

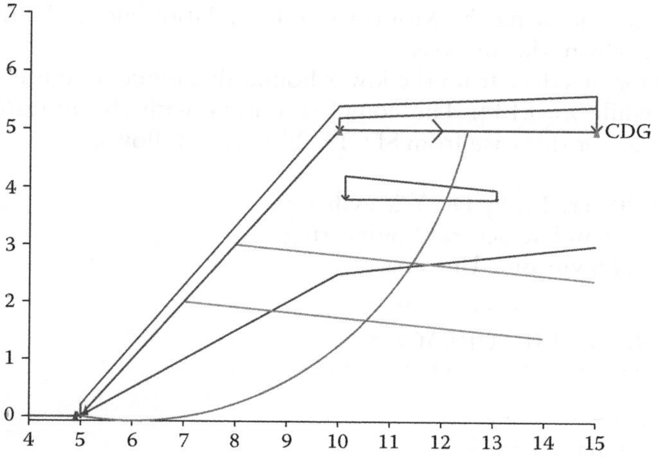

Figure A.3 A simple slope with two soil nails, three surface loads and one underground trapezoidal vertical load, two soil nails and a water table.

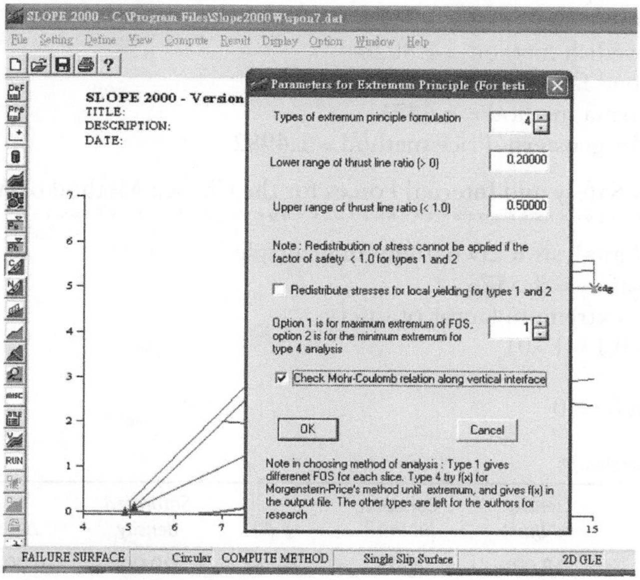

Figure A.4 Parameters for extremum principle.

checkbox for checking the Mohr–Coulomb relation along the interfaces and then perform the analysis.

The factor of safety from the lower bound theorem/extremum principle is 1.558 while $\lambda=0.134$. The complete output with the *interslice force function f*(*x*) for this case from SLOPE 2000 is as follows:

SLOPE 2000 ver. 2.3 by Dr. Y.M. Cheng
Dept. of Civil and Structural Engineering
Hong Kong Polytechnic University

****

* SINGLE SLIP SURFACE *

****

= =

= Basic data =

= =

Density of water: 9.81 (kN/m^3)
Tolerance in analysis: 0.00050
Slip surface: Circular
Number of slices: 10

****

FOS for Bishop method = 1.5298
FOS for Janbu simplified = 1.4514
FOS for Swedish method = 1.3020
FOS for Load factor method = 1.6211
FOS for Sarma method = 1.6471
FOS for Morgenstern–Price method = 1.4982

* Factor of Safety and Internal Forces for the Chosen Method of Analysis *
****

Method of analysis = 2D extremum principle
Factor of safety = 1.5576
Maximum extremum factor of safety
Lambda = 0.134E+01

Material type = 0

** Soil properties **

Soil name	*Density (kN/m^3)*	*Cohesion (kN/m^2)*	*ϕ (°)*	*Saturated density*	*ΔC*	*Δϕ*
CDG	18.00	5.00	36.00	20.00	0.00	0.00

```
==============================
= Soil profile coordinates =
==============================
```

X/Y Coor (m)	*4.00*	*5.00*	*10.00*	*15.00*
CDG	0.00	0.00	5.00	5.00
Water (m)	0.00	0.00	2.50	3.00

```
====================================
= = Slip surface coordinates = =
====================================
```

No.	*X (m)*	*Y (m)*	*Line of thrust (m)*
1	5.000	0.000	0.000
2	5.625	−0.066	−0.066
3	6.250	−0.073	0.945
4	6.875	−0.021	1.352
5	7.500	0.091	1.580
6	8.125	0.266	1.808
7	8.750	0.510	2.043
8	9.375	0.831	2.290
9	10.000	1.243	2.571
10	11.250	2.455	3.459
11	12.500	5.000	5.000

Centre of circle (*X*,*Y*) = (6.011, 6.608)
Radius of circle = 6.685

```
=======================
= Surface load =
=======================
```

No.	*StartX (m)*	*EndX (m)*	*VPress (kN/m²)*	*VPress2 (kN/m²)*	*HPress (kN/m²)*	*HPress2 (kN/m²)*	*Depth1 (m)*	*Depth2 (m)*
1	5.000	15.000	5.000	15.000	0.000	0.000		
2	10.000	15.000	5.000	10.000	0.000	0.000		
3	12.000	12.000	0.000	0.000	10.000	20.000		
4	10.120	13.068	10.000	4.000	0.000	0.000	3.802	3.802

============================
= Soil nail information =
============================

Diameter of grout hole = 0.07500
Soil/grout bond stress factor of safety = 2.000
Bond load determined from Hong Kong practice by the overburden stresses

Number of rows of soil nails: 2
Soil nail horizontal spacing: 1.000
All the loads are defined per nail

	Nail head coordinates		*Nail angle*	*Bond strength*	*Length*	*Tensile strength*	*Actual load*
No.	*X (m)*	*Y (m)*	*(°)*	*(kN)*	*(m)*	*(kN)*	*(kN)*
1	7.000	2.000	5.00		7.00		12.24
2	8.000	3.000	5.00		7.00		10.40

	Nail/slip coordinates		*Ungrout length*	*Bond length*	*Nail spacing*
No.	*X (m)*	*Y (m)*	*(m)*	*(m)*	*(m)*
1	10.467	1.697	0.00	3.519	1.000
2	11.373	2.705	0.00	3.615	1.000

Nail load applied at the ground surface

===========================
Slice details
===========================

No.	*Weight (kN)*	*Surcharge load (kN)*	*Horizontal load (kN)*	*Base angle (°)*	*Base length (m)*	*Base pore pressure (kPa)*	*Base friction tan(φ)*	*Base cohesion (kPa)*
1	4.12	3.32	0.00	−6.01	0.628	1.86	0.727	5.000
2	12.00	3.71	0.00	−0.63	0.625	5.28	0.727	5.000
3	19.14	4.10	0.00	4.73	0.627	8.12	0.727	5.000
4	25.54	4.49	0.00	10.15	0.635	10.39	0.727	5.000
5	31.17	4.88	0.00	15.65	0.649	12.04	0.727	5.000
6	35.97	5.27	0.00	21.31	0.671	13.06	0.727	5.000
7	39.86	5.66	0.00	27.20	0.703	13.35	0.727	5.000
8	42.70	6.05	0.00	33.42	0.749	12.82	0.727	5.000
9	72.67	30.32	0.00	44.11	1.741	7.00	0.727	5.000
10	28.64	28.09	0.00	63.84	2.835	0.06	0.727	5.000

No.	Base normal (kN)	Base shear (kN)
1	52.455	25.941
2	11.536	5.849
3	7.892	3.318
4	17.313	7.038
5	67.494	29.920
6	−2.152	−2.936
7	70.049	30.555
8	21.101	7.770
9	101.881	47.430
10	46.455	30.693

Nail	Loads at slice	
Slice	Horizontal (kN)	Vertical (kN)
1	0.00	0.00
2	0.00	0.00
3	0.00	0.00
4	12.19	1.07
5	10.36	0.91
6	0.00	0.00
7	0.00	0.00
8	0.00	0.00
9	0.00	0.00
10	0.00	0.00

Interface number	Interface length l(i,i + 1)	Interface friction tan(ϕ) (m)	Interface cohesion c(i,i + 1) (kPa)	Interface normal force E(i,i + 1) (kN)	Interface shear force X(i,i + 1) (kN)	f(x)
1	0.69	0.73	5.00	31.29	−42.01	1.000
2	1.32	0.73	5.00	37.26	−37.77	0.755
3	1.90	0.73	5.00	39.92	−22.67	0.423
4	2.41	0.73	5.00	55.99	−9.85	0.131
5	2.86	0.73	5.00	76.96	−45.96	0.445
6	3.24	0.73	5.00	75.01	−1.64	0.016
7	3.54	0.73	5.00	70.17	−32.39	0.344
8	3.76	0.73	5.00	65.03	−5.52	0.063
9	2.54	0.73	5.00	28.16	−8.70	0.230

Total weight of the soil mass = 311.81

To search for the critical failure surface corresponding to the problem in Figure A.3, define the left and right search range as (4.0, 6.0) and (10.0, 15.0) as shown in Figure A.5. This means that the left exit end will be controlled within (4.0, 6.0) and the right exit end will be controlled within (10.0, 15.0). Choose the shape of the failure surface from the failure surface option under the define menu (default to non-circular in SLOPE 2000), then proceed to the selection of the method of analysis as shown in Figure A.6. The minimum factor of safety 1.4 for the Spencer method is obtained as shown in Figure A.7, and the default tolerance in locating the critical failure surface is 0.0001, which can be adjusted in the default if necessary. SLOPE 2000 is the only program at present for which a tolerance in the optimization search can be defined.

SLOPE 2000 is robust and has been used in many countries (Hong Kong, Taiwan, China, Italy, the United States, Finland, Syria, Argentina). This program has been used in China for many major national projects and a simplified Chinese version is available. The interface is identical to the English one except that the words are all in simplified Chinese.

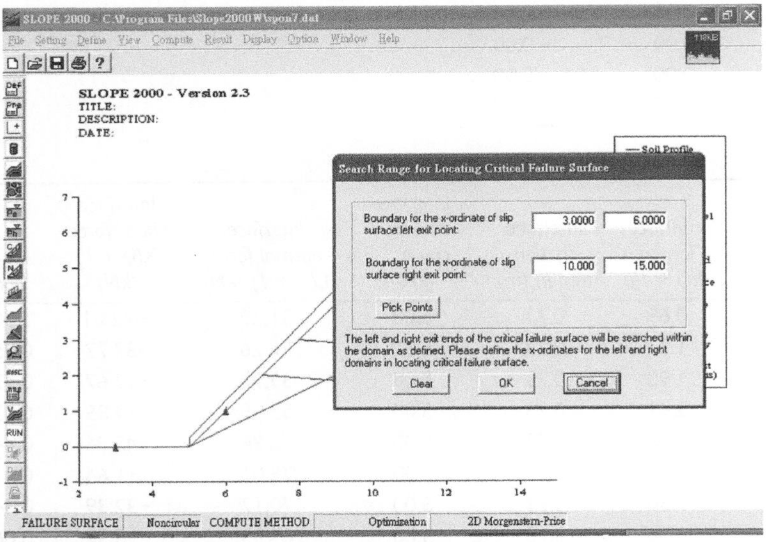

Figure A.5 Defining the search range for optimization analysis.

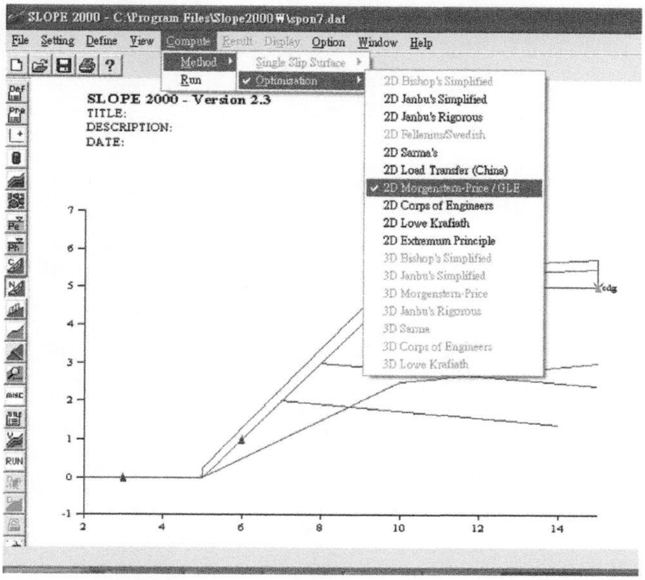

Figure A.6 Choose the stability method for optimization analysis.

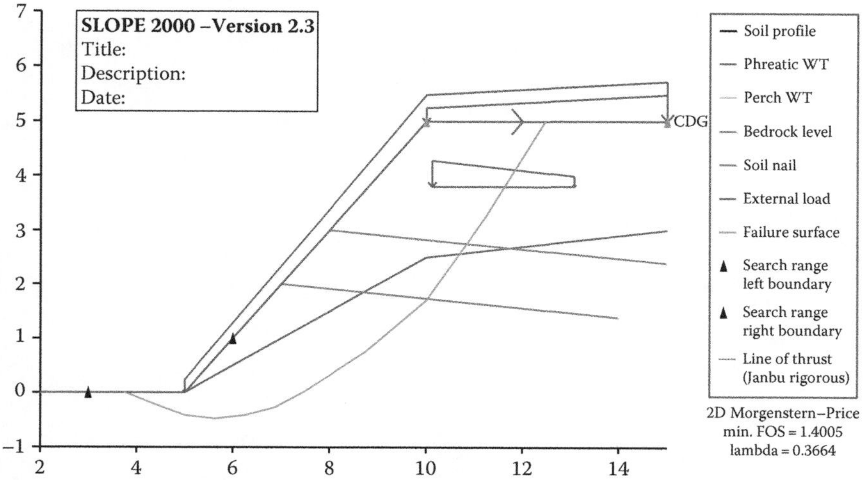

Figure A.7 The critical failure surface with the minimum factor of safety corresponding to Figure A.6.

References

Abramson L.W., Lee T.S., Sharma S., and Boyce G.M. (2002) *Slope Stability and Stabilization Methods*, 2nd edn., New York: John Wiley.

ACADS. (1989) *Soil Slope Stability Programs Reviews*, Publication No. U255, Melbourne, Victoria, Australia: ACADS.

Anderheggen E. and Knopfel H. (1972) Finite element limit analysis using linear programming, *International Journal of Solids and Structures*, 8(12), 1413–1431.

Arai K. and Tagyo K. (1985) Determination of noncircular slip surfaces giving the minimum factor of safety in slope stability analysis, *Soils and Foundations*, 25, 43–51.

Assadi A. and Sloan S.W. (1990) Undrained stability of shallow square tunnel, *Journal of the Geotechnical Engineering Division, ASCE*, 117(8), 1152–1173.

Ausilio E., Conte E., and Dente G. (2001) Stability analysis of slopes reinforced with piles, *Computers and Geotechnics*, 28(8), 591–611.

Azzouz A.S. and Baligh M.M. (1983) Loaded areas on cohesive slopes, *Journal of Geotechnical Engineering, ASCE*, 109, 709–729.

Baker R. (1980) Determination of the critical slip surface in slope stability computations, *International Journal of Numerical and Analytical Methods in Geomechanics*, 4, 333–359.

Baker R. (2003) Sufficient conditions for existence of physically significant solutions in limiting equilibrium slope stability analysis, *International Journal of Solids and Structures*, 40(13–14), 3717–3735.

Baker R. (2005) Variational slope stability analysis of materials with non-linear failure criterion, *Electronic Journal of Geotechnical Engineering*, 10, Bundle A.

Baker R. and Garber M. (1977) Variational approach to slope stability, in: *Proceedings of the Ninth International Conference on Soil Mechanics and Foundations Engineering*, Tokyo, Japan, Vol. 2, pp. 9–12.

Baker R. and Garber M. (1978) Theoretical analysis of the stability of slopes, *Geotechnique*, 28, 395–411.

Baker R. and Leshchinsky D. (2001) Spatial distribution of safety factors, *Journal of Geotechnical and Geoenvironmental Engineering*, 127(2), 135–145.

Baligh M.M. and Azzouz A.S. (1975) End effects on the stability of cohesive soils, *Journal of the Geotechnical Engineering Division, ASCE*, 101(GT11), 1105–1117.

Bishop A.W. (1955) The use of the slip circle in the stability analysis of earth slopes, *Geotechnique*, 5(1), 7–17.

Bishop A.W. (1972) Shear strength parameters for undisturbed and remoulded soils specimens, in: R.H.G. Parry (ed.), *Stress–Strain Behaviour of Soils*, pp. 3–58, London, U.K.: Foulis.

Bolton H.P.J., Heymann G., and Groenwold A. (2003) Global search for critical failure surface in slope stability analysis, *Engineering Optimization*, 35(1), 51–65.

Bolton M.D. and Lau C.K. (1993) Vertical bearing capacity factors for circular and strip footings on a Mohr–Coulomb soil, *Canadian Geotechnical Journal*, 30(6), 1024–1033.

Booker J.R. and Zheng X. (2000) Application of the theory of classical plasticity to the analysis of the stress distribution in wedges of a perfectly frictional material, in: M. Zaman, G. Gioda, and J.R. Booker (eds.), *Modelling in Geomechanics*, pp. 329–358, New York: John Wiley.

Bottero A., Negre R., Pastor J., and Turgeman S. (1980) Finite element method and limit analysis theory for soil mechanics problems, *Computer Methods in Applied Mechanics and Engineering*, 22, 131–149.

Brent R.P. (1973) *Algorithms for Minimization without Derivatives*, Englewood Cliffs, NJ: Prentice-Hall.

Buildings Department. (2009a) *Technical Memorandum for Supervision Plans*, Hong Kong: HKSAR Government.

Buildings Department. (2009b) *Code of Practice for Site Supervision*, Hong Kong: HKSAR Government.

Byrne R.J., Cotton D., Porterfield J., Wolschlag C., and Ueblacker, G. (1996) Manual for design and construction monitoring of soil nail walls, Federal Highway Administration (FHWA) Report No. FHWA-SA-96-069, Washington, DC.

Cai F. and Ugai K. (2000) Numerical analysis of the stability of a slope reinforced with piles, *Soils and Foundations*, 40(1), 73–84.

Castilo E. and Luceno A. (1980) Evaluation of variational methods in slope analysis, in: *Proceedings of International Symposium on Landslides*, New Delhi, India, Vol. 1, pp. 255–258.

Castilo E. and Luceno A. (1982) A critical analysis of some variational methods in slope stability analysis, *International Journal for Numerical and Analytical Methods in Geomechanics*, 6, 195–209.

Cavounidis S. (1987) On the ratio of factors of safety in slope stability analyses, *Geotechnique*, 37(2), 207–210.

Celestino T.B. and Duncan J.M. (1981) Simplified search for non-circular slip surface, in: *Proceedings of 10th International Conference on Soil Mechanics and Foundation Engineering*, Stockholm, Sweden, pp. 391–394.

Chan P.S. (1999) Distinct element slope stability analysis, Master thesis, Hong Kong Polytechnic University, Hong Kong.

Chang C.S. (1992) Discrete element method for slope stability analysis, *Journal of Geotechnical Engineering, ASCE*, 118(12), 1889–1905.

Chang M. (2002) A 3D slope stability analysis method assuming parallel lines of intersection and differential straining of block contacts, *Canadian Geotechnical Journal*, 39(4), 799–811.

Chatterjee A. and Siarry P. (2006) Nonlinear inertia weight variation for dynamic adaptation in particle swarm optimization, *Computers and Operations Research*, 33, 859–871.

Chen J. (2004) Slope stability analysis using rigid elements, PhD thesis, Hong Kong Polytechnic University, Hong Kong.

Chen L.T. and Poulos H.G. (1997) Piles subjected to lateral soil movements, *Journal of Geotechnical and Geoenvironmental Engineering*, 123(9), 802–811.

Chen R.H. and Chameau J.L. (1982) Three-dimensional limit equilibrium analysis of slopes, *Geotechnique*, 32(1), 31–40.

Chen W.F. (1975) *Limit Analysis and Soil Plasticity*, Burlington, VT: Elsevier.

Chen W.F. and Giger M.W. (1971) Limit analysis of stability of slopes, *Journal of the Soil Mechanics and Foundations Division, Proceedings of the American Society of Civil Engineers*, 97, 19–26.

Chen Z. (1998) On Pan's principles of rock and soil stability analysis, *Journal of Tsinghua University (Sci & Tech)*, 38, 1–4 (in Chinese).

Chen Z. and Morgenstern N.R. (1983) Extensions to generalized method of slices for stability analysis, *Canadian Geotechnical Journal*, 20(1), 104–109.

Chen Z. and Shao C. (1983) Evaluation of minimum factor of safety in slope stability analysis, *Canadian Geotechnical Journal*, 25(4), 735–748.

Chen Z.Y., Wang J., Wang Y.J., Yin J.H., and Haberfield C. (2001b) A three-dimensional slope stability analysis method using the upper bound theorem. Part II: Numerical approaches, applications and extensions, *International Journal of Rock Mechanics and Mining Sciences*, 38(3), 379–397.

Chen Z.Y., Wang X.G., Haberfield C., Yin J.H., and Wang Y.J. (2001a) A three-dimensional slope stability analysis method using the upper bound theorem. Part I: Theory and methods, *International Journal of Rock Mechanics and Mining Sciences*, 38(3), 369–378.

Cheng Y.M. (1998) Advancement and improvement in discontinuous deformation analysis, *Computers and Geotechnics*, 22(2), 153–163.

Cheng Y.M. (2002) Slip line solution and limit analysis for lateral earth pressure problem, in: *Proceedings of the Ninth Conference on Computing in Civil and Building Engineering*, April 3–5, Taipei, Taiwan, pp. 311–314.

Cheng Y.M. (2003) Locations of critical failure surface and some further studies on slope stability analysis, *Computers and Geotechnics*, 30(3), 255–267.

Cheng Y.M. (2007a) Global optimization analysis of slope stability by simulated annealing with dynamic bounds and Dirac function, *Engineering Optimization*, 39(1), 17–32.

Cheng Y.M. (2007b) Slope 2000, User Guide, http://www.cse.polyu.edu.hk/~ceymcheng/

Cheng Y.M. and Au S.K. (2005) Slip line solution of bearing capacity problems with inclined ground, *Canadian Geotechnical Journal*, 42, 1232–1241.

Cheng Y.M., Au S.K., and Lai X.L. (2013a) An innovative Geonail System for soft ground stabilization, *Soils and Foundation*, 53(2), 282–298.

Cheng Y.M., Hu Y.Y., and Wei W.B. (2007c) General axisymmetric active earth pressure by method of characteristic – Theory and numerical formulation, *Journal of Geomechanics, ASCE*, 7(1), 1–15.

Cheng Y.M., Lansivaara T., Baker R., and Li N. (2013b) The use of internal and external variables and extremum principle in limit equilibrium formulations with application to bearing capacity and slope stability problems, *Soils and Foundation*, 53(1), 130–143.

Cheng Y.M., Lansivaara T., and Siu J. (2008a) Impact of convergence on slope stability analysis and design, *Computers and Geotechnics*, 35(1), 105–113.

Cheng Y.M., Lansivaara T., and Wei W.B. (2007a) Two-dimensional slope stability analysis by limit equilibrium and strength reduction methods, *Computers and Geotechnics*, 34, 137–150.

Cheng Y.M., Li D.Z., Lee Y.Y., and Au S.K. (2013c) Solution of some engineering partial differential equations governed by the minimal of a functional by global optimization method, *Journal of Mechanics*, 29, 507–516.

Cheng Y.M, Li D.Z., Li L., Sun Y.J., Baker R., and Yang Y. (2011b) Limit equilibrium method based on approximate lower bound method with variable factor of safety that can consider residual strength, *Computers and Geotechnics*, 38, 628–637.

Cheng Y.M., Li L., and Chi S.C. (2007b) Studies on six heuristic global optimization methods in the location of critical slip surface for soil slopes, *Computers and Geotechnics*, 34(6), 462–484.

Cheng Y.M., Li L., Chi S.C., and Wei W.B. (2007d) Particle swarm optimization algorithm for location of a critical non-circular failure surface in two-dimensional slope stability analysis, *Computers and Geotechnics*, 34(2), 92–103.

Cheng Y.M., Li L., Lansivaara T., Chi S.C., and Sun Y.J. (2008c) Minimization of factor of safety using different slip surface generation methods and an improved harmony search minimization algorithm, *Engineering Optimization*, 42(2), 95–115.

Cheng Y.M., Li L., Sun Y.J., and Au S.K. (2012) A coupled particle swarm and harmony search optimization algorithm for difficult geotechnical problems, *Structural and Multidisciplinary Optimization*, 45, 489–501.

Cheng Y.M., Liang L., Chi S.C., and Wei W.B. (2008b) Determination of the critical slip surface using artificial fish swarms algorithm, *Journal of Geotechnical and Geoenvironmental Engineering*, 134(2), 244–251.

Cheng Y.M., Liu H.T., Wei W.B., and Au S.K. (2005) Location of critical three-dimensional non-spherical failure surface by NURBS functions and ellipsoid with applications to highway slopes, *Computers and Geotechnics*, 32(6), 387–399.

Cheng Y.M., Yeung A.T., Tham L.G., Au S.K., So T.C., Choi Y.K., and Chen J. (2009) New soil nail material-pilot study of grouted GFRP pipe nails in Korea and Hong Kong, *Journal of Civil Engineering Materials, ASCE*, 21(3), 93–102.

Cheng Y.M. and Yip C.J. (2007) Three-dimensional asymmetrical slope stability analysis – Extension of Bishop's, Janbu's, and Morgenstern–Price's techniques, *Journal of Geotechnical and Geoenvironmental Engineering*, 133(12), 1544–1555.

Cheng Y.M., Zhao Z.H., and Sun Y.J. (2010) Evaluation of interslice force function and discussion on convergence in slope stability analysis by the lower bound method, *Journal of Geotechnical and Geoenvironmental Engineering, ASCE*, 136(8), 1103–1113.

Cheng Y.M. and Zhu L.J. (2005) Unified formulation for two dimensional slope stability analysis and limitations in factor of safety determination, *Soils and Foundations*, 44(6), 121–128.

Chow Y.K. (1996) Analysis of piles used for slope stabilization, *International Journal for Numerical and Analytical Methods in Geomechanics*, 20, 635–646.

Chugh A.K. (1986) Variable factor of safety in slope stability analysis. *Geotechnique*, 36, 57–64.
Chun R.H. and Chameau J.L. (1982) Three-dimensional limit equilibrium analysis of slopes, *Geotechnique*, 32(1), 31–40.
Civil Engineering Development Department (CEDD). (2005) *When Hillsides Collapse – A Century of Landslides in Hong Kong*, Hong Kong: CEDD, Hong Kong SAR Government.
Clouterre. (1991) Soil nailing recommendations 1991 for design, calculating, constructing and inspecting earth support system using soil nailing, *Recommendations Clouterre, FHWA-SA-93-026, Federal Highway Administration, US.*
Collin A. (1846) *Recherches Expérimentales sur les Glissements Spontanés des Terrains Argileux, accompagnées de Considerations sur Quelques Principes de la Méchanique Terrestre*, Paris, France: Carilian-Goeury and Dalmont.
Collins I.F. (1974) A note on the interpretation of Coulomb's analysis of the thrust on a rough retaining wall in terms of the limit theorems of limit plasticity, *Geotechnique*, 24, 106–108.
Cox A.D. (1962) Axially-symmetric plastic deformation in soils – II, indentation of ponderable soils, *International Journal of Mechanical Sciences*, 4, 371–380.
Cundall P.A. (1988) Formulation of a three-dimensional distinct element model – Part I: A scheme to detect and research contacts in a system composed of many polyhedral blocks, *International Journal of Rock Mechanics Mining Science and Geomechanics*, 25(3), 107–116.
Cundall P.A. and Strack O.D.L. (1979) A discrete numerical model for granular assemblies, *Geotechnique*, 29(1), 47–65.
D'Appolonia E., Alperstein R., and D'Appolonia D.J. (1967) Behavior of a colluvial slope, *Journal of Soil Mechanics and Foundations Division, ASCE*, 93(SM4), 447–473.
David F.R. (2001) *An Introduction to NURBS: With Historical Perspective*, San Francisco, CA: Morgan Kaufmann Publishers.
Dawson E., Motamed F., Nesarajah S., and Roth W. (2000) Geotechnical stability analysis by strength reduction. Slope stability 2000, in: *Proceedings of Sessions of Geo-Denver 2000*, Denver, CO, August 5–8, 2000, pp. 99–113.
Dawson E.M., Roth W.H., and Drescher A. (1999) Slope stability analysis by strength reduction, *Geotechnique*, 49(6), 835–840.
De Beer E.E. and Wallays M. (1970) Stabilization of a slope in schist by means of bored piles reinforced with steel beams, in: *Proceedings of Second International Congress Rock on Mechanics*, Beograd, Serbia, Vol. 3, pp. 361–369.
Dolan C.W. (1993) FRP development in the United States, in: A. Nanni (ed.), *Fiber-Reinforced Plastic (FRP) Reinforcement for Concrete Structures: Properties and Applications*, pp. 129–163, Amsterdam, the Netherlands: Elsevier.
Donald I. and Chen Z.Y. (1997) Slope stability analysis by the upper bound approach: Fundamentals and methods, *Canadian Geotechnical Journal*, 34(6), 853–862.
Donald I.B. and Giam S.K. (1988) Application of the nodal displacement method to slope stability analysis, in: *Proceedings of the Fifth Australia–New Zealand Conference on Geomechanics*, Sydney, New South Wales, Australia, pp. 456–460.
Dorigo M. (1992) Optimization, learning and natural algorithms, PhD thesis, Department of Electronics, Politecnico di Milano, Milano, Italy.

Dowling N.E. (1999) *Mechanical Behavior of Materials: Engineering Methods for Deformation, Fracture and Fatigue*, 2nd edn., Upper Saddle River, NJ: Prentice Hall.

Drescher A. and Detournay E. (1993) Limit load in translational failure mechanisms for associative and non-associative materials, *Geotechnique*, 43(3), 443–456.

Drucker, D.C., Greenberg, W., and Prager, W. (1951) The safety factor of an elastic plastic body in plane strain, *Transactions of the ASME, Journal of Applied Mechanics*, 73, 371.

Drucker D.C. and Prager W. (1952) Soil mechanics and plastic analysis or limit design, *Quarterly of Applied Mathematics*, 10, 157–165.

Duncan J.M. (1996) State of the art: Limit equilibrium and finite element analysis of slopes, *Journal of Geotechnical Engineering, ASCE*, 122(7), 577–596.

Duncan J.M. and Wright S.G. (2005) *Soil Strength and Slope Stability*, Hoboken, NJ: John Wiley.

Elias V. and Juran I. (1991) *Soil Nailing for Stabilization of Highway Slopes and Excavations*, FHWA-RD-89-198, McLean, VA: United States Federal Highway Administration.

Endicott L.J. (1982) Analysis of piezometer data and rainfall records to determine groundwater conditions, *Hong Kong Engineer*, 10(9), 53–56.

Espinoza R.D. and Bourdeau P.L. (1994) Unified formulation for analysis of slopes with general slip surface, *Journal of Geotechnical Engineering, ASCE*, 120(7), 1185–1204.

Eyre D. and Lewis D.A. (1987) Soil corrosivity assessment (Contractor Report 54), Transport and Road Research Laboratory, Crowthorne, U.K.

Fan K., Fredlund D.G., and Wilson G.W. (1986) An interslice force function for limit equilibrium slope stability analysis, *Canadian Geotechnical Journal*, 23(3), 287–296.

Farzaneh O. and Askari F. (2003) Three-dimensional analysis of nonhomogeneous slopes, *Journal of Geotechnical and Geoenvironmental Engineering*, 129(2), 137–145.

FHWA. (1996) Geotechnical Engineering Circular No. 2 – Earth retaining systems, U.S. Department of Transport, Report FHWA-SA-96-038, Washington, DC.

FHWA. (1999) *Manual for Design and Construction of Soil Nail Walls*, Washington, DC: U.S. Department of Transport.

Fellenius W. (1918) Kaj-och jordrasen i Goteborg, *Teknisk Tidskrift V.U.*, 48, 17–19.

Fellenius W. (1927) *Erdstatische Berechnungen mit Reibung und Kohasion* (in German), Berlin, Germany: Ernst.

Fellenius W. (1936) Calculation of the stability of earth dams, in: *Transactions of the Second Congress on Large Dams, International Commission on Large Dams of the World Power Conference*, Washington, DC, 4, 445–462.

Fisher R.A. and Yates F. (1963) *Statistical Tables for Biological, Agricultural and Medical Reasearch*, 6th edn., Edinburgh, Scotland: Oliver & Boyd.

Fredlund D.G. and Krahn J. (1977) Comparison of slope stability methods of analysis, *Canadian Geotechnical Journal*, 14(3), 429–439.

Fredlund D.G. and Krahn J. (1984) Analytical methods for slope analysis, in: *Proceedings of the Fourth International Symposium on Landslides*, Toronto, Ontario, Canada, pp. 229–250.

Fredlund D.G., Zhang Z.M., and Lam L. (1992) Effect of axis on moment equilibrium in slope stability analysis, *Canadian Geotechnical Journal*, 29(3), 456–465.

Frohlich O.K. (1953) The factor of safety with respect to sliding of a mass of soil along the arc of a logarithmic spiral, in: *Proceedings of the Third International Conference on Soil Mechanics and Foundation Engineering*, Zurich, Switzerland, Vol. 2, pp. 230–233.

Fukuoka M. (1977) The effects of horizontal loads on piles due to landslides, in: *Proceedings of the 10th Specialty Session, Ninth International Conference Soil Mechanics and Foundation Engineering*, Tokyo, Japan, pp. 27–42.

Gassler, G. (1993) The first two field tests in the history of soil nailing on nailed walls pushed to failure, in: *Soil Reinforcement: Full Scale Experiments of the 80's. Presses de l'ecole Nationale des Ponts et Chaussees, CEEC*, Paris, France, pp. 7–34.

Geem Z.W., Kim J.H., and Loganathan G.V. (2001) A new heuristic optimization algorithm: Harmony search, *Simulation*, 76(2), 60–68.

Geotechnical Control Office. (1984) *Geotechnical Manual for Slopes*, 2nd edn., Hong Kong: HKSAR Government.

Geotechnical Control Office. (1993) *Guide to Retaining Wall Design – Geoguide 1*, 2nd edn., Hong Kong: HKSAR Government.

Geotechnical Engineering Office. (1976) Report on the slope failures at Sau Mau Ping, August, Hong Kong: Hong Kong Government.

Geotechnical Engineering Office. (1984) *Geotechnical Manual of Slopes*, Hong Kong: Hong Kong Government, the HKSAR Government.

Geotechnical Engineering Office. (1995) *Guide to Slope Maintenance* (*Geoguide 5*), Hong Kong: CEDD, HKSAR Government.

Geotechnical Engineering Office. (1996a) Report on the Fei Tsui Road landslide of 13 August 1995, The HKSAR Government, Hong Kong.

Geotechnical Engineering Office. (1996b) Report on the Shum Wan Road landslide of 13 August 1995, the HKSAR Government, Hong Kong.

Geotechnical Engineering Office. (1996c) Investigation of some major slope failures between 1992 and 1995, GEO Report No. 52, The HKSAR Government, Hong Kong.

Geotechnical Engineering Office. (1998) Report on the Ching Cheung Road Landslide of 3 August 1997, GEO Report No. 78, The HKSAR Government, Hong Kong.

Geotechnical Engineering Office. (2001) Regional variation in extreme rainfall values, GEO Report No. 115, The HKSAR Government, Hong Kong.

Geotechnical Engineering Office. (2002) *Application of Prescriptive Measures to Slopes and Retaining Walls*, 2nd edn., GEO Report No. 56, Hong Kong.

Geotechnical Engineering Office. (2003) Enhancing the reliability and resolution of engineered soil cut slope, Technical Guidance Note No. 11, Hong Kong.

Geotechnical Engineering Office. (2004a) Guidelines for classification of consequence-to-life category for slope features, GEO Technical Guidance Note No. 15, Hong Kong.

Geotechnical Engineering Office. (2004b) Review of cases with problems encountered during soil nail construction (1993–2003), GEO Technical Note TB1/2004, Hong Kong.

Geotechnical Engineering Office. (2007) Engineering geological practice in Hong Kong, GEO Publication No. 1/2007, the HKSAR Government, Hong Kong.

Geotechnical Engineering Office. (2008) *Guide to Soil Nail Design and Construction* (*Geoguide* 7), Hong Kong: HKSAR Government.

Geotechnical Engineering Office. (2011) Technical guidelines on landscape treatment for slopes, GEO Publication No. 1/2011, Hong Kong.

Geotechnical Engineering Office and Hong Kong Institution of Engineers. (2011) *Design of Soil Nails for Upgrading Loose Fill Slopes*, Hong Kong: the HKSAR Government.

Giger M.W. and Krizek R.J. (1975) Stability analysis of vertical cut with variable conner angle, *Soils and Foundations*, 15(2), 63–71.

Glover F. (1989) Tabu search – Part I, *ORSA Journal on Computing*, 1(3), 190–206.

Glover F. (1990) Tabu search – Part II, *ORSA Journal on Computing*, 2(1), 4–32.

Greco V.R. (1996) Efficient Monte Carlo technique for locating critical slip surface, *Journal of Geotechnical Engineering, ASCE*, 122, 517–525.

Greenwood J.R. (1987) Effective stress stability analysis. Discussion in *Ninth European Conference on Soil Mechanics and Foundations*, Dublin, Ireland, September 1987, Vol. 3, *Post Conference Proceedings*, Balkema, 1989, pp. 1082–1108.

Griffiths D.V. and Lane P.A. (1999) Slope stability analysis by finite elements, *Geotechnique*, 49(3), 387–403.

Haefeli R. (1948) The stability of slopes acted upon by parallel seepage, in: *Proceedings of the Second International Conference on Soil Mechanics and Foundation Engineering*, Rotterdam, the Netherlands, Vol. 1, pp. 57–62.

Hassiotis S., Chameau J.L., and Gunaratne, M. (1997) Design method for stabilization of slopes with piles, *Journal of Geotechnical and Geoenvironmental Engineering*, 123(4), 314–323.

Hazzard J.F., Maxwell S.C., and Young R.P. (1998) Micromechanical modelling of acoustic emissions, in: *Proceedings of ISRM/SPE Rock Mechanics is Petroleum Engineering, Eurock 98*, Trondheim, Norway, pp. 519–526, SPE 47320.

Hoek E. and Bray J.W. (1981) *Rock Slope Engineering*, 3rd edn., London, U.K.: The Institute of Mining and Metallurgy.

Holland J.H. (1975) *Adaptation in Natural and Artificial Systems*, Ann Arbor, MI: University of Michigan Press.

Hong Kong Geological Survey. (1987) 1:20,000 Solid and Superficial Geology Map, Hong Kong.

Hong Kong Government. (1972) Final Report of the Commission of Enquiry into the Rainstorm Disasters, Hong Kong.

Hovland H.J. (1977) Three-dimensional slope stability analysis method, *Journal of the Geotechnical Engineering Division, ASCE*, 103(GT9), 971–986.

Huang C.C. and Tsai C.C. (2000) New method for 3D and asymmetrical slope stability analysis, *Journal of Geotechnical and Geoenvironmental Engineering, ASCE*, 126(10), 917–927.

Huang C.C., Tsai C.C., and Chen Y.H. (2002) Generalized method for three-dimensional slope stability analysis, *Journal of Geotechnical and Geoenvironmental Engineering, ASCE*, 128(10), 836–848.

Hungr O. (1987) An extension of Bishop's simplified method of slope stability analysis to three dimensions, *Geotechnique*, 37(1), 113–117.

Hungr O. (1994) A general limit equilibrium model for three-dimensional slope stability analysis: Discussion, *Canadian Geotechnical Journal*, 31(5), 793–795.

Hungr O., Salgado F.M., and Byrne P.M. (1989) Evaluation of a three-dimensional method of slope stability analysis, *Canadian Geotechnical Journal*, 26(4), 679–686.

Itasca. (1999) PFC2D/3D – Particle flow code in 2 dimensions, Minneapolis, MN: Itasca Consulting Group Inc.

Ito T. and Matsui T. (1975) Methods to estimate lateral force acting on stabilizing piles, *Soils and Foundations*, 15(4), 43–60.

Ito T., Matsui T., and Hong W.P. (1981) Design method for stabilizing piles against landslide–one row of piles, *Soils and Foundations*, 21(1), 21–37.

Ito T., Matsui T., and Hong W.P. (1982) Extended design method for multi-row stabilising piles against landslide, *Soils and Foundations*, 22(1), 1–13.

Izbicki R.J. (1981) Limit plasticity approach to slope stability problems, *Journal of the Geotechnical Engineering Division, Proceedings of the American Society of Civil Engineers*, 107, 228–233.

Janbu, N. (1957) Earth pressure and bearing capacity by generalized procedure of slices, in: *Proceedings of the Fourth International Conference on Soil Mechanics*, London, U.K., pp. 207–212.

Janbu N. (1973) Slope stability computations, in: R.C. Hirschfield and S.J. Poulos (eds.), *Embankment-Dam Engineering*, pp. 47–86, New York: John Wiley.

Janbu N., Bjerrum L., and Kjaernsli B. (1956) *Soil Mechanics Applied to Some Engineering Problems*, Norway: Norwegian Geotechnical Institute, Publ. No. 16.

Jiang J.C. and Yamagami T. (1999) Determination of the sliding direction in three-dimensional slope stability analysis, in: *Proceedings of 44th Symposium on Geotechnical Engineering*, Reno, NV, pp. 193–200.

Jiao J.J., Ding G.P., and Leung C.M. (2006) Confined groundwater near the rock-head in igneous rocks in the Mid-Levels area, Hong Kong, China, *Engineering Geology*, 84(3–4), 207–219.

Juran I., Beech J., and Delaure E. (1984) Experimental study of the behavior of nailed soil retaining structures on reduced scale models, in: *Proceedings of the International Conference on In-situ Soil and Rock Reinforcements*, Paris, France.

Karel K. (1977a) Application of energy method, *Journal of the Geotechnical Engineering Division, Proceedings of the American Society of Civil Engineers*, 103, 381–397.

Karel K. (1977b) Energy method for soil stability analyses, *Journal of the Geotechnical Engineering Division, Proceedings of the American Society of Civil Engineers*, 103, 431–445.

Kawai T. (1977) New discrete structural models and generalization of the method of limit analysis, in: P.G. Bergan et al. (eds.), *Finite Elements in Nonlinear Mechanics*, pp. 885–906, Trondheim, Norway: NIT.

Kennedy J. and Eberhart R. (1995) Particle swarm optimization, in: *Proceedings of the IEEE International Conference on Neural Networks*, Perth, Western Australia, Australia, pp. 1942–1948.

King G.J.W. (1989) Revision of effective stress method of slices, *Geotechnique*, 39(3), 497–502.

Kirkpatrick S., Gelatt Jr. C.D., and Vecchi M.P. (1983) Optimization by simulated annealing, *Science*, 220, 671–680.

Krahn J. (2003) The 2001 R.M. Hardy lecture: The limits of limit equilibrium analyses, *Canadian Geotechnical Journal*, 40(3), 643–660.

Kwong J.S.M. (1991) Shaft stability, PhD thesis, King's College London, London, U.K.

Lam L. and Fredlund D.G. (1993) A general limit equilibrium model for three-dimensional slope stability analysis, *Canadian Geotechnical Journal*, 30(6), 905–919.

Lam L. and Fredlund D.G. (1994) A general limit equilibrium model for three-dimensional slope stability analysis: Reply, *Canadian Geotechnical Journal*, 31(5), 795–796.

Lam L., Fredlund D.G., and Barbour S.L. (1987) Transient seepage model for saturated–unsaturated soil systems: A geotechnical engineering approach, *Canadian Geotechnical Journal*, 24(4), 565–580.

Lau J.C.W. and Lau C.K. (1998) Risk management in design-and-build in foundation works, in: *Half-Day Seminar on Foundation Design and Construction in Difficult Ground*, Jointly Organized by the Building and Geotechnical Divisions, HKIE, Hong Kong, 20 February.

Law K.T. and Lumb P. (1978) A limit equilibrium analysis of progressive failure in the stability of slopes, *Canadian Geotechnical Journal*, 15(1), 113–122.

Lee C.Y., Hull T.S., and Poulos H.G. (1995) Simplified pile-slope stability analysis, *Computers and Geotechnics*, 17(1), 1–16.

Lee C.Y., Poulos, H.G., and Hull, T.S. (1991) Effect of seafloor instability on offshore pile foundations, *Canadian Geotechnical Journal*, 28(5), 729–737.

Lee K.S. and Geem Z.W. (2005) A new meta-heuristic algorithm for continuous engineering optimization: Harmony search theory and practice, *Computer Methods in Applied Mechanics and Engineering*, 194, 3902–3933.

Les P. and Wayne T. (1997) *The NURBS Book*, 2nd edn., Berlin, Germany: Springer Verlag.

Leshchinsky D., and Baker R. (1986) Three-dimensional analysis for slope stability analysis, *International Journal for Numerical and Analytical Methods in Geomechanics*, 9, 199–223.

Leshchinsky D., Baker R., and Silver M.L. (1985) Three-dimensional analysis of slope stability, *International Journal for Numerical and Analytical Methods in Geomechanics*, 9(3), 199–223.

Leshchinsky D. and Huang C.C. (1992) Generalized three-dimensional slope stability analysis, *Journal of Geotechnical Engineering, ASCE*, 118, 1559–1576.

Li N. (2013) Failure mechanism of slope under several conditions by two-dimensional and three-dimensional distinct element analysis, M.Phil. Thesis, Department of Civil and Environmental Engineering, Hong Kong Polytechnic University.

Li L., Chi S.C., and Lin G. (2005) The genetic algorithm incorporating harmony procedure and its application to the search for the non-circular critical slip surface for soil slopes, *Journal of Hydraulic Engineering*, 36, 913–918 (in Chinese).

Low B.K., Gilbert R.B., and Wright S.G. (1998) Slope reliability analysis using generalized method of slices, *Journal of Geotechnical and Geoenvironmental Engineering*, 124(4), 350–362.

Lowe J. and Karafiath L. (1960) Stability of Earth dams upon drawdown, in: *Proceedings of the First Pan-American Conference on Soil Mechanics and Foundation Engineering*, Mexico City, Vol. 2, pp. 537–552.

Lumb P. (1962) Effect of rain storms on slope stability, in: *Proceedings of the Symposium on Hong Kong Soils*, Hong Kong, pp. 73–87.

Lumb P. (1975) Slope failures in Hong Kong, *Quarterly Journal of Engineering Geology*, 8, 31–65. (Abstract published in Geotechnical Abstracts, 1975, no. GA 99.20.)

Lyamin A.V. and Sloan S.W. (1997) A comparison of linear and nonlinear programming formulations for lower bound limit analysis, in: S. Pietruszczak and G.N. Pande (eds.), *Proceedings of the Sixth International Symposium on Numerical Models in Geomechanics*, Balkema, Rotterdam, the Netherlands, pp. 367–373.

Lyamin A.V. and Sloan S.W. (2002b) Upper bound limit analysis using linear finite elements and non-linear programming, *International Journal for Numerical and Analytical Methods in Geomechanics*, 26(2), 181–216.

Lysmer J. (1970) Limit analysis of plane problems in soil mechanics, *Journal of the Soil Mechanics and Foundations Division, Proceedings of the American Society of Civil Engineering*, 96, 1311–1334.

Malkawi A.I.H., Hassan W.F., and Sarma S.K. (2001) Global search method for locating general slip surface using Monte Carlo techniques, *Journal of Geotechnical and Geoenvironmental Engineering*, 127, 688–698.

Martin C.M. (2004) *User Guide for ABC – Analysis of Bearing Capacity (v1.0)*, Oxford, U.K.: Department of Engineering Science, University of Oxford.

Martin R.P. (2000) Geological input to slope engineering in Hong Kong, Engineering Geology HK, November, IMM (HK Branch), pp. 117–138.

Matsui T. and San K.C. (1992) Finite element slope stability analysis by shear strength reduction technique, *Soils and Foundations*, 32(1), 59–70.

Michalowski R.L. (1989) Three-dimensional analysis of locally loaded slopes, *Geotechnique*, 39(1), 27–38.

Michalowski R.L. (1995) Slope stability analysis: A kinematical approach, *Geotechnique*, 45(2), 283–293.

Michalowski R.L. (1997) An estimate of the influence of soil weight on bearing capacity using limit analysis, *Soils and Foundations*, 37(4), 57–64.

Morgenstern N.R. (1992) The evaluation of slope stability – A 25-year perspective, in: *Stability and Performance of Slopes and Embankments – II*, Geotechnical Special Publication No. 31, ASCE, New York.

Morgenstern N.R. and Price V.E. (1965) The analysis of stability of general slip surface, *Geotechnique*, 15(1), 79–93.

Morrison I.M. and Greenwood J.R. (1989) Assumptions in simplified slope stability analysis by the method of slices, *Geotechnique*, 39(3), 503–509.

Mroz Z. and Drescher A. (1969) Limit plasticity approach to some cases of flow of bulk solids, *Journal of Engineering for Industry, Transactions of the American Society of Mechanical Engineers*, 51, 357–364.

Nash D. (1987) A comparative review of limit equilibrium methods of stability analysis, in: M.G. Anderson and K.S. Richards (eds.), *Slope Stability*, pp. 11–75, New York: John Wiley & Sons.

Naylor D.J. (1982) Finite elements and slope stability, in: *Numerical Methods in Geomechanics, Proceedings of the NATO Advanced Study Institute*, Lisbon, Portugal, 1981, pp. 229–244.

Nguyen V.U. (1985) Determination of critical slope failure surfaces, *Journal of Geotechnical Engineering, ASCE*, 111(2), 238–250.

Ourique C.O., Biscaia E.C., and Pinto J.C. (2002) The use of particle swarm optimization for dynamical analysis in chemical processes, *Computers and Chemical Engineering*, 26(12), 1783–1793.

Pan J. (1980) *Analysis of Stability and Landslide for Structures*, Beijing, China: Hydraulic Press, pp. 25–28 (in Chinese).

Peterson P. and Kwong H. (1981) A design rain storm profile for Hong Kong, Royal Observatory, HKSAR, Technical Note No. 58, Hong Kong.

Petterson K.E. (1955) The early history of circular sliding surfaces, *Geotechnique*, 5, 275–296.

Poulos, H.G. (1973) Analysis of piles in soil undergoing lateral movement, *Journal of the Soil Mechanics and Foundations Division, ASCE*, Vol. 99(SM5), 391–406.

Poulos, H.G. (1995) Design of reinforcing piles to increase slope stability, *Canadian Geotechnical Journal*, 32(5), 808–818.

Poulos, H.G. and Davis, E.H. (1980) *Pile Foundation Analysis and Design*, New York: John Wiley & Sons.

Prandtl L. (1920) Über die Härte plastischer Köroer, *Nachr. Kgl. Ges. Wiss.* Göttingen, Math. Phys. Klasse, pp. 74–85.

Pun W.K. and Shiu Y.K. (2007) Design practice and technical developments of soil nailing in Hong Kong, *HKIE Geotechnical Division Annual Seminar*, Hong Kong, pp. 192–212.

Pun W.K. and Yeo K.C. (1995) Report on the investigation of the 23 July and 7 August 1994 landslides at Milestone 14½ Castle Peak Road, GEO Report No. 52, Geotechnical Engineering Office, Hong Kong, pp. 83–97.

Qian L.X. and Zhang X. (1995) Rigid finite element method and its applications in engineering, *Acta Mechanica Sinica* (English edition), 11(1), 44–50.

Reese L.C., Wang S.T., and Fouse J.L. (1992) Use of drilled shafts in stabilising a slope, in: R.B. Seed and R.W. Boulanger (eds.) *Stability and Performance of Slopes and Embankments – II*, Vol. 2, pp. 1318–1332, New York: American Society of Civil Engineers.

Revilla J. and Castillo E. (1977) The calculus of variations applied to stability of slopes, *Geotechnique*, 27, 1–11.

Rocscience. (2006) Slide Verification Manual I, Rocscience Inc., US.

Ruegger R. and Flum D. (2001) Slope stabilization with high-performance steel wire meshes in combination with nails and anchors, in: *International Symposium, Earth Reinforcement*, IS Kyushu, Fukuoka, Japan, November, pp. 14–16.

Saeterbo Glamen M.G., Nordal S., and Emdal A. (2004) Slope stability evaluations using the finite element method, NGM 2004, *XIV Nordic Geotechnical Meeting*, Finland, Vol. 1, pp. A49–A61.

Salman A., Ahmad I., and Madani S.A. (2002) Particle swarm optimization for task assignment problem, *Microprocessors and Microsystems*, 26, 363–371.

Sarma S.K. (1973) Stability analysis of embankments and slopes, *Geotechnique* 23(3), 423–433.

Sarma S.K. (1979) Stability analysis of embankments and slopes, *Journal of the Soil Mechanics and Foundations Division*, 105(GT12), 1511–1522.

Sarma S.K. (1987) A note on the stability of slopes, *Geotechnique*, 37(1), 107–111.

Sarma S.K. and Tan D. (2006) Determination of critical slip surface in slope analysis, *Geotechnique*, 56(8), 539–550.

Schlosser F. (1982) Behaviour and design of soil nailing, in: *Proceedings of the Symposium on Recent Development Techniques*, Bangkok, Thailand, pp. 399–413.

Schlosser F. and Unterreiner P. (1991) Soil nailing in France: Research and practice, *C.R. 70th Congres Annuel TRB*, Washington, DC, pp. 72–79.

Shen C.K., Bang S., and Herrman L.R. (1981) Ground movement analysis of earth support system, *Journal of Geotechnical Engineering, ASCE*, 107(12), 1610–1624.

Shi G.H. (1996) Manifold method, in: *Proceedings of First International Forum on DDA and Simulations of Discontinuous Media*, Berkeley, CA, pp. 52–204.

Shield R.T. (1955) On the plastic flow of metals under conditions of axial symmetry, *Proceedings of the Royal Society of London Series A*, 233, 267–287.

Shiu Y.K. and Cheung W.M. (2003) Long-term durability of steel soil nails, GEO Report No. 135, Geotechnical Engineering Office, Civil Engineering and Development Department, the Government of Hong Kong Special Administrative Region, China.

Shukha R. and Baker R. (2003) Mesh geometry effects on slope stability calculation by FLAC strength reduction method – linear and non-linear failure criteria, in: *Third International Conference on FLAC and Numerical Modeling in Geomechanics*, Sudbury, Ontario, Canada, pp. 109–116.

Sieniutycz S. and Farkas H. (2005) *Variational and Extremum Principles in Macroscopic Systems*, Amsterdam, the Netherlands: Elsevier.

Sloan S.W. (1988a) Lower bound limit analysis using finite elements and linear programming, *International Journal for Numerical and Analytical Methods in Geomechanics*, 12(1), 61–77.

Sloan S.W. (1988b) A steepest edge active set algorithm for solving sparse linear programming problem, *International Journal for Numerical Methods in Engineering*, 26(12), 2671–2685.

Sloan S.W. (1989) Upper bound limit analysis using finite elements and linear programming, *International Journal for Numerical and Analytical Methods in Geomechanics*, 13(3), 263–282.

Sloan S.W. and Kleeman P.W. (1995) Upper bound limit analysis using discontinuous velocity fields, *Computer Methods in Applied Mechanics and Engineering*, 127(1–4), 293–314.

Sokolovskii V.V. (1965) *Statics of Soil Media* (trans. D.H. Jones and A.N. Scholfield), London, U.K.: Butterworths Scientific.

Song E. (1997) Finite element analysis of safety factor for soil structures, *Chinese Journal of Geotechnical Engineering*, 19(2), 1–7 (in Chinese).

Spencer E. (1967) A method of analysis of the stability of embankments assuming parallel inter-slice forces, *Geotechnique*, 17, 11–26.

Stark T.D. and Eid H.T. (1998) Performance of three-dimensional slope stability methods in practice, *Journal of Geotechnical and Geoenvironmental Engineering, ASCE*, 124(11), 1049–1060.

Stocker M.F., Korber G.W., Gassler G., and Gudehes G. (1979) Soil nailing, in: *Proceedings of the International Conference on Reinforcement des sols*, Paris, France, pp. 469–474.

Sultan H.A. and Seed H.B. (1967) Stability analysis of sloping core earth dam, *Journal of the Soil Mechanics and Foundation Engineering, ASCE*, 93, 45–67.

Sun. (2013) Studies of slope stability problems by limit equilibrium method, strength reduction method and distinct element method, Ph.D. Thesis, Department of Civil and Environmental Engineering, Hong Kong Polytechnic University.

Taylor D.W. (1937) Stability of earth slopes, *Journal of the Boston Society of Civil Engineers*, 24, 197–246.

Taylor D.W. (1948) *Fundamentals of Soil Mechanics*, New York: John Wiley.

Thomaz J.E. and Lovell C.W. (1988) Three-dimensional slope stability analysis with random generation of surfaces, in: *Proceedings of the Fifth International Symposium on Landslides*, Rotterdam, the Netherlands, Vol. 1, pp. 777–781.

Thornton C. (1997) Coefficient of restitution for collinear collisions of elastic perfectly plastic spheres, *Journal of Applied Mechanics*, 64, 383–386.

Turnbull W. J. and Hvorslev M. L. (1967) Special problems in slope stability, *Journal of the Soil Mechanics and Foundation Engineering, ASCE*, 93(4), 499–528.

Ugai K. (1985) Three-dimensional stability analysis of vertical cohesive slopes, *Soils and Foundations*, 25, 41–48.

Ugai K. and Leshchinsky D. (1995) Three-dimensional limit equilibrium and finite element analysis: A comparison of results, *Soils and Foundations*, 35(4), 1–7.

Wang M.C., Wu A.H., and Scheessele D.J. (1979) Stress and deformation in single piles due to lateral movement of surrounding soils, in: R. Lundgren, (ed.), *Behavior of Deep Foundations, ASTM 670*, pp. 578–591, Philadelphia, PA: American Society for Testing and Materials.

Wang Y.J. (2001) Stability analysis of slopes and footings considering different dilation angles of geomaterial, PhD thesis, Department of Civil and Structural Engineering, Hong Kong Polytechnic University, Hong Kong.

Wang J., Liu Z., and Lu P. (2008) Electricity load forecasting based on adaptive quantum-behaved particle swarm optimization and support vector machines on global level, *Proceedings of the 2008 International Symposium on Computational Intelligence and Design*, pp. 233–236, Wuhan, China.

Watkins A.T. and Powell G.E. (1992), Soil nailing to existing slopes as landslip preventive works, *Hong Kong Engineer*, 20(3), 20–27.

Wei W.B. and Cheng Y.M. (2009a) Soil nailed slope by strength reduction and limit equilibrium methods, *Computers and Geotechnics*, 37, 602–618.

Wei W.B. and Cheng Y.M. (2009b) Strength reduction analysis for slope reinforced with one row of piles, *Computers and Geotechnics*, 36, 1176–1185.

Whitman R.V. and Bailey W.A. (1967) Use of computers for slope stability analyses, *Journal of the Soil Mechanics and Foundation Division, ASCE*, 93(4), 475–498.

Won J., You K., Jeong S., and Kim S. (2005) Coupled effects in stability analysis of pile-slope systems, *Computers and Geotechnics*, 32(4), 304–315.

Wong H.N. (2001) Recent advances in slope engineering in Hong Kong, in: *Proceedings of the 14th Southeast Asian Geotechnical Conference*, Hong Kong, Vol. 1, pp. 641–659.

Works Bureau. (1999) WBTC No. 13/99, Geotechnical Manual for Slopes – Guidance on Interpretation and Updating, Hong Kong SAR Government, Hong Kong.

Wu L.Y. and Tsai Y.F. (2005) Variational stability analysis of cohesive slope by applying boundary integral equation method, *Journal of Mechanics*, 21, 187–195.

Xing Z. (1988) Three-dimensional stability analysis of concave slopes in plan view, *Journal of Geotechnical Engineering, ASCE*, 114(6), 658–671.

Yamagami T. and Jiang J.C. (1996) Determination of the sliding direction in three-dimensional slope stability analysis, in: *Proceedings of the 2nd International Conference on Soft Soil Engineering*, Nanjing, China, May, Vol. 1, pp. 567–572.

Yamagami T. and Jiang J.C. (1997) A search for the critical slip surface in three-dimensional slope stability analysis, *Soils and Foundations*, 37(3), 1–16.

Yamagami T., Jiang J.C., and Ueno K. (2000) A limit equilibrium stability analysis of slopes with stabilizing piles, in: D.V. Griffiths, G.A. Fenton, and T.R. Martin (eds.), *Slope Stability 2000: Proceeding of Sessions of Geo-Denver 2000*, August 5–8, 2000, Denver, CO, Sponsored by the Geo-Institute of the ASCE, pp. 343–354.

Yeung A.T., Cheng Y.M., Lau C.K., Mak L.M., Yu R.S.M., Choi Y.K., and Kim J.H. (2005) An innovative Korean system of pressure-grouted soil nailing as a slope stabilization measure, *HKIE Geotechnical Division, 25th Annual Seminar*, Hong Kong, 4 May, pp. 43–49.

Yeung A.T., Cheng Y.M., Tham L.G., Au A.S.K., So S.T.C., and Choi Y.K. (2007) Field evaluation of a glass-fiber soil reinforcement system, *Journal of Performance of Constructed Facilities, ASCE*, 21(1), 26–34.

Yin J.H., Ding X.L., Yang Y.W., Lau C.K., Huang D.F., and Chen Y.Q. (2002) An integrated system for slope monitoring and warning in Hong Kong, in: *Proceedings of Advanced Building Technology 2002*, 4–6 December, Hong Kong.

Yin P.Y. (2004) A discrete particle swarm algorithm for optimal polygonal approximation of digital curves, *Journal of Visual Communication and Image Representation*, 15(2), 241–260.

Yu H.S., Salgado R., Sloan S.W., and Kim J.M. (1998) Limit analysis versus limit equilibrium for slope stability, *Journal of Geotechnical and Geoenvironmental Engineering, ASCE*, 124(1), 1–11.

Yu H.S., Sloan S.W., and Kleeman P.W. (1994) A quadratic element for upper bound limit analysis, *Engineering Computations*, 11, 195–212.

Zhang X. (1988) Three-dimensional stability analysis of concave slopes in plan view, *Journal of Geotechnical Engineering*, 114, 658–671.

Zhang X. (1999) Slope stability analysis based on the rigid finite element method, *Geotechnique*, 49(5), 585–593.

Zhang X. and Qian L.X. (1993) Rigid finite element and limit analysis, *Acta Mechanica Sinica* (English edition), 9(2), 156–162.

Zheng H. (2012) A three-dimensional rigorous method for stability analysis of landslides, *Engineering Geology*, 145, 30–40.

Zheng H., Liu D.F., and Li C.G. (2008) On the assessment of failure in slope stability analysis by the finite element method, *Rock Mechanics and Rock Engineering*, 41(4), 629–639.

Zheng H., Tham L.G., and Liu D.F. (2006) On two definitions of the factor of safety commonly used in finite element slope stability analysis, *Computers and Geotechnics*, 33, 188–195.

Zheng H., Zhou C.B., and Liu D.F. (2009) A robust solution procedure for the rigorous methods of slices, *Soils and Foundations*, 49(4), 537–544.

Zheng Y.R., Zhao S.Y., Kong W.X., and Deng C.J. (2005) Geotechnical engineering limit analysis using finite element method, *Rock and Soil Mechanics*, 26(1), 163–168.

Zhu D.Y., Lee C.F., Qian Q.H., and Chen G.R. (2005) A concise algorithm for computing the factor of safety using the Morgenstern–Price method, *Canadian Geotechnical Journal*, 42(1), 272–278.

Zhu D.Y., Lee C.F., Qian Q.H., Zou Z.S., and Sun F. (2001) A new procedure for computing the factor of safety using the Morgenstern–Price's method, *Canadian Geotechnical Journal*, 38(4), 882–888.

Zhu D.Y., Lee C.F., and Jiang H.D. (2003) Generalised framework of limit equilibrium methods for slope stability analysis, *Geotechnique*, 53(4), 377–395.

Zhuo J.S. and Zhang Q. (2000) *Interface Element Method for Discontinuous Mechanics Problems* (in Chinese), Beijing, China: Science Press.

Zienkiewicz O.C., Humpheson C., and Lewis R.W. (1975) Associated and non-associated visco-plasticity and plasticity in soil mechanics. *Geotechnique*, 25(4), 671–689.

Zolfaghari A.R., Heath A.C., and McCombie P.F. (2005) Simple genetic algorithm search for critical non-circular failure surface in slope stability analysis, *Computers and Geotechnics*, 32(3), 139–152.

Index

D

E

F

G

I

J

M

T

U

For Product Safety Concerns and Information please contact our EU representative GPSR@taylorandfrancis.com Taylor & Francis Verlag GmbH, Kaufingerstraße 24, 80331 München, Germany

Printed and bound by CPI Group (UK) Ltd, Croydon, CR0 4YY
01/05/2025
01858497-0001